GANGS
IN AMERICA'S
COMMUNITIES

To my wife and LCP, Karen Q. Howell
Our beloved daughter, Megan Q. Howell

In Memoriam
Walter B. Miller (1920–2004)
Gang Researcher Extraordinaire
Posthumously published book:
City Gangs
http://gangresearch.asu.edu/walter_miller_library/walter-b.-miller-book

GANGS
IN AMERICA'S
COMMUNITIES

JAMES C. HOWELL

Los Angeles | London | New Delhi
Singapore | Washington DC

Los Angeles | London | New Delhi
Singapore | Washington DC

FOR INFORMATION:

SAGE Publications, Inc.
2455 Teller Road
Thousand Oaks, California 91320
E-mail: order@sagepub.com

SAGE Publications Ltd.
1 Oliver's Yard
55 City Road
London EC1Y 1SP
United Kingdom

SAGE Publications India Pvt. Ltd.
B 1/I 1 Mohan Cooperative Industrial Area
Mathura Road, New Delhi 110 044
India

SAGE Publications Asia-Pacific Pte. Ltd.
33 Pekin Street #02-01
Far East Square
Singapore 048763

Acquisitions Editor: Jerry Westby
Editorial Assistant: Erim Sarbuland
Production Editor: Karen Wiley
Copy Editor: Gretchen Treadwell
Typesetter: C&M Digitals (P) Ltd.
Proofreader: Christine Dahlin
Indexer: Wendy Allex
Cover Designer: Gail Buschman
Marketing Manager: Erica DeLuca
Permissions Editor: Karen Ehrmann

Printed in the United States of America

Library of Congress Cataloging-in-Publication Data

Howell, James C.

Gangs in America's communities / James C. Howell.

p. cm.
Includes bibliographical references and index.

ISBN 978-1-4129-7953-5 (pbk.: alk. paper)

1. Gangs—United States. 2. Violence—United States—Prevention. I. Title.

HV6439.U5H679 2012
364.106′60973—dc23   2011037034

This book is printed on acid-free paper.

11 12 13 14 15 10 9 8 7 6 5 4 3 2 1

Contents

Detailed Contents

Preface

Street gangs are at times perplexing to everyone and fighting them often is considered a futile exercise. The two main purposes of this book are, first, to demonstrate that the essential features of street gangs can be understood despite their highly varied and sometimes enigmatic public presence, and second, that some gang prevention, control strategies, and programs are effective, in contradiction of widespread proclamations that nothing works.

Street gangs are not well understood largely because they are at once shrouded in myths (some of which they create themselves), folklore, urban legends, media exaggerations, popular misconceptions, and international intrigue often associated with them. Taking a historical approach to the emergence of gangs in the United States, the book uncovers their origins and traces their development, first, in the Northeast region of the United States; next, in the Midwest; then in the West region; and last, in the Southern region. The author analyzes the key historical events that produced waves of gang growth in the respective regions. These trends are brought up to date with 14 years of annual national survey data showing a marked increase in gang activity since the beginning of the new millennium. The book also examines gang trends along the U.S.–Mexico border, and in Central America, Canada, and Europe, along with an assessment of the threat of such highly publicized gangs as Mara Salvatrucha (MS-13), 18th Street gangs, and prison gangs such as the Mexican Mafia.

American gang history also serves as an excellent backdrop for reviews of myths about gangs, theories of gang formation, and various ways of defining and classifying gangs. Gangs emerged in the United States in a rainbow of colors, beginning with White ones, that reflect both outside immigration and internal migration patterns. Understanding the history of evolving gangs in America also engenders a stark realization that gang joining is often a logical choice for marginalized youth who have been relegated to the fringes of society, where powerlessness is commonly felt. Social and economic conditions in inner-city areas, organized crime, and deviancy centers foster widespread criminal activity where ganging together for safety is an understandable response.

The text explains how youngsters who are making the transition from childhood to adolescence form new gangs. These *starter gangs* often form somewhat spontaneously by authority-rejecting children and adolescents who have been alienated from families and schools. Finding themselves spending a great deal of time on the street, they may form gangs with other socially marginalized youths and look to each other for protection and street socialization. Although most youths who join are on average in a gang for less than 1 year, some of these gangs increase their criminal activity, especially when conflict with other street groups solidifies them, becoming a formidable force in the streets. Girls often are active participants in youth gangs, and they commit very similar crimes to boys. Interestingly, research on younger gangs shows that the most criminally active ones tend to be gender balanced.

To be sure, there is a harsh, cold reality about street gangs in major cities that we ignore at our own peril. Many of the gangs incubated in the most poverty-stricken zones of very large American cities begin as the youngest cliques or sets of well-established gangs, in systematic age-graded succession. These gangs can dominate inner-city streets and create a feudal-like territory that often leads to ongoing gang wars for turf, dominance, and physical prowess—typically in very small gang *set spaces*.

Cities with populations in excess of 100,000 persons are home to the overwhelming majority of dangerous gangs representing the bulk of gang members in the entire country, particularly older, more violent gangs with mainly young adult participants. Two-thirds of these cities consistently experience large numbers of gang-related homicides and other gang-related violence, mayhem, intimidation, and pervasive fear. Case studies illustrate that cities have gang-problem histories much like individuals' careers in crime. The author and colleagues have identified common gang-history patterns among groups of cities as these unfolded across 14 years of annual national survey data. Very large cities with long histories of gang problems tend to display relatively stable patterns of serious gang activity; in contrast, small cities, towns, and counties fluctuate in seriousness and gangs may actually dissolve in many of these places.

Preventing gangs from forming and eliminating established gangs altogether is virtually impossible, when they are rooted in the cracks of our society. But the exceedingly good news is that gang crime can be reduced—even among some of the worst gangs—and communities can be made safe from the social destruction that often follows in their wake. Although there is no quick fix, no magic bullet, several steps can be taken to bring measurable relief. But to expect dramatic results would be naïve, given the community conditions in which gangs thrive and that well-established street gangs place unusual demands on their members including an oath of loyalty, a code of secrecy, penalties for violating gang behavioral codes, and unequivocal promises of protection.

The main implication is that communities must organize themselves better than the gangs and present a more formidable front. Once communities make a commitment to this end, they are in an excellent position to undertake strategic planning toward overcoming the gangs. Each community needs to assess its own gang activity, prepare a strategic plan that fits its specific gang problem, and develop a continuum of programs and activities that parallels youths' gang involvement over time. *Prevention* programs are needed to target children and early adolescents at risk of gang involvement, in order to reduce the number of youths that join gangs. *Intervention* programs and strategies are needed to provide necessary sanctions and services for slightly older youth who are actively involved in gangs in order to separate them from gangs. And law enforcement *suppression* strategies are needed to target the most violent gangs and older, criminally active gang members. Each of these components helps make the others more effective, provided that evidence-based services and strategies are incorporated in the continuum. The final chapters provide ample examples of these and link readers to online resources for more detailed information. Students and community stakeholders should then have the capacity to use these electronic resources to assess gang problems and actively assist or guide the mapping of a strategic plan in a given neighborhood or community.

Acknowledgments

My journey in gang research began in earnest 15 years ago, when I was privileged to begin working at the newly created National Youth Gang Center (recently renamed the National Gang Center). This center was created by the federal Office of Juvenile Justice and Delinquency Prevention (OJJDP) in 1994 at the Institute for Intergovernmental Research (IIR) in Tallahassee, Florida, with three main goals in mind. First, this center was expected to standardize and conduct annual measurements of gang activity across the United States and undertake a program of research. Second, it would review and identify research-based programs that work to prevent gang involvement, reduce gang crime, and control gangs. Third, it would provide training and technical assistance on what works. This book draws upon much of the work in which I have been involved at the National Gang Center.[1]

First and foremost, I must express deep appreciation to Emory Williams for hiring me, and along with him, Doug Bodrero, Bruce Buckley, Gina Hartsfield, Clay Jester, John P. Moore, and other Institute for Intergovernmental Research officials for creating a stellar organization in which gang research, program development, and training could thrive. Colonel (Ret.) John P. Moore inspired my work with his extraordinary knowledge of gangs and gang intelligence. Most important, he has masterfully directed the National Gang Center from the beginning and the recent expansion in achieving its goals and other equally notable accomplishments; he also has been a valuable collaborator on many survey and research products, and an invaluable critic of my work.

Walter Miller first stimulated my attention to youth gangs, beginning in 1974, when I first went to work at the U.S. Department of Justice. His pioneering research prompted the U.S. Congress to create a program of gang research and program development at the federal Office of Juvenile Justice and Delinquency Prevention, and to authorize the establishment of a national gang center. It was he who envisioned the potential value of a National Youth Gang Survey. I also am deeply indebted to several other eminent gang researchers. Joan Moore and Diego Vigil educated me about Chicano gangs and the history of Mexican American gangs. Finn Esbensen's pioneering student surveys have provided me and the gang field extraordinary insights into youth gangs. Irving Spergel, Jim Short, and Ron Huff constantly tutored me on gang research and important research questions. I also acknowledge Rolf Loeber (Pittsburgh), Terry Thornberry (Rochester), and David Huizinga (Denver) as the directors of the three major U.S. longitudinal studies of delinquency. The OJJDP launched these studies 23 years ago in its Program of Research on the Causes and Correlates of Delinquency. A few years later, OJJDP embedded studies of gang members in these large representative samples, and more than 20 research reports have been published on gang members in these three landmark studies. I draw on a number of them in this book. Researchers in a fourth longitudinal study, the Seattle Social Development Project, undertook a similar line of research under the leadership of Karl Hill, which also has produced very noteworthy findings (see Further Reading in Chapter 5 for a listing of reports on each of these four projects).

I have benefited from both short- and long-term relationships with a number of colleagues that enriched this book, including Dave Barciz, Becky and Richard Block, Beth Bjerregaard, Jim Burch, Dave Curry, Scott Decker, Arlen Egley Jr., Finn Esbensen, David Farrington, Victor Gonzalez, Rachel Gordon, Karl Hill, Ron Huff, Lorine Hughes, David Huizinga, Chuck Katz, Marion Kelly, Mac Klein, Marv Krohn, Mark Lipsey, Alan Lizotte, Rolf and Magda Loeber, Jim Lynch, Rebecca Petersen, Jim Short, Irv Spergel, Terry Thornberry, Deborah Weisel, and Susan Whitten. I must give special recognition to John J. Wilson, with whom I coauthored *A Comprehensive Strategy for Serious, Violent and Chronic Juvenile Offenders* in 1993. This collaboration broadened my perspective of solutions to juvenile delinquency and gang problems that could encompass counties, cities, states, and counties.

The following colleagues and SAGE reviewers kindly provided valuable comments on draft chapters: Beth Bjerregaard, Brenda Chaney, Dave Curry, Scott Decker, Finn Esbensen, Adrianne Freng, Erika Gebo, Camille Gibson, Rachel Gordon, Laura Hansen, Megan Howell, Ron Huff, Lorine Hughes, Robert B. Jenkot, Soraya Kawucha, John Moore, Joan Moore, Robert Lombardo, Dan Okada, Dana Peterson, Becky Petersen, Andrew Papachristos, Pamela Preston-Black, David Pyrooz, Irving Spergel, Diego Vigil, and Elvira M. White.

I also have learned a great deal about working with gang members and gang programming from my colleagues at the National Gang Center and other experts with whom we regularly work, including Michelle Arciaga, Van Dougherty, Arlen Egley, Erica Fearby-Jones, Victor Gonzalez, Candice Kane, Kim Porter, and Steve Ray. Along with Col. Moore, Van Dougherty deserves special recognition for her contributions to development and refinement of the Comprehensive Gang Model and for conceiving the ideal administrative structure for it (Figure 10.2). Henry Kahn, David Kolar, Karen Ohman, and Jim Bachemin at the IIR have also supported my work in important ways.

I am indebted to several close colleagues for invaluable contributions to this book. George Tita, Beth Griffiths, and Arlen Egley performed the innovative analyses for our publication, *U.S. Gang Problem Trends and Seriousness, 1996–2009*, which I summarize in Chapters 7 and 8. John P. Moore collaborated with me on the initial report on *History of Street Gangs in the United States*, which I expanded in Chapter 1. Arlen Egley Jr., extraordinary administrator of the National Youth Gang Survey, also prepared all of the figures and data tables on the NYGS that appear throughout this book. Dave Curry also played an instrumental role in developing the NYGS and analyses; however, the author is responsible for any errors in presentations of the survey results. He and I also collaborated on a comprehensive review of risk factors for gang membership, annual NYGS fact sheets, other survey reports, and numerous American Society of Criminology presentations, several of which are referenced herein. Finn Esbensen generously provided tables and research reports from his two pioneering GREAT studies. Colleagues at the U.S. Department of Justice have constantly supported my work including Jim Burch, James Chavis, Catherine Doyle, Barbara Kelley, Dennis Mondoro, Stephanie Rapp, Jeff Slowikowski, and Phelan Wyrick.

I must acknowledge my North Carolina colleagues, Megan Howell, Billy Lassiter, Susan Whitten, Nancy Hodges, Dave Barciz, Deborah Weisel, and Fran Cook for their leadership in statewide youth gang programming in North Carolina.

To my long-time friend and historian extraordinaire, Larry P. Riggs, I extend heartfelt thanks for constantly encouraging me in this endeavor and for insisting that I pursue the historical development of street gangs.

Suzanne Sinclair, librarian at the Health Sciences Library, First Health of the Carolinas, in Pinehurst, North Carolina, greatly facilitated my research by obtaining articles and chapters promptly from other sources.

Dr. Walter B. Miller passed away in 2004 without seeing the publication of his monumental gang book. It centered on a long-term study of 21 street gangs in Boston and a successful program to address them that he largely developed in the early 1950s and meticulously evaluated, the Midcity Project. His renowned scholarship is now available in book form. Unfortunately, the publication of Dr. Miller's book was not completed in time for inclusion in this book. Fortunately for readers, however, it is online and accessible free of charge thanks to the largess of Scott Decker, Foundation Professor and Director, School of Criminology and Criminal Justice, Arizona State University. Dr. Decker masterfully organized Miller's epic work for publication and it can be accessed at http://gangresearch.asu.edu/walter_miller_library/walter-b.-miller-book. Dr. Miller is widely recognized as one of the six most distinguished early gang scholars, along with Malcolm Klein, Joan Moore, Jim Short, Irving Spergel, and Diego Vigil.

I would be inexcusably remiss if I did not recognize important collaborators and supporters in the preparation of this book at SAGE. Jerry Westby, acquisitions editor, envisioned what this book could become. Erim Sarbuland, editorial assistant, constantly provided support. Karen Wiley, production editor, brought the project to fruition. Megan Koraly worked hard to secure permissions. The copy editor Gretchen Treadwell masterfully reorganized disjointed sections and often magically turned my ordinary prose into clarity. Readers are indebted to her.

Note

1. The National Gang Center was supported by Cooperative Agreement No. 2010-GP-BX-K076, awarded by the Bureau of Justice Assistance, and Grant No. 2007-JV-FX-0008 and Cooperative Agreement No. 2011-MU-MU-K001, awarded by the Office of Juvenile Justice and Delinquency Prevention, Office of Justice Programs. The opinions, findings, and conclusions or recommendations expressed in this publication are those of the author and do not necessarily reflect the views of the U.S. Department of Justice.

History of Gangs in the United States

Introduction

A widely respected chronicler of British crime, Luke Pike (1873), reported the first active gangs in Western civilization. While Pike documented the existence of gangs of highway robbers in England during the 17th century, it does not appear that these gangs had the features of modern-day, serious street gangs. Later in the 1600s, London was "terrorized by a series of organized gangs calling themselves the Mims, Hectors, Bugles, Dead Boys [and they] fought pitched battles among themselves dressed with colored ribbons to distinguish the different factions" (Pearson, 1983, p. 188). According to Sante (1991), the history of street gangs in the United States began with their emergence on the East Coast around 1783, as the American Revolution ended. These gangs emerged in rapidly growing eastern U.S. cities, out of the conditions created in large part by multiple waves of large-scale immigration and urban overcrowding.

This chapter examines the emergence of gang activity in four major U.S. regions, as classified by the U.S. Census Bureau: the Northeast, Midwest, West, and South. The purpose of this regional focus is to develop a better understanding of the origins of gang activity and to examine regional migration and cultural influences on gangs themselves. Unlike the South, in the Northeast, Midwest, and West regions, major phases characterize gang emergence. Table 1.1 displays these phases.

Table 1.1 Key Timelines in U.S. Street Gang History

Northeast Region (mainly New York City)

First period: 1783–1850s

- The first ganglike groups emerged immediately after the American Revolution ended, in 1783, among the White European immigrants (mainly English, Germans, and Irish).
- Serious ganging in New York City commenced around 1820.
- The first well-organized gang formed in 1826 in the back room of Rosanna Peers's greengrocery.

Second period: 1860–1920s

- The Chinese began setting up their own highly structured tongs around 1860.
- The arrival of the Poles, Italians, Austrians, and other peoples in the period 1890 to 1930 created even worse slum conditions.
- The first U.S. police war on gangs occurred in New York City in 1915–1916.

Third period: 1930s–1980s

- Beginning in the 1930s, the most intensive gang activity in New York City shifted from downtown (Manhattan) to both northern (East Harlem and the Bronx) and southeastern (Brooklyn) locations in the metropolitan area.
- More fighting gangs emerged after the arrival of African American migrants from the South and Latino immigrant groups (from Latin America, the Caribbean, and Puerto Rico) in the 1930s and 1940s.
- Black gangs appeared by the 1950s.
- In the late 1950s, a "slum clearance" project moved several thousand poor Puerto Rican and African American families into high-rise public housing in East Harlem.
- During the 1980s, new Asian and non–Puerto Rican Latino immigrants populated gangs.

Midwest Region (mainly Chicago)

First period: 1860s–1920s

- Chicago's first street gangs developed among White immigrants along ethnic lines in the 1860s—particularly Irish, German, and Lithuanian people.
- In the 1920s, gangs became entrenched in the patronage networks operated by ward politicians, with notorious criminals and rum-runners, the most notable of which was the Al Capone gang.

Second period: 1940s–1970s

- Following the 1919 race riot, Black males formed gangs to confront hostile White gang members who were terrorizing the African American communities.
- Mexican American gangs likely formed in the 1950s, if not earlier.
- Chicago's largely African American gang problem exploded in the 1960s, with more gangs and more violence.
- Public housing high-rises became gang incubators and drug turf battlegrounds beginning in the 1970s.

West Region (mainly Los Angeles)

First period: 1890s–1920s

- Ganglike groups of Mexican descent appeared in the West region in the 1890s.
- The first Los Angeles Mexican American gangs likely formed in the 1920s.

Second period: 1940s–1950s

- Two events in the 1940s stimulated growth of Mexican American gangs in the West: the Sleepy Lagoon murder and the zoot suit riots.
- Mexican immigration accelerated in the early 1950s.

Third period: 1950s–1980s

- By the 1950s, African American gangs in Los Angeles were beginning to assume a street gang presence.
- African American gangs were well established by the 1960s in low-income housing projects.
- Mexican American gangs steadily grew following the Vietnam War, the War on Poverty, and the Chicano movement of the 1960s and 1970s.
- By the 1970s, street gangs had emerged in most populated areas across California.
- In the 1980s, the gang culture melded with crack cocaine dealing and consumption in the African American ghettos.

South Region

First period: 1970s–1990s

- Gang activity likely did not emerge in the southern states prior to the 1970s.
- As of 1980, only Miami and San Antonio were considered to have a moderately serious gang problem.
- Several southern states saw sharp increases in gang activity in multiple cities and counties by 1995.
- Before the end of the 20th century, the South region matched the other major regions in the prevalence of gang activity.
- Houston emerged in the past decade as a major gang center.

The Influence of Population Migration Patterns on Gang Emergence

Three large groups of early immigrants populated the Northeast and Midwest regions of the United States. According to Pincus and Ehrlich (1999), so-called old immigrants first came, predominantly from England and English territories, and also Dutch, German, Swedish, and Scandinavian peoples. In the second large wave, from 1865 to 1890, approximately 11 million immigrants arrived from mainly northern and western regions of Europe, especially Great Britain, Germany, and Scandinavia (Denmark, Norway, Sweden). The third group of immigrants from countries of southern and eastern Europe—the Poles, Italians, Austrians, and many others—another 11 million or so, arrived from 1890 to 1930. Largely consisting of low-skilled, low-wage laborers, not unexpectedly, the three large immigrant surges overwhelmed the housing and welfare capacity of the young northeastern and midwestern cities, contributing directly to slum conditions and the accompanying crime problems, gangs included. Street gangs emerged from similar conditions of social disorganization in Chicago.

In contrast, gangs initially grew out of the preexisting Mexican culture in the West region, and subsequent Mexican migrations continuously fueled their growth. Immigrant groups along the trail from Mexico to Los Angeles initially populated El Paso, Albuquerque, and Los Angeles. The migrants brought an embryo, or pregang, subculture called *pachuco* (J. W. Moore, 1978; Vigil, 1988, 1990) that quite likely began forming in El Paso (J. W. Moore, 2007a). Gang emergence in the South does not appear to have been grounded in any of the preexisting conditions found in the Northeast, Midwest, and West.

Street Gang Emergence in the Northeast

New York City's Ellis Island was the major port of entry to the United States. Dutch immigrants arrived first, in the early 1600s, and as Bourgois (2003) reports, they promptly stole Manhattan Island from the indigenous people who inhabited the island and hunted and fished there. The Lower East Side of the city—particularly around the Five Points—later fell victim to rapid Irish immigration and ensuing political, economic, and social disorganization (Riis, 1902/1969). Bourgois (2003) also identifies Irish and Italian immigrants as early European settlers in East Harlem. Virtually all of the Puerto Ricans arrived there much later, mainly in the two decades following World War II.

Street gangs on the East Coast developed in three phases. The first phase began after the American Revolution. These ganglike groups were not seasoned criminals—only youth fighting over local turf. The beginning of serious ganging in New York City, the second phase, commenced a few years later, around 1820, when immigration began to pick up (Pincus & Ehrlich, 1999). A third wave of gang activity ensued in the 1930s and 1940s when Latino and Black populations began to arrive in large numbers. Soon, according to Gannon (1967), more than two-thirds of the New York gangs were Puerto Rican or Black.

First Period of New York City Gang Growth

Three developments in particular appear to have contributed to the emergence of New York City's street gangs: (1) social disorganization in slum areas, (2) the establishment of greengrocery stores, and (3) the involvement of politicians in street gangs. The isolation and marginalization of early immigrants in the rapidly growing New York City may have prompted them to establish what Ley (1975) describes as "a small secure area where group control [could] be maximized against the flux and uncertainty of the . . . city" (pp. 252–253). Conflict was therefore imminent, and street gangs grew in such environments, largely motivated by a desire to exercise some power or control over a chaotic environment. Nevertheless, these first gangs were largely inconsequential.

The serious street gangs that first drove clearly defined stakes in the streets of New York in the late 18th century grew out of a second development, the establishment of greengrocery speakeasies that sold vegetables. However, in most cases, vegetable sales were nothing more than a front for the back room in which "the fiery liquor of the period" was sold at lower prices than in the respectable saloons (Asbury, 1927). As Adamson (1998) and Sante (1991) report, many of the older gang members were employed, mostly as common laborers including bouncers in saloons and dance halls, dockers, carpenters, sail makers, and shipbuilders. "They engaged in violence, but violence was a normal part of their always-contested environment; turf warfare was a condition of the neighborhood" (Sante, 1991, p. 198). Barroom brawling was a common denominator. "The majority of dives featured one or another of a variation of the basic setup: bar, dance floor, private boxes, prostitution, robbery" (p. 112). Hagan (1995) asserts that these "deviance service centers" catered to the demand for illicit sex, alcohol, guns, protection, and even murder for hire.

The first gang with a definite, acknowledged leadership—named the Forty Thieves and made up largely of local thieves, pickpockets, and thugs—formed around 1826 in the back room of Rosanna Peers's greengrocery (Haskins, 1974). The second gang that formed in the area, the Kerryonians, named themselves after the county in Ireland from which they originated. Other similar gangs quickly formed in the Five Points area, including the Chichesters, Roach Guards, Plug Uglies (named after their large plug hats), Shirt Tails (distinguished by wearing their shirts outside their trousers), and Dead Rabbits. Haskins (1974) explains how the Dead Rabbits were so-named when, "at one of the gang's stormy meetings someone threw a dead rabbit into the center of the room. One of the squabbling factions accepted this as an omen and its members withdrew, forming an independent gang and calling themselves the Dead Rabbits" (p. 27). Their battle symbol was a dead rabbit impaled on a pike. Haskins notes that in Five Points jargon, a rabbit was a super tough guy.

The Irish Bowery Boys soon formed in a nearby area known as the Bowery. Battles between the Bowery Boys and Dead Rabbits (claiming more than 1,000 members each) were legendary, each of which was supported by smaller gangs they had spawned. These gangs out-manned police and both the National Guard and the regular army were summoned on occasion to quell the fights. Altogether, they waged as many as 200 battles with Five Points gangs over a span of 10 years, beginning in 1834 (Asbury, 1927). A 2002 movie, *Gangs of New York*, vividly depicted these gangs, albeit with some exaggerations and distorted history.

Shrewd politicians immediately recognized the potential asset that the street gangs might represent (Haskins, 1974). In the early 1830s, several politicians (ward and district leaders) bought grocery stores in Five Points and the saloons and dance halls in the Bowery, the gathering places for the gangs. In return for their assured protection of the gangs' meeting places, and financial rewards offered to the gangs for their loyalty, gang leaders returned the favor by taking care of jobs like blackjacking political opponents and scaring unsupporting voters away from the polls. "Nearly every shrewd ward and district leader had at least one gang working for him" (p. 32).

Sante (1991) suggests that the gangs were permitted to thrive, to kill each other and drink themselves to death, by authorities who willingly allowed the gangs to act as principal agents of natural selection in the slums of New York City. "The gangs repaid this courtesy by demonstrating their mingled respect and derision for the world outside their turf through parody: parody of order, parody of law, parody of commerce, parody of progress" (p. 235).

Second Period of New York City Gang Growth

For 20 years following the Civil War, corruption and vice were rampant in New York City. Haskins (1974) pinpoints a governmental and political organization, Tammany Hall, at the center of much of the corruption—even aiding and abetting gang activity. Needless to say, gang membership grew enormously during this period:

> By 1855 it was estimated that there were at least 30,000 who owed allegiance to the gang leaders and, through them, to politicians of various factions. At every election the gangs burned ballot boxes, beat up ordinary citizens trying to exercise their right to vote, and themselves voted many times over. (pp. 34–35)

Police were powerless to arrest gang members; they were also beaten when they attempted to guard polling places. By 1857, the city police force was so corrupt that the New York State legislature intervened and replaced it.

In the first decade of the 1900s, another 8.8 million immigrants reached the United States (Pincus & Ehrlich, 1999). The arrival of the Poles, Italians, Austrians, and other peoples in New York City during the period from 1890 to 1930 created even worse slum conditions (Riis, 1902/1969). Inundated with immigrants, New York City could not provide enough homes for the enormous influx. Tenement houses were created as a temporary solution that became permanent. This ethnic succession continued into the 20th century. Sante (1991) elaborates on how economic conditions and the American drive for upward mobility particularly influenced this process. It was the harshness of economic conditions that mainly accounted for the overcrowding of Manhattan's neighborhoods and the succession of resident groups. As families bettered themselves economically, they would move to better living conditions. This signified advancement as assimilated Americans. In turn, newer immigrants would occupy the lower rung in society that advancing families vacated. "Meanwhile, the physical fabric of the original slums rotted away even as they continued to be overpopulated, and nobody cared very much, not even the inhabitants, as long as they thought they had a chance to move elsewhere" (p. 22).

Gangs and other criminal groups were virtually unfettered from forging their own wedges in the social and physical disorder. In fact, Asbury (1927) reports that around 1913, "there were more gangs in New York than at any other period in the history of the metropolis" (p. 360). This shifted by 1916, when, according to Haskins (1974), police "smashed" most of the large gangs in the first U.S. war on gangs. Police beat and arrested gang members, and criminal courts imprisoned more than 200 of the most important gang leaders. The gang action then moved to Chinatown and other outlying areas of New York City.

The Chinese began setting up their own highly structured tongs around 1860. Chin (1996, 2000) suggests that these eventually put the street gangs to shame in running a criminal operation that controlled opium distribution, gambling, and political patronage. By the 1870s, highly organized Chinatown gangs had established a notable street presence as a result of their active involvement in extortion, robbery, debt collection, and protection of Asian-owned vice businesses. Sante (1991) describes how the Chinese immigrants established this influence early with no opposition, merging "the functions, resources, and techniques of politicians, police, financiers, and gangsters (p. 226).

Philadelphia and Boston could also lay claim to having substantial street gangs before the Civil War. Adamson (2000) asserts that Philadelphia's *Public Ledger* identified nearly 50 Philadelphia gangs between 1840 and 1870. These gangs persisted. Davis and Haller (1973) document how Blumin, a *New York Tribune* reporter, described the northern suburbs of Philadelphia in 1948 and 1949 as swarming with gangs. However, these gangs do not appear to have been well organized and certainly not as ferocious as the New York gangs.

Third Period of New York City Gang Growth

Beginning in the 1930s, the most intensive gang activity in New York City shifted from downtown (Manhattan) to both northern (Harlem and the Bronx) and southeastern (Brooklyn) locations in the metropolitan area. German and Irish Catholics already populated the northern areas of the city, and rural southern Italians arrived at the turn of the century, to face ethnic hostility from English Americans and the Irish in particular. Gangs were visible in East Harlem and that area would be a gang hot spot, although when they formed there is uncertain. Bourgois (2003) suggests that it was quite likely by the early 1900s, growing out of ethnic Irish and Italian clashes. Soon more African Americans would arrive, in the Great Migration of Blacks from the rural South northward between 1910 and 1930, making up 14% of New York City's population by the end of that period. By the time of World War II, Harlem was one of the first Black ghettos in America, and "the area could not have been riper for the sprouting of street gangs" (Haskins, 1974, p. 80).

More fighting gangs took root after the arrival of Latinos (from Central America, South America, and the Caribbean) in the 1930s and 1940s, who settled in areas of the city populated by European Americans—particularly in East Harlem, the South Bronx, and Brooklyn. Bourgois (2003) describes how the largest new group of immigrants, 1.5 million Puerto Ricans, "were wrenched from sugar cane fields, shantytowns, and highland villages to be confined to New York City tenements and later to high-rise public housing projects in the two decades following World War II" (p. 51).

By the 1940s, three-way race riots involved Italian Americans, Puerto Ricans, and African Americans in Harlem (Bourgois, 2003).[1] Ethnic succession was a key precipitating factor. With the massive influx of Puerto Ricans, the previously fashionable countryside retreat for wealthy New Yorkers called East Harlem, later dubbed Italian Harlem, would soon be known as El Barrio, and informally as Spanish Harlem in reference to Puerto Ricans' Spanish culture and language heritage. (*El Barrio* in New York City now refers specifically to East Harlem.)

In East Harlem, adolescent gang fights between Italians and Puerto Ricans were depicted vividly in Leonard Bernstein's classic musical *West Side Story*. Gang members appeared younger than in the past, and primarily non-White (Black, Mexican American, and Latino).[2] Haskins (1974) contends the new gangs were more organized, better armed, and often involved in drug activity. Following another wave of Black migration beginning in the 1950s, some Black gangs were very prevalent in East Harlem and other segregated communities in New York City. By 1950 there were 800,000 Blacks in the city (Haskins, 1974).

Overall, by this time, Gannon (1967) reports more than two-thirds of the New York gangs were Puerto Rican or Black. During this same period, a surging Hispanic/Latino population succeeded Whites across New York City, creating a preponderance of both all-minority and multiethnic neighborhoods (Lobo et al., 2002). More serious gang fights ensued in the late 1950s, following a "slum clearance" project that moved several thousand poor Puerto Rican and African American families into high-rise public housing in East Harlem, making it what Bourgois (2003) calls "one of the most concentrated foci of dislocated poverty and anomic infrastructure in all of New York City" (p. 65).

Throughout the 1940s to 1970s, a mixture of youth gangs remained in both the northern and southern areas of New York City. In the southernmost sections—Brooklyn, the South Bronx, and Chinatown—a variety of gangs had emerged (M. L. Sullivan, 1993). Originally composed of Italians and Irish, the newer gangs in the South Bronx and Chinatown were populated by much older members involved in organized crime, and the Brooklyn (Sunset Park, largely poor Puerto Rican Latino) gangs were classic fighting youth gangs that drew widespread media attention for the large number of murders. The 1970s and 1980s brought another large wave of migrants to the United States, around 7 million people, respectively (Pincus & Ehrlich, 1999). Sullivan (1993) asserts that during the 1980s, many of the new immigrants into Brooklyn were Asian and non–Puerto Rican Latinos, especially Dominicans followed by Central and South Americans. The newer Hispanic groups began to succeed Puerto Ricans. "In fact, by the late 1990s, Hispanics had replaced Blacks as the largest minority group in the city" (Lobo et al., 2002, p. 704). Overall, according to Bourgois (2003), 51% of Harlem residents were Latino/Puerto Rican, 39% were African Americans, and the remainders (only 10%) were Whites and other groups.

Modern-Day Eastern Gangs

New York City is no longer the epicenter of serious street gang activity in the Northeast, as was the case in the early 1900s. In the 1980s and 1990s, the New York Police Department (NYPD) successfully thwarted drug trafficking and associated violence among New York City's gangs (Hagedorn & Rauch, 2007). Nevertheless, many gangs remain active in the city. In the 1990s, post–World War II urban renewal, slum clearances, and ethnic migration pitted gangs of African American, Puerto Rican, and Euro-American youth against each other in battles in New York City to dominate changing neighborhoods and to establish and maintain their turf and honor (Schneider, 1999). The Federal Bureau of Investigation (FBI, 2009) shows a similar trend evident in the broader Northeast region, with increasing gang-related violence as a result of competition among gangs for control of territories.

Gradually, gang activity in this region expanded to include other New England states, particularly Pennsylvania, New Jersey, and Connecticut. According to the FBI's (2009) intelligence reports, "the most significant gangs operating in the East Region are Crip, Latin King, MS-13, Ñeta, and United Blood Nation" (p. 16).

Street Gang Emergence in the Midwest

Chicago emerged as an industrial hub between the Civil War and the end of the 19th century. The city's capacity to produce gangs was enhanced when it recruited a massive labor force from the peasantry of southern and eastern Europe, becoming what Finestone (1976) calls "a latter-day tower of Babel" (p. 6). Gangs that flourished in Chicago grew mainly from the same immigrant groups that populated the early serious street gangs of New York City. Polish and Italian gangs were the most numerous early gangs in Chicago.

First Period of Chicago Gang Growth

Chicago's street gangs developed among White immigrants along ethnic lines, mainly Irish, German, and Lithuanian (Thrasher, 1927/2000). While these were merely mischievous groups at first, by the 1860s, more menacing gangs of Irish and German youth had clubrooms in the basement of saloons. By the 1880s, large Irish gangs (e.g., the Dukies and the Shielders) exerted a powerful influence on the streets around the stockyards—robbing men leaving work, fighting among themselves, and terrorizing the German, Jewish, and Polish immigrants who settled there from the 1870s to the 1890s. These gangs fought constantly among themselves, but they occasionally united to battle nearby Black gangs. The Black immigrants had arrived following the U.S. Civil War, to escape the misery of Jim Crow laws and the sharecropper's life in the southern states. But serious Black gangs likely did not appear until the 1920s, and Perkins (1987) asserts that "the impact of Black street gangs on the Black community was minimal, at best, prior to the 1940s" (pp. 19, 25).

Soon street gangs became entrenched in the patronage networks operated by ward politicians (Adamson, 2000), and the city's gangs "thrived on political corruption" (Moore, 1998, p. 76). Diamond (2005) further asserts that they stuffed ballot boxes and intimidated potential voters to ensure the elections of their political patrons. Moreover, Arredondo (2004) claims the reigning Irish gang, Ragen's Colts, in marking the racialized boundaries of "their" space, attacked both Mexican American and Black youth. Other White gangs also patrolled that area. "Reportedly, young Irish men, particularly on the east side of the yards, applied violent tactics similar to those of Ragen's Colts, waylaying Mexicans and beating them up" (p. 406).

Notorious criminals and rum-runners headed some of the early Chicago gangs (Thrasher, 1927/2000) and organized crime groups were very prevalent. Peterson (1963) identifies the Al Capone gang as most notable, while McKay describes how street gangs thrived "in the very shadow of these institutions" (p. 36).

In his 1927 book, Thrasher plotted on a map of the city the location of the 1,313 early gangs (with some 25,000 members) that he found in Chicago. This exercise revealed Chicago's "gangland." The heyday of Chicago's White ethnic gangs came to an end soon after Thrasher's research was completed, however. As Moore (1998) explains, "The gangs of the 1920s were largely a one-generation immigrant ghetto phenomenon" (p. 68).

Second Period of Chicago Gang Growth

Mexican American and Black gangs became prominent in the second period of Chicago gang growth. None of the Chicago gangs that Thrasher (1927/2000) classified in the 1920s were of Mexican descent and only 7% (63 gangs) were Black. This rapidly changed, beginning in the 1940s, after massive migration of both groups into Chicago.

Although some Mexican Americans already had a continuous presence in Chicago (Valdés, 1999), the first major wave of Mexican migration occurred during the years 1919 to 1939. According to Arredondo (2004), it was instigated by the revolutionary period in Mexico and new employment opportunities in Chicago, particularly in the meatpacking and steel industries. Soon, Mexican immigrants spread into two Chicago communities that had long been settled by the Irish, Germans, Czechs, and Poles (Pilsen and Little Village), wherein Spergel (2007) suggests Mexican American gangs grew to join the ranks of the most violent gangs in the city. By the 1940s, the Mexican migration into Chicago had swelled, and it reached 56,000 by 1960, prompting residents to dub the city as the "Mexico of the Midwest." Partly in response to what Diamond (2005) tags as growing racial and ethnic violence, Black, Puerto Rican, and Mexican American gangs proliferated in the late 1950s.

Between 1910 and 1930, during the Great Migration of more than a million Blacks from the rural South to the urban North for jobs, Chicago gained almost 200,000 Black residents (Marks, 1985; B. Miller, 2008), giving the city an enormous urban Black population—along with New York City, Cleveland, Detroit, Philadelphia, and other Northeast and Midwest cities. Cureton (2009) traces the origin of Chicago's serious Black street gangs to this peoples' segregated residency in inner-city areas, beginning in the early 1900s. Black gangs likely formed to counter the aggressive White youth, but these relatively unorganized groups at first were no match for the well-organized, all-White gangs that were centered in their athletic clubs. Perkins (1987) directly attributes the race riot of 1919—in which Black males united to confront hostile White gang members who were terrorizing the Black community—to gang formation. Diamond (2005) also believes Black, Puerto Rican, and Mexican American gangs alike proliferated in the late 1950s partly in response to growing racial and ethnic violence.

From 1940 to 1950, the Chicago Black population nearly doubled, from 278,000 to nearly 500,000 (B. Miller, 2008). Most of the immigrant Blacks in Chicago settled in the area known as the Black Belt (a 30-block stretch of dilapidated housing along State Street on the south side), where abject poverty was concentrated. To alleviate the housing shortage and better the lives of poor city residents, from 1955 to 1968 the Chicago Housing Authority (CHA) constructed 21,000 low-income family apartments (90% of which were in high-rise buildings), the best known of which is Robert Taylor Homes. Venkatesh (2000, 2008) reports that this complex consisted of 28 16-story buildings in uniform groups of two and three along a 2-mile stretch from the industrial area near downtown into the ghetto. Believed to be the biggest public housing project in the world, Robert Taylor Homes housed approximately 30,000 poor people, and 90% of the adults reported welfare as their sole source of support. In time, the public housing high-rises became gang incubators and drug turf battlegrounds. This setting not only provided a strong base for gangs, but also brought them into regular and direct contact. Gangs grew stronger in the buildings, and also in several instances took control of them, literally turning them into high-rise forts. Venkatesh (2008) notes, "Most remarkably, law enforcement officials deemed Robert Taylor Homes too dangerous to patrol" (p. 36) and this particular high-rise was widely recognized as "the hub of Chicago's gang and drug problem" (p. 37). Gang wars erupted, and Chicago's largely Black gang problem "exploded" in the 1960s, a period of increased gang "expansion and turbulence" in Chicago (Perkins, 1987, p. 74).

Cureton (2009) identifies three major Black street gang organizations formed in Chicago during the latter years of 1950s and early 1960s: the Devil's Disciples, P-Stones, and Vice Lords. Two of these gangs, the Vice Lords and the Black P-Stone Nation/Black Stone Rangers, were created in the Illinois State Reformatory School at Saint Charles (p. 353). Established in 1960, the Devil's Disciples gang splintered into three warring factions between 1960 and 1973: the Black Disciples, Black Gangster Disciples, and Gangster Disciples (p. 354). Gang wars occurred frequently among these large gangs in the late 1960s (Block, 1977; Block & Block, 1993). Perkins describes some of the implications:

> By the early sixties, Chicago's Black street gangs had grown to such proportions that they not only posed a threat to themselves but to the Black community as well. . . . [These gangs] were being perceived as predators who preyed on whomever they felt infringed on their lust for power [and] they turned to more criminal activities' and the control of turf became their number one priority. . . . by controlling turf, gangs were able to exercise their muscles to extort monies from businesses and intimate the Black community. (p. 32)

Perkins adds that those who dared resist often found themselves in confrontational situations that usually led to intimidation, threats, or bodily harm. "In fact, in some communities, it got to the point where being a gang member was the safest thing to do" (p. 32).

Racial unrest also contributed to rapid gang growth in Chicago. "The Civil Rights Movement was advocating nonviolence, racial pride, and unity. But black students who were having nonviolent demonstrations in the South had little influence on black street gang members [in Chicago] who were having their own distinctly more violent demonstrations" (Perkins, 1987, p. 29). The rise of the Black Panthers instilled Black pride, and their demise stirred resentment. Diamond (2001) describes how the Black gangs that were prevalent in Chicago in the 1960s "lived and acted in a world that overlapped with that of other youths [and the gang members] were surely participants in a street culture" that promoted racial empowerment and racial unity (p. 677). The youth subculture was a ready source of distinctive clothing, hairstyles, music, and other symbols including clenched fists.

Modern-Day Midwest Gangs

In the mid-1970s, Latino gangs, Black gangs, and Caucasian gangs in Chicago formed loose alliances such as the People and the Folk. The remaining gangs were independents and were not aligned with either of these groups. Until recent years, these alliances were respectfully maintained on Chicago's streets and the People and the Folk were strong rivals. "Now, although street gangs still align themselves with the People and the Folk, law enforcement agencies all seem to agree that these alliances mean little" (Chicago Crime Commission, 2006, p. 11). Nevertheless, Cureton (2009) explains, "the Chicago style of gangsterism stretches to Gary, Indiana, and Milwaukee, Wisconsin, where alliances are fragile enough to promote interracial mistrust and solid enough to fuel feuds lasting for decades" (p. 354). But Maxson (1998) contends that few, if any, of these situations can be attributed to gang expansion.

In Focus 1.1
Gang Names and Alliances

Chicago Gang Alliances and Supergangs

During the 1960s, a pattern of gang branches became popular in some cities, whereby a number of gangs adopted a variant of a common gang name. In Chicago in the 1960s, about 10 local gangs used the Vicelord name, including the California Lords, War Lords, Fifth Avenue Lords, and Maniac Lords. These gangs claimed to be part of a common organization—the Vicelord Nation—related to one another by ties of alliance and capable of engaging in centrally directed activity (Keiser, 1969).

In the mid-1970s, Latino gangs, Black gangs, and Caucasian gangs in Chicago formed two major alliances, the People and the Folk. The remainder were independents and not aligned with either the People or Folk. According to the Chicago Crime Commission (2009), the People and the Folk were formed in the penitentiary system by incarcerated gang members seeking protection through coalition building. The two alliances apparently carved out turf boundaries similar to agreements among modern nations. "Until recent years, these alliances were respectfully maintained on Chicago's streets and the People and the Folk were strong rivals....Now, although street gangs still align themselves with the People and the Folk, law enforcement agencies all seem to agree that these alliances mean little" (p. 11).

In the 1980s, Chicago's African American and Latino gangs claimed affiliations of various sorts including *families, coalitions, confederations, nations,* and *supergangs*. These were not intentional expansions, however. Rather, Venkatesh (2000) says "they followed from incidental contacts in prison and juvenile detention facilities, non-neighborhood recruitment, family movement, and seasonal residence patterns" (p. 216). Based on his research, Venkatesh notes that available data do not support the existence of well-coordinated supergangs in Chicago.

Los Angeles Gang Alliances

From the 1960s onward, the pattern of claiming a federated relationship with other gangs grew in popularity in Los Angeles. The most prominent of these were the Crips and Bloods—two rival gangs originally formed in Los Angeles—with locality designations reflecting neighborhoods in that city (e.g., Hoover Crips, East Side 40th Street Gangster Crips, Hacienda Village Bloods, and 42nd Street Piru Bloods). Many of the Bloods and Crips gangs or "sets" regarded one another as mortal enemies and engaged in a continuing blood feud. In succeeding years, hundreds of gangs across America adopted the Bloods and Crips names. A 1994 survey counted more than 1,100 gangs in 115 cities throughout the nation with Bloods or Crips in their names.

Sureños and Norteños

In the 1960s, Al Valdez (2007) reports a rivalry that developed in California's prison system between northern and southern gang members. Many Southern California gangsters aligned with La Eme, a southern-based Mexican prison gang. Southern California gangsters began using terms like *Sureño* (meaning southerner, often expressed as *Sur*) to refer to themselves, along with the 13th letter of the alphabet, *M*, which is *Eme* in Spanish. La Eme's recruitment efforts prompted the formation of Nuestra Familia, a second Mexican prison-based gang that recruited members from Northern California. Northern California street and prison gang members began to use the number 14 (representing the 14th letter, *N*) to identify with *Norteño* (the Spanish word for northerner). "Rival Southern California Hispanic street gangs then had one thing in common, they were enemies with anyone from Northern California. This rivalry would unite them in jail and in state prisons. The same was true for Northern Hispanic street gang members, except their enemy was any gang member from the south" (p. 139). The north–south barrier is north of the city of San Jose. From time to time, members of these gang alliances, and sets claiming allegiances, have been documented in other regions of the United States.

Primary sources: People and Folk: W.B. Miller, 2001, pp. 43–44; Sureños and Norteños: Al Valdez, 2007, pp. 137–143.

Traditional Chicago gangs still have the strongest presence in the Midwest region. The most recent chapter in Chicago gang history is the proliferation of gangs to suburban areas. By 2006, the Chicago Crime Commission (2006) reports 19 gang turfs were scattered around Chicago, throughout Cook County. Other cities in this region that have extensive gang activity include Cleveland, Detroit, Joliet, Kansas City, Minneapolis, Omaha, and St. Louis. According to the FBI (2009), Mexican American gangs with ties to the Southern California–based Mexican Mafia (La Eme) prison gang have established a presence in the central region and are attempting to expand their influence there. Hispanic Sureños' 13 members have also been reported in the region.

Street Gang Emergence in the West

The existence of the Mexican population in the United States dates back to the 16th century, when people of Indian, Spanish, Mexican, and Anglo backgrounds inhabited a broad region that was then northern Mexico and is currently the American South-west, encompassing parts of present-day Arizona, Colorado, Nevada, New Mexico, and Utah. Telles and Ortiz (2008) note that Mexicans constitute the largest contemporary immigrant group and also largest immigrant group in American history.

Ganglike groups are said to have first appeared in the West region as early as the 1890s (Redfield, 1941; Rubel, 1965). Both J. W. Moore (1978, 1991) and Vigil (1990, 1998), widely recognized experts on Mexican American gang origins, suggest that the precursors of urban gangs in the West region were the *palomilla* (meaning literally, flock of doves). These are best described as small groups of young Mexican men that formed out of what Rubel (1965) calls a "male cohorting tradition," first reported in south Texas in the early 1900s. These nascent gangs, according to Vigil (1998), grew within Mexican culture along the immigration trail that originated in Mexico and continued along a route through El Paso and Albuquerque, and onward to Los Angeles (Vigil, 1998). Seemingly coalesced under urban social pressures, the first Mexican Los Angeles gangs, which Bogardus called "boy gangs" in 1926, clearly were patterned after the palomilla (Bogardus, 1926, J. W. Moore, 1978; Vigil, 1990, 1998).

Mexican immigration was greatly accelerated as a result of the Mexican Revolution (1910–1920), Mexico's new rail system, and the labor needs of the Southwest and the Midwest. Telles and Ortiz (2008) documented these three factors combined to draw 700,000 legal Mexican immigrants to the United States from 1911 to 1930. The trail from Mexico to Los Angeles would come to resemble a well-traveled road, with a mul-tigeneration tradition of migration to and from Mexico and the United States. In fact, Vigil (1998) contends that events leading to the presence of Mexican street gangs in Los Angeles and the entire West region began long before the first gangs appeared there. Following the end of the war between the United States and Mexico, under the Treaty of Guadalupe Hidalgo (in 1848), the Mexican government ceded a large south-western region to the U.S. Mexican citizens in the area now known as California, Nevada, Utah, Arizona, and Texas; residents in parts of New Mexico and Colorado also became naturalized U.S. citizens. Even though they were naturalized citizens, the Mexicans became alienated in their own homeland. Valdez (2007) explains:

> Many Mexican street gang members felt—and still feel—that the United States stole this part of their country from their ancestors….They often were treated as second-class citizens and were told to go back to their home, Mexico. In their mind, they were home, but now [their homeland] was part of the United States because of the annexation. They were in a country where they were not wanted, but they could not return to Mexico because of their new status with the United States. (p. 94)

The unwelcoming surroundings prompted the Mexican Americans to form barrios and rural *colonias* (colonies) to protect and maintain ethnic traditions (Vigil, 1998).

First Period of Los Angeles Gang Growth

Two forces served to incubate street gangs of Mexican origin in Los Angeles and in other Western cities: physical and cultural "marginalization" (Vigil, 1988, 2002, 2008).

The barrios in which the earliest and most firmly established gangs developed were well-demarcated settlements of Mexican immigrants. J. W. Moore (1993) asserts that the Mexican American gangs in the barrios (neighborhoods) of East Los Angeles typically formed in adolescent friendship groups in the 1930s and 1940s, although Bogardus (1926) suggests the first ones appeared in the 1920s. Moore (1993) and Vigil (1993) both believe conflict with groups of youth in other barrios, school officials, police, and other authorities solidified them as highly visible groups. This intense bonding to barrios and gangs is unique to Los Angeles and other southwestern cities. Vigil and Long (1990) explain: "Each new wave of immigrants has settled in or near existing barrios and created new ones, [providing] a new generation of poorly schooled and partially acculturated youths from which the gangs draw their membership" (p. 56). Thus, isolationism and stigmatization were major contributing factors to gang growth and expansion.

In Focus 1.2
Gaining Admission to Mexican American Gangs

Admission of nonresidents into gangs can occur thorough kinship, alliance in fights, extensions of barrio boundaries, and forming branches.

Kinship. Gang membership is readily extended to relatives who live outside the barrio. For *Chicano* gang boys, a *homeboy* (fellow gang member) is the equivalent of a *carnal* (blood) brother. In the time-honored Mexican cultural norm, the gang takes on kinlike characteristics, especially mutual obligations among gang members.

Alliance. The defining characteristic of barrio gangs is fighting, particularly with another gang. "In essence, the gang is a group of boys who are allied in fights, and boys from other communities can be pressed into service" (Moore et al., 1983, p. 186).

Expansion of boundaries. Gang boundaries may extend into several barrios when members of multiple gangs live within them. Expansion typically works as follows: As a gang recruits one or two boys who live a block or so outside one of its boundaries, it becomes more difficult for a rival gang to defend that area. Thus the gang begins to claim that area as its own. Eventually, the rival gang ceases to claim its original turf and the successful recruiting gang has expanded into the area.

Forming branches. Extension from the home barrio by forming branches can mean forming *klikas* in noncontiguous areas. This typically happens when new Chicanos move into the general area.

Regardless of the method of bringing nonresidents into gangs, Moore and colleagues assert "expansion feeds upon itself" because "expansion through kinship and alliances enhances the feelings of mutual attachment and the fighting strength of the gang and lets it prosper, thus legitimating more non-residents and further expansion of territory" (p. 189).

Source: Reprinted from Moore, J.W., Vigil, D., & Garcia, R. (1983). Residence and territoriality in Chicano gangs. *Social Problems, 31,* 182–194. Copyright © 1983 the Society for the Study of Social Problems Inc. Reprinted with permission.

This history is not all that distinguishes Mexican American gangs from those in New York and Chicago. Another main difference Adamson (1998, 2000) notes is that gangs in these two cities emanated from conflicts with other racial/ethnic groups, whereas the first Mexican American gangs drew enormous strength from their own ethnic history. A second main difference is that the Mexican American gangs in the West region did not grow out of severe social disorganization, as appeared to be the case in the evolutionary history of New York and Chicago gangs. Generally speaking, Moore and Pinderhughes (1993) emphasize that poverty did not become as concentrated in Mexican American neighborhoods in Los Angeles as in Chicago or New York City. But, W. J. Wilson (1987) adds that later in Chicago, abject poverty and racism undergirded Mexican American gangs.

Second Period of Los Angeles Gang Growth

Following a hiatus during the Great Depression, Mexican immigration accelerated again, beginning in the early 1950s, bringing what Telles and Ortiz (2008) pinpoint as almost 1.4 million more persons by 1980. The Mexican-origin population in the United States grew from 2.5 million to 8.7 million during this period. The Los Angeles area received the most Mexican immigrants. Indeed, "Los Angeles has long been the Latino 'capital' of the United States, housing more people of Mexican descent than most cities in Mexico" (Moore & Vigil, 1993, p. 27).

Two other historic events proved pivotal in the growth of Mexican American gangs in the West: the Sleepy Lagoon murder and the zoot suit riots. Sleepy Lagoon was a popular swimming hole in what is now East Los Angeles. A Mexican youngster was killed there in 1942, and members of the 38th Street Mexican American gang were arrested and charged with murder by the Los Angeles Police Department. Unfortunately, the criminal trial resembled a "kangaroo court," in which five of the gang members were convicted and sentenced to prison. Al Valdez (2007) details the magnitude of this event: "Mexican street gangs changed forever because of these convictions. The jail sentences also acted as a glue to unite the Mexican community in a common cause, a fight against class distinction based on prejudice and racism, a fight against the establishment" (p. 98). The 38th Street gang members' cause continued in prison. They maintained their dignity and "demonstrated a type of gang pride and resolve never seen before. These behaviors also elevated the incarcerated 38th Street gang members to folk hero status in the Mexican community. The street gang members especially held them in high esteem" (p. 99).

The zoot suit riots had a similar unifying effect for Mexican Americans and fueled gang recruitment. Zoot suits were a fashionable clothing trend in the late 1920s and popularized in the nightclubs of Harlem. The exaggerated zoot suit included an oversized jacket with wide lapels and shoulders, and baggy pants that narrowed at the ankles, typically accompanied by a wide-brimmed hat. The style traveled west and south into Mexico and, most likely, was introduced into California via the El Paso Mexican street gang population. According to Katz and Webb (2006), by 1943, the Anglo community, the police, and the media began to view the zoot suiters as a savage group that presumably had attacked vulnerable White women and was also said to be responsible for several local homicides. Vigil (2002) elaborates that military personnel on leave and citizen mobs chased and beat anyone wearing a zoot suit—Chicano and Black youth alike—during a 5-day riotous period.

Third Period of Los Angeles Gang Growth

In this third stage, the development of Black gangs in Los Angeles follows a pattern that is similar to the emergence of Black gangs in Chicago. As in Chicago, Harrison (1999) shows a pattern of south-to-north Black migration in the 1950s, 1960s, and 1970s. Cureton (2009) believes this subsequently fueled the growth of Black gangs in Los Angeles, stating, "Southern Blacks were simply looking for a better life, and the West was considered the land of prosperity because of employment opportunities in factories" (p. 355). Instead, institutional inequality (in housing, education, and employment) and restrictive housing covenants legalized in the 1920s rendered much of Los Angeles off-limits to most minorities (Alonso, 2004; Cureton, 2009). Black residents challenged these covenants, leading to violent clashes between White social clubs and clusters of Black youth. Cureton explains, "Fear of attack from Whites was widespread and this intimidation led to the early formation of Black social street clubs aimed at protecting Black youths against persistent White violence directed at the Black community" (p. 664).

Alonso (2004) documents Black gang formation in Los Angeles principally in two phases: in the late 1940s and in the 1970s. Several observers report that Black Los Angeles gangs formed in the late 1940s, as a defensive response to White violence in the schools. Vigil (2002) reports the first racial gang wars to have occurred "at Manual Arts High in 1946, at Canoga Park High in 1947, and at John Adams Junior High in 1949" (p. 68). Quite likely, many of these Black gangs and others initially formed in the marginal areas of communities, typically close to Whites, which permitted the Black gangs to draw more members. "Black youths were thus able to vie with whites over the social space of the schools and entertainment areas" (p. 68). In the second phase, the effects of residential segregation (particularly in public housing projects), police brutality, and racially motivated violence in the aftermath of the 1960s civil rights conflicts "created a breeding ground for gang formation in the early 1970s" (Alonso, 2004, p. 659).

Unlike the Mexican American gangs, which were located in geographically restricted barrios, territorial boundaries were less important to the Black gangs, allowing them to encompass a wider area, creating gangs that were more confederations than single entities (Vigil, 2002). Thus, it is not surprising that the gangs that grew in the 1950s and 1960s were far more serious gangs than the earlier ones. Vigil (1988) explains that beginning as early as 1940, low-income housing projects helped to curb social problems for impoverished Los Angeles families, but these large-scale settlements also contributed to gang growth among Black and Mexican youths alike. Five such projects in East Los Angeles have become barrios in their own right. But Black gangs appear to have evolved principally out of Black–White racial conflicts, and Black clubs that played a central role in developing resistance strategies to counter White intimidation. "As white clubs began to fade from the scene, eventually the black clubs, which were first organized as protectors of the community, began to engage in conflicts with other black clubs. Black gang activity [soon] represented a significant proportion of gang incidents" across Los Angeles (Alonso, 2004, p. 665).

Cureton (2009) attributes the Black civil rights movement (1955 to 1965) to an underclass-specific, socially disorganized, and isolated Black community. Alonso (2004) elaborates:

> The end of the 1960s was the last chapter of the political, social, and civil rights movement by Black groups in LA, and a turning point away from the development of positive Black identity in the city....[But the] deeply racialized context coincided with the resurgence of new emerging street groups [between 1970 and 1972]. (p. 668)

According to Davis (2006), this occurred in large part because of poverty and high unemployment rates that were most prevalent amongst Black youth.

The emergence of a wide variety of street groups also expanded the base of Black gangs into two camps, Crips and Bloods, yet there are competing accounts of how Bloods and Crips gangs formed. Prominent among these is Cureton's (2009) research, indicating former Black Panther president Bunchy Carter and Raymond Washington formed the Crips in 1969 out of disappointment with the failure of the Black Panther Party to achieve its goals. According to Cureton, the Crips originally were organized to be a community self-help association; however, following Carter's death, the Crips's leadership shifted its focus to "drug (marijuana, PCP, and heroin) and gun (Uzi, AK-47, and Colt AR-15 assault rifles) sales that involved much violence and crime" (p. 356). Street gang feuds soon erupted. Neighborhood groups who opposed the Crips formed an umbrella organization to unify themselves.

Regardless of the disputed formation, the Crips and Bloods began to emulate the territory-marking practices that the early Los Angeles Mexican American gangs developed (Valdez, 2007). Crips wore blue clothing; the Bloods chose red. Both the Bloods and the Crips drew large memberships in the public housing projects built in the 1950s. Valdez (2002) reports Blacks made up nearly 95% of the membership of these two gangs, whose presence, according to Alonso (2004), quickly spread into other areas of South Los Angeles, including Compton and Inglewood. "Crip identity took over the streets of South L.A. and swept Southside schools in an epidemic of gang shootings and street fights by 1972," first involving 18 Black gangs, that multiplied to 60 by 1978 and to 270 throughout Los Angeles County by the 1990s (p. 669).

Mexican American gangs also steadily grew in number during this period, fueled by three historical developments: the Vietnam War, the War on Poverty, and the Chicano movement of the 1960s and 1970s (Acuna, 1981). Vigil (1990) contends that the Vietnam War depleted the barrios of a generation of positive role models. The ending of the War on Poverty eliminated jobs and increased marginalization. Perhaps most important, the Chicano civil rights movement "brought attention to the overall plight of the Mexican-American people, particularly long-suffering barrio populations" (p. 126). Moore and Vigil (1993) assert that its climax occurred with regionwide participation in the Chicano Moratorium march in East Los Angeles on August 29, 1970: "The event ended in death and destruction: a crowd dispersed by police and tear gas reacted by looting and destroying stores in the commercial section of East Los Angeles" (p. 37). In the meantime, major demographic shifts occurred throughout the greater Los Angeles area as another surge of Mexican immigrants that arrived in the 1960s joined the other Latino groups that began migrating to Los Angeles in the late 1970s:

> These first-generation residents have replaced heretofore Black and third-generation Latino ghettos and barrios, respectively; established new barrios in multistoried downtown apartments while making over the downtown shopping district; and generally transformed the spatial and social fabric of greater Los Angeles. (Vigil, 1990, pp. 126–127)

The Los Angeles gang culture soon began to draw the attention of youth in nearby cities. By the 1970s, according to W. B. Miller (1982/1992), street gangs had emerged in most populated areas across California. The Bloods became particularly strong in the Black communities in South Central Los Angeles—especially in places on its periphery such as Compton—and in outlying communities such as Pacoima, Pasadena, and Pomona (Alonso, 2004; Vigil, 2002). By 1972, Vigil (2002) reports there were 18 Crips and Bloods gangs in Los Angeles, and these were the largest of the more than 500 active gangs in the city in the 1970s.

Modern-Day Western Gangs

Southern California has produced four gang forms that have gained national prominence in the past two decades: (1) the traditional Black Bloods and Crips, (2) a mixture of prison gangs, (3) the highly publicized Mexican American 18th Street gang and Salvadorian Mara Salvatrucha gang—both of which are viewed by the media and federal agencies to be transnational gangs, and (4) Asian gangs.

By the 1980s, Black gangs had become "a major street force" (Vigil, 2002, p. xvi), and the gangs and other street groups melded the gang culture with crack dealing and consumption in the ghettos of South Central Los Angeles in the 1980s (Cockburn & St. Clair, 1998). This development expanded the visibility of both Black and Mexican American street gangs and quickly drew media (Reeves & Campbell, 1994) and police attention (U. S. Government Accountability Office, 1996). Today, this visibility has led, according to Valdez (2007), to many West Coast Black street gang members affiliating themselves with the Bloods or Crips. Many other gangs and naïve youth across the America mimic them and adopt their symbols and other elements of their gang culture. This phenomenon is equally pronounced—if not more so—among Mexican American gangs (Martinez, Rodriguez, & Rodriguez, 1998; Vigil, 2002).

Several prison gangs have gained notoriety in the West region in the past two decades. The Mexican Mafia is one of the oldest such gangs, having formed in California prisons in the late 1950s (FBI, 2009). (See Chapter 7 for a brief history of this gang and others in the West and Southwest regions.) Altogether, the Mexican Mafia is said to control approximately 50,000 to 75,000 California Sureños gang members and associates (GAO, 2010). The significant street and prison gangs operating in the West and Pacific regions (particularly California, Nevada, and Hawaii) are La Eme, 18th Street, Mara Salvatrucha, and Nuestra Familia (FBI, 2009). The most prominent Mexican American gangs among these are 18th Street, La Eme, and Nuestra Familia. The FBI considers the California-based Mexican Mafia (La Eme) to be one of seven major prison gangs. (The others are Aryan Brotherhood, Barrio Azteca, Black Guerrilla Family, Hermanos de Pistoleros Latinos, Mexikanemi—also known as Texas Mexican Mafia or Emi—and Ñeta.) The Government Accountability Office (GAO, 2010) classifies the main criminal activities of prison gangs as assault, carjacking, homicide, robbery, distributing illegal drugs within the prison system and on the streets, and extortion of drug distributors outside prison. Decker (2007) describes the incidence of prison crimes connected to the outside: "Many incidents in prison [are] linked to the street and many incidents of street violence [are] linked to prison violence" (p. 399).

Los Angeles gang culture produced two gangs that have been called *transnational gangs* and no other street gangs exceed them in generating widespread public fear. These are the notorious 18th Street Mexican American gang and Mara Salvatrucha, a Salvadorian Los Angeles gang. There is no single definition of a transnational gang. Franco (2008b) cites one or more of the following characteristics in various definitions:

- Such gangs are criminally active and operational in more than one country.
- Criminal activities gang members commit in one country are planned, directed, and controlled by gang leaders in another country.
- Such gangs tend to be mobile and adapt to new areas.
- The criminal activities of such gangs tend to be sophisticated and transcend borders. (p. 2)

For a gang to be considered transnational, Franco (2008b) suggests that it should have more than one of the preceding characteristics; however, this rule is not followed in much of the literature and media coverage, which characterizes gangs as transnational merely because they are present in more than one country. Both the 18th Street Mexican American gang and Mara Salvatrucha have been said to be involved to some extent with major drug cartels in Central America and Mexico, but more likely only along the U.S.–Mexico border. (These gangs are described in more detail in Chapter 7.)

The West region is also known for its Asian gangs that grew there in the 1990s and first decade of the 21st century among Filipinos, Koreans, Samoans, and Southeast Asians (Cambodians, Thais, and Vietnamese) (Valdez, 2007; Vigil, 2002; Vigil & Yun, 1990). Among these immigrant groups, the Vietnamese gangs seem to have drawn the most attention, because of their nonterritorial style, avoidance of monikers, and fluid structure (incessant changing membership). Vigil (2002) notes that these features contrast them sharply with typical Black and Mexican American gangs, while Al Valdez (2007) explains they also can be highly mobile, using their networked connections with one another to execute multicity and multistate robberies.

Street Gang Emergence in the South

The broad South region emerged much later than other regions as an important gang territory. First, it lacked a central large city that could have provided a springboard for gang growth. For many years, as Telles and Ortiz (2008) note, San Antonio was the only large city, but it was too isolated to extend its gang influence. Second, the early immigrant groups were dispersed across the area. Hence, W. B. Miller (1982/1992) concludes gang activity likely did not emerge in the southern states prior to the 1970s. Toward the end of that decade, only six southern cities reported gang activity— Dallas, Texas; Durham, North Carolina; Fort Worth, Texas; New Orleans, Louisiana; Miami, Florida; and San Antonio, Texas. Among these cities, only Miami and San Antonio were considered to have a moderately serious gang problem at that time. Actually, Dallas, San Antonio, St. Louis, Fort Worth, and Miami reported a greater problem with disruptive local groups than gangs in the 1970s.

However, before the end of the 20th century, Miller (1982/1992) claims the South region matched the other major regions in the prevalence of gang activity. Several southern states saw sharp increases in the number of new gang counties by 1995: Florida (23%), South Carolina (15%), Alabama (12%), and Texas (8%). From the 1970s through 1995, the South region led the nation in the number of new gang cities, a 32% increase, versus increases of 26% in the Midwest, 6% in the Northeast, and 3% in the West. In addition, gang activity emerged in multiple cities in a number of southern counties by 1995, including Dallas County, Texas (18 cities); Broward County, Florida (15 cities); Palm Beach County, Florida (11 cities); Dade County, Florida (8 cities); and St. Louis County, Missouri (6 cities).

Texas is particularly noted for its prison gangs. Vogel (2007) asserts the two best-known Mexican prison gangs are the Texas Syndicate and Mexican Mafia, and have been for decades. A third major Texas prison gang, Tango Blast, was established inside Texas's state prisons during the early 1990s to shield inmates from other prison gangs (FBI, 2009; Vogel, 2007). A unique feature of Tango Blast is its lateral organizational structure, which is divided up by cities, or hometowns, for expansive protection. The Houstone Tango Blast prison gang is the overall fastest growing gang in Houston, with a current membership of 2,759 documented members, and its membership extends into 12 of 17 nearby counties (Houston Intelligence Support Center, 2010).

Houston emerged in the past decade as a major gang center in the South. In addition to Tango Blast and Houstone Tango Blast, the Latin Disciples is another regional gang that operates in Houston, along with more than 20 local gangs that present a significant threat, according to the Houston Multi-Agency Gang Task Force (n.d.). Aside from Houstone Tango Blast, the Houston Intelligence Support Center (2010) identifies the next largest gangs in the city as the 52 Hoover Crips, 59 Bounty Hunters, Southwest Cholos, Bloods (general), Mara Salvatrucha, 59 Piru, Treetop Piru, and La Primera—each with more than 300 members.

Another Wave of Immigrant Groups

The Immigration and Nationality Act of 1965 ended the national quotas on foreigners in the United States. This led to a shift in immigration to the states, from European origins to Central and South America and Asia (Bankston, 1998). The next 25 years brought in many groups of Asians (Cambodians, Filipinos, Koreans, Samoans, Thais,

Vietnamese, and others) and Latin Americans (Colombians, Cubans, Dominicans, Ecuadorians, Mexicans, Panamanians, Puerto Ricans, and others) (W. Miller, 2001)— altogether about 16.6 million people of all nationalities (Pincus & Ehrlich, 1999). Native American gangs also would emerge much later (Bell & Lim, 2005; Major et al., 2004). In 1975, Miller concluded the majority of gangs in America were no longer White, with various European backgrounds. By the late 1980s, the children of many American-born or Americanized parents among the new immigrants, dubbed "the new second genera- tion" of the post-1960s immigrant groups (principally Asian and Latin Americans) had reached adolescence or young adulthood (Portes & Rumbaut, 2005; Portes & Zhou, 1993), and many of them joined gangs. Studies show that because of the successful assimilation of early European migrant groups into American society, gangs virtually disappeared by the third generation (Telles & Ortiz, 2008; Waters, 1999).

This has not been the case with Mexican Americans. With each generation, famil- iarity with the gang lifestyle increased and thus gang involvement grew, at least through the fourth generation (Telles & Ortiz, 2008). Overall, the fact that emergence of Black gangs occurred more rapidly in those three regions than in the South leads Alba and Ne (2003) to strongly suggest that failed or prohibited assimilation of Black families into American society was a main factor contributing to gang involvement. Continuous immigration has produced more hybrid gangs. Greater prevalence of *ethnic churning*— the process of racial/ethnic transition in a neighborhood and changes in the propor- tions of each racial/ethnic group (Pastor, Sadd, & Hipp, 2001)—surely contributes to more multiracial gangs that have been reported recently (Esbensen et al., 2008).

Hybrid Gangs

By the 1980s, law enforcement began reporting *hybrid gangs*. These are distin- guished in particular by racial/ethnic mixing (Valdez, 2007; Starbuck, Howell, & Lindquist 2001). No doubt, as Telles and Ortiz (2008) note, immigrant groups that came to the United States in the past 40 years have contributed significantly to racial/ ethnic mixtures seen in street gangs, with each successive generation. In a recent mul- ticity survey, Esbensen and fellow researchers (2008) found that about one-quarter of White and Black youth said their gangs had members with other racial/ethnic origins.

Hybrid gangs were first documented by David Starbuck, then a sergeant in the Kansas City Missouri Police Department Gang Unit, in the Midwest in the early 1980s (Starbuck et al., 2001). Once he began observing hybrid gangs in Kansas City, Starbuck came to the realization that the city was "a textbook example of a locality experiencing gang migration. Located in almost the geographical center of the continental United States, Kansas City [soon] had approximately 5,000 documented gang members and affiliates and numerous Chicago- and California-style gangs in the metropolitan area" (personal communication, May 5, 2009). Motivated to come to Kansas City for drug traf- ficking opportunities, they brought with them the gang culture from their cities of origin:

> Now hundreds of cities and towns across the United States report the presence of gangs that bear names of old time large LA gangs such as Rollin 60s Crip, Inglewood Family Gangster Bloods etc., or Chicago origin gangs such as Latin Kings and Gangster Disciples.

However, a large percentage of these gangs have little or no real connection to the original gangs and often put their own variations into the way they operate. (personal communication, May 5, 2009)

By the latter part of the 1990s, a *hybrid gang culture* was evident in a number of jurisdictions across the United States (Howell, Egley, & Gleason, 2002; Starbuck et al., 2001). Indeed, as Maxson (1998) reports, this is largely attributable to widespread population movement of families for social reasons. Egley, Howell, and Major (2006) find that a majority of law enforcement agencies report hybrid gangs across all population sizes, indicating a nationwide prevalence of these types of gangs. In addition, Al Valdez (2007) notes certain large-scale gangs even have a mixture of race/ethnicities such as the Los Angeles 18th Street gang. Chettleburgh (2003) concludes hybrid gangs are also common in Canada. Furthermore, Van Gemert, Lien, and Peterson (2008) document hybrid gangs in some European countries, with racial and ethnic migration along with diffusion of the gang culture, also a very important factor.

Jankowski's (1991) participant observation study of 37 gangs in New York, Boston, and Los Angeles over an 11-year period (1978–1989) found distinctive gangs of a wide variety of ethnicities including Black, Jamaican, Puerto Rican, Dominican, Mexican American, Central American, and Irish. Several of the gangs were composed of ethnic mixtures: Blacks with Puerto Ricans, Mexican Americans with Central Americans, and Irish with other Whites.

In sum, the U.S. hybrid gangs tend to have the following nontraditional features:

- They may or may not have an allegiance to a traditional gang color. In fact, much of the hybrid gang graffiti in the United States is a composite of multiple gangs with conflicting symbols. For example, Crip gang graffiti painted in red (the color the rival Blood gang uses) would be unheard of in California but have appeared elsewhere in the hybrid gang culture.
- Local gangs may adopt the symbols of large gangs in more than one city. For example, a locally based gang named after the Los Angeles Bloods may also use symbols from the Chicago People Nation, such as five-pointed stars and downward-pointed pitchforks.
- Gang members may switch their affiliation from one gang to another, and more than once, and existing gangs may change their names or suddenly merge with other gangs to form new ones.
- Although many gangs continue to be based on race/ethnicity, many of them are increasingly diverse. Esbensen and colleagues (2008) support this in their multicity U.S. study, in which 28% of White gang members and 27% of Black gang members said most or all of their gang members were not of the same race/ethnicity. A much smaller proportion of Mexican Americans (only 10%) said their gang membership was diverse in this respect.
- Southern California hybrid gangs, according to Al Valdez (2007), are characterized by mixed gender and mixed race/ethnicity. They are also less territorial and apt to not reside in the same area or even the same city. Their gang colors are nontraditional and they tend to wear a variety of tattoos.

The early American gangs were very homogeneous with respect to race/ethnicity of members and gang culture itself. This changed over time, owing to three main factors.

First, continuous immigration of peoples of varying nationalities naturally instigated some mixing. Second, the mobility of gangs themselves promoted diversity of gang culture, signs, symbols, clothing, and so on. Third, diffusion of the gang culture in popular media contributed to more cosmopolitan gangs.

Concluding Observations

Important differences in the history of gang emergence are apparent in the four major U.S. gang regions. First, the timing differed. Serious gangs first emerged on the East Coast in the 1820s, led by New York City. A half-century passed before gangs emerged in the Midwest (Chicago), while West (Los Angeles) regions saw significant gang development a full century later than New York City. The South would not experience significant gang problems for another half-century, in the 1970s. Second, the racial or ethnic composition of gangs in each region varied over time. In both New York and Chicago, the earliest gangs arose in concert with external migration of European origins—the traditional classic ethnics of the period from 1783 to 1860 (particularly German, French, British, Scandinavian). Other groups of White ethnics soon arrived during 1880 to 1920—mainly Irish, Italians, Jews, and Poles—and the second-generation youth were most susceptible to gang involvement. The latter nationalities almost exclusively populated the early serious street gangs of New York and Chicago. By the 1960s and 1970s, the gang composition had changed dramatically in both of these cities, with a far greater proportion of Black and Latino members. A new wave of immigrants, principally Asians and Latinos, was welcomed into the United States in the mid-1990s by permissive immigration policies. The story of gang involvement among their future generations is yet to be told because gang involvement tends to increase in each successive generation of immigrants.

The West region gang history contrasts sharply with that in the Northeast and Midwest. Western gangs never had a White ethnic history. Instead, for at least half a century, virtually all of the gangs were of Mexican descent. In contrast with New York and Chicago, street gangs in the West region appear to have emerged from aggressive groups of young Mexican men, age-graded Mexican *palomilla* cohorts, and nascent gang forms. These groups developed into boy gangs that were attached to barrios in Mexico and also in Los Angeles. In this region, a youth subculture that grew among the *cholo* (marginalized) youth provided the street lifestyle that supported gang formation. Extreme poverty appears to have been less important than cultural pride that arose as a result of extreme social and cultural isolation. This national pride has long been a characteristic feature of the Chicano gangs in the United States. However, the failed or blocked assimilation of Mexican American families into American society is likely the most predominant factor contributing to gang involvement.

Each of the three major gang regions also saw a pronounced second wave of Black gang development as a result of internal migration. However, it appears that the impact of this population shift from South to Northeast, Midwest, and West on gang emergence differs among the regions. Notably, Black gangs that developed in conjunction with this migration do not appear to have gained the foothold in New York City that they gained in the Midwest and West. Factors that might account for this difference are not readily apparent.

It also appears certain that the formation of civil rights and social movements in the 1960s had long-term impacts on street gangs in each of the regions. Several Black gang historians demonstrate that the resurgence of new emerging street groups in Chicago and Los Angeles coincided with the political, social, and civil rights movements, particularly the Black Panther Party and the U.S. Organization (both of which were racially motivated entities). But in Chicago, the ambivalence of civil rights leaders toward street gangs sent a mixed signal.

In Los Angeles, White street clubs that intimidated Black youth directly led to formation of similar Black clubs. It also is apparent that Los Angeles gangs were strengthened by the Chicano civil rights movement that drew attention to the overall plight of the long-suffering Mexican American people in barrios in multiple cities.

A common denominator of gang growth in New York City, Chicago, and Los Angeles is the policy of concentrating poverty in the high-rise public housing units—a remarkable urban planning blunder. In major cities, a high-rise public housing settings provided gangs with cohesion because it was a clearly identified and secure home base. But this policy was carried out far more widely in Chicago and Los Angeles than in New York City, and stimulated gang growth much more in the "Windy City" than either in New York or Los Angeles

Another key cross-region difference is that the early Mexican American gangs in the West region were not only populated by waves of newly arriving immigrants, but also by families with gang-ready youths. In the first phase of cultural diffusion, when they arrived in the United States, street gangs were already present in the barrios into which they moved. In the second phase, gang culture in Mexico was enriched by reverse migration. Children often came to the United States, stayed for a period, and returned home, having learned a gang culture. In turn, they introduced American gang lifestyle to younger youths in Mexico and Central America, so that in the third phase, the next generation of immigrants arrived in the states fully prepared for active gang involvement. To this day, gang culture in the West region is continually reinforced with wave after wave of immigrants from Mexico and Central America, suggesting an important point: regions' and cities' gang dynamics clearly differ in some respects; sweeping generalizations are ill-advised.

DISCUSSION TOPICS

1. Why do street gangs exist? What explanations does history suggest?

2. What role did racial/ethnic conflict, organized crime, and political corruption play in the development of street gangs? Which of these factors was more important in each region?

3. Why are immigration and internal migration important?

4. Why is the South region so different in its street gang history?

5. Why was high-rise public housing such an important contributor to street gang problems? To emergence or seriousness?

RECOMMENDATIONS FOR FURTHER READING

Conditions That Give Rise to Gangs

Anbinder, T. (2001). *Five points*. New York: Free Press.

Bernard, W. (1949). *Jailbait*. New York: Greenberg.

Hayden, T. (2005). *Street wars: Gangs and the future of violence*. New York: New Press.

Howell, J. C. (1998). Youth gangs: An overview. *Juvenile Justice Bulletin. Youth Gang Series*. Washington, DC: U.S. Department of Justice, Office of Juvenile Justice and Delinquency Prevention.

Short, J. F., Jr., & Hughes, L. A. (Eds.). (2006b). *Studying youth gangs*. Lanham, MD: AltaMira Press.

Riis, J. A. (1892). *Children of the poor*. New York: Charles Scribner's Sons.

Riis, J. A. (1902/1969). *The battle with the slum*. Montclair, NJ: Paterson Smith.

Ro, R. (1996). *Gangsta: Merchandizing the rhymes of violence*. New York: St. Martin's Press.

West, C. (1993). *Race matters*. Boston: Beacon.

New York City Gang History

Bourgois, P. (2003). *In search of respect: Selling crack in El Barrio* (2nd ed.). New York: Cambridge University Press.

Puffer, J. A. (1912). *The boy and his gang*. Boston: Houghton Mifflin.

Sante, L. (1991). *Low life: Lures and snares of old New York*. New York: Vintage Books.

Chicago Gang History

Arredondo, G. F. (2004). Navigating ethno-racial currents: Mexicans in Chicago, 1919–1939. *Journal of Urban History, 30*, 399–427.

Chicago Crime Commission. (1995). *Gangs: Public enemy number one, 75 years of fighting crime in Chicagoland*. Chicago: Chicago Crime Commission.

Cureton, S. R. (2009). Something wicked this way comes: A historical account of Black gangsterism offers wisdom and warning for African American leadership. *Journal of Black Studies, 40*, 347–361.

Dawley, D. (1992). *A nation of lords: The autobiography of the vice lords*. (2nd ed.). Prospect Heights, IL: Waveland Press.

Keiser, R. L. (1969). *The Vice Lords: Warriors of the street*. New York: Holt, Rinehart & Winston.

Lombardo, R. M. (1994). The Social Organization of Organized Crime in Chicago. *Journal of Contemporary Criminal Justice, 10*, 290–313.

Lombardo, R. M. (2002). The Black Hand: Terror by letter in Chicago. *Journal of Contemporary Criminal Justice, 18*, 394–409.

Perkins, U. E. (1987). *Explosion of Chicago's Black street gangs: 1900 to the present*. Chicago: Third World Press.

Los Angeles Gang History

Alonso, A. A. (2004). Racialized identities and the formation of black gangs in Los Angeles. *Urban Geography, 25*, 658–674.

Davis, M. (2006). *City of quartz: Excavating the future in Los Angeles* (2nd ed.). New York: Verso.

Martinez, R., Rodriguez, J., & Rodriguez, L. (1998). *East Side stories: Gang life in East L.A.* New York: PowerHouse.

Moore, J. W. (1978). *Homeboys: Gangs, drugs and prison in the barrios of Los Angeles*. Philadelphia: Temple University Press.

Moore, J. W. (1991). *Going down to the barrio: Homeboys and homegirls in change*. Philadelphia: Temple University Press.

Valdez, A. (2007). *Gangs: A guide to understanding street gangs* (5th ed.). San Clemente, CA: LawTech Publishing.

Vigil, J. D. (2002). *A rainbow of gangs: Street cultures in the mega-city.* Austin: University of Texas Press.

Vigil, J. D. (2007). *The projects: Gang and non-gang families in East Los Angeles.* Thousand Oaks, CA: Sage.

Major Gangs in all Regions

Federal Bureau of Investigation. (2009). *National Gang Threat Assessment: 2009.* Washington, DC: U.S. Department of Justice, Federal Bureau of Investigation.

Notes

1. *Ethnicity* merely reflects cultural differences, whereas *race* overlaps with ethnicity and "refers to a group that is defined as culturally or physically distinct and, furthermore, ranked on a social hierarchy of worth and desirability" (Telles & Ortiz, 2008, p. 23). Mexican Americans have an ambiguous status. "Although Mexican Americans are often referred to as an ethnic group and not as a race, they were referred to as the latter in earlier times and arguably continue to be referred to and treated as such in societal interactions today" (p. 24).

2. In deference to Black scholars who commonly use this term, we also do. Persons in the United States of Mexican descent who have established citizenship prefer to be called Mexican Americans (Telles & Ortiz, 2008).

2

Myths and Realities of Youth Gangs

Introduction

Because of their criminal activities and deliberate efforts to control the streets, gangs can engender enormous fear in community adults and youth (Lane & Meeker, 2000, 2003). But two groups in particular have a tendency to exaggerate the nature and seriousness of gangs: the broadcast media and the gangs themselves (Esbensen & Tusinski, 2007; Thompson, Young, & Burns, 2000). Moore (1993) explains, "Most typically, [the media stereotype] is that gangs are composed of late-adolescent males, who are violent, drug- and alcohol-soaked, sexually hyperactive, unpredictable, confrontational, drug-dealing criminals. . . . They are demonic, and all the worse for being in a group" (p. 28). In many communities, when gangs are enshrouded in images such as this, the determination of appropriate community responses can be thwarted.

This chapter presents several popular gang myths along with research that substantiates realities that contradict the myths, or at least brings them into serious question. Technically speaking, *myths* refer to beliefs that are strongly held and convenient to believe but are based on little actual information; they are not necessarily false (Bernard, 1992). Beliefs that are unequivocally false are properly labeled *fallacies*. Although useful, such a clear-cut distinction often cannot be made in reference to gangs because, depending on how they are defined, at least one exception may be found to every myth, thus the more inclusive term is used herein.

Felson (2006) argues that the gangs themselves complicate community action by creating myths as part of what he calls their *big gang theory*. The process often transpires as follows: Youths sometimes feel that they need protection on the streets in their communities. The gang provides this service. However, few gangs are nasty enough to be particularly effective in protecting youths. Hence, they need to appear more dangerous than they actually are to provide maximum protection. Felson observed that gangs use a ploy found in nature to maximize the protection they seek to provide. In order to scare off threatening predators, some harmless animals and insects will mimic a more dangerous member of their species. In turn, predators learn to avoid all species—both harmless and dangerous—that look alike. For example, Felson notes that the coral snake, an extremely dangerous viper, is mimicked by the scarlet king snake, which is often called the "false coral snake" because of its similar colors and patterns. Although the latter snake is not venomous at all, it scares off potential predators by virtue of its appearance.

In the late 1980s, Skolnick's (1989, 1990) interviews of imprisoned gang members from the California Crips and Bloods further reinforced the big gang theory and is a prime example of Felson's (2006) observation that gangs can grossly exaggerate their nastiness. The gang members said they were transforming themselves into formal criminal organizations to profit from the "crack cocaine epidemic."[1] They also claimed they were expanding their criminal operations across the country. Crime and gang "reporting waves" supported various myths. In short, the gangs' involvement in cocaine drug trafficking connections in California was overstated (Klein, Maxson, & Cunningham, 1991; Maxson, 1998).

The California gang members' story influenced public perceptions of gangs via broadcast media in several ways. The myth of formal organization is particularly important. The notion that gangs were becoming huge powerful criminal organizations—much like highly structured corporations—became widely accepted. This feature of the big gang theory was also promoted at every level of government, including the United States Congress, and in the executive branch of the federal government (see Howell, 2009). The story became gospel, and it was elaborated to embrace the notion that Southern California and Chicago gangs formed alliances in their respective regions and expanded across the United States. How adolescents and uneducated young adults could manage to execute such an operation never was explained.

Felson (2006) suggests that gangs use the same strategy, providing signals for local gang members for making their gangs resemble truly dangerous big city gangs. These standardized signals or symbols typically consist of hand signs, colors, graffiti, clothes, and language content. Gang members can display these scary signals at will to create a more menacing image. Employing a famous gang name will help them intimidate others. Once enough people believe their overblown dangerous image, it becomes accepted as reality.

In the late 1980s, a panel of professionals knowledgeable of youth gangs convened in Washington, DC, to assess the gangs–drugs situation. The panel concluded that juveniles and gangs were integrally involved in drug trafficking. "Juvenile involvement in gang activities is not new, but it has new and alarming ramifications due largely to competition for the lucrative illegal drug trade. This competition has led to increased recruitment of juveniles, indiscriminate killings, and the spread of gangs into suburban areas and small and midsize cities across the country" (McKinney, 1988, p. 1). Another conclusion drawn from the expert panel was that "recent—and frequent—news reports of gang violence are not simply media hype. Gang violence, often driven by the illegal drug trade, is real, and so are its victims" (V. Speirs, as cited in McKinney, 1988, p. 1).

A subsequent national conference concluded that "it is well known that gang members are key players in the illegal drug trade," and that "there is clear evidence . . . that the demand for drugs, especially crack cocaine, has led to the migration of Los Angeles gang members across the country" (Bryant, 1989, pp. 2–3). The threat drug-trafficking youth gangs represented to the nation seemed apparent. "The fierce circle of drugs, profits, and violence threatens the freedom and public safety of citizens from coast to coast. It holds in its grip large jurisdictions and small ones, urban areas and rural ones" (T. Donahue, as cited in Bryant, 1989, p. 1).

Consideration of Key Myths About Gangs

Misrepresentations of gangs in the print media have been well documented in four analyses covering articles published over the past four decades (Best & Hutchinson, 1996; Esbensen & Tusinski, 2007; W. Miller 1974a; Thompson, Young, & Burns, 2000). As Bjerregaard (2003) notes, legislators also sometimes foster overreactions to gangs with very broad laws that prescribe severe penalties for any type of gang involvement. Almost invariably though, newspaper accounts, popular magazine articles, and electronic media broadcasts on youth gangs contain at least one myth or fallacy. First, the leading newsweeklies and most major newspapers consider "gangs" to be a monolithic phenomenon and do not describe the diversity among distinctively different types of gangs, such as prison gangs versus drug gangs and youth gangs. Second, the demographic image of gang members as exclusively males and racial or ethnic minorities is perpetuated. Third, news outlets portray gangs as an urban problem that has spread to new areas, as part of a conspiracy to establish satellite sects across the country. Fourth, most gangs are characterized as hierarchical organizations with established leaders and operating rules. Fifth, the pervasiveness of violence is exaggerated. And the members themselves are prone to overstatements, for example, always claiming they were victorious in fights (Klein, 1995; Valdez, 2007).

Youth gang members, broadcast media, and politicians are not the only ones to promulgate misleading information about gangs—so do criminologists on occasion. A noteworthy example is the case in which two criminologists, Katz and Jackson-Jacobs (2004), suggested that gangs do not increase a youth's involvement in criminality; gangs do not raise the level of violence in some cities; and gang youths could not account for more violence in a city than other delinquent groups. For a brief period, some believed these well-intentioned scholars. But Krohn and Thornberry's (2008) comprehensive review of research on these topics corrected Katz and Jackson-Jacobs' understatements.

Myth I: Gangs Have a Formal Organization

A key premise of the big gang theory is that modern-day gangs are highly organized and function in a ruthless manner much like organized crime groups or drug cartels. A main reason why a gang appears to be more menacing than a mere collection or group of lawbreakers is that the term *gang* implies that its members are organized, commit crimes in groups, and are thus resolutely committed to violence and mayhem (McCorkle & Miethe, 2002).

Reality. A few street gangs have evolved into highly organized, entrepreneurial adult criminal organizations (Coughlin & Venkatesh, 2003; Papachristos, 2001, 2004). However, studies in a growing number of cities show that gangs are far less organized than expected. "Gangs," says Klein (1995), "are not committees, ball teams, task forces, production teams, or research teams. . . . They do not gather to achieve a common, agreed-upon end" (p. 80). In fact, very few youth gangs could meet the essential criteria for classification as "organized crime" (Decker, Bynum, & Weisel, 1998; Klein, 1995). As Klein notes (2004), "Organized crime groups such as drug cartels must have strong leadership, codes of loyalty, severe sanctions for failure to abide by these codes, and a level of entrepreneurial expertise that enables them to accumulate and invest proceeds from drug sales" (pp. 57–59). Such criminal gangs and organized crime networks are often highly structured.

In Focus 2.1
Cities That Have Gangs With Low Levels of Organization

- Chicago, Illinois (Decker, Bynum, & Weisel, 1998)
- Denver, Colorado (Esbensen, Huizinga, & Weiher, 1993)
- Detroit (Bynum & Varano, 2003)
- Cleveland and Columbus, Ohio (Huff, 1989, 1996, 1998; J. A. Miller, 2001)
- Kansas City, Missouri (Fleisher, 1998)
- Milwaukee, Wisconsin (Hagedorn, 1988)
- Las Vegas and Reno, Nevada (McCorkle & Miethe, 2002; Miethe & McCorkle, 1997)
- Los Angeles (J. W. Moore, 1978, 1991; Vigil, 1988, 2002)
- New York City (Kontos, Brotherton, & Barrios, 2003; M. L. Sullivan, 2006)
- Phoenix, Arizona (Zatz, 1987; Zatz & Portillos, 2000)
- San Diego, California (Decker et al., 1998; Sanders, 1994)
- San Francisco, California (Waldorf, 1993)
- Seattle, Washington (Fleisher, 1995)
- St. Louis, Missouri (Decker & Curry, 2000; Decker & Van Winkle, 1996; Monti, 1993)
- Washington, DC (McGuire, 2007)

In contrast, street gangs are generally loosely organized groups that are constantly changing—consolidating, reorganizing, and splintering (Monti, 1993; Weisel, 2002a, 2002b). Chapters 3 and 8 reference various street gang structures, and none of these resembles a corporate structure; typically only an informal division of labor with "shot callers" who play a key leadership role, and these may change from one gang activity to another. Tita, Cohen, and Endberg (2005) contend that gangs' public image and reputations are very large, yet their set spaces are very small, typically much smaller than neighborhoods or even census tracts—even, as Block (2000) notes, for very violent Chicago gangs.

Myth 2: Some Gangs Have 20,000 Members

Along with the big gang theory, gangs constantly exaggerate gang membership in interviews and in media reports. In particular, Perkins (1987) reports claims of 20,000 membership levels among some Black gangs in Chicago were made in the 1960s, when they called themselves "nations." In 2007, Al Valdez found some estimates of the Los Angeles–based 18th Street gang's membership are as high as 20,000 in California, other states, and a few other countries. The supposed rivalry between *supergangs* over drug turf is another component of the big gang theory, featuring the Crips and the Bloods in Los Angeles, and the People and the Folks in Chicago.

Reality. In reference to the large estimates of Black gang "nations" in Chicago, Perkins (1987) insists that "these figures were never substantiated and, no doubt, were exaggerated" (p. 64). Estimated membership in these popular gangs is based on the assumption that groups carrying the same name are somehow affiliated even though they may exist thousands of miles apart, in remote areas. Moreover, in each of these gangs, local sets or sections constantly fight with one another.

The same observation can be made of present-day Black Crip and Blood gangs and the 18th Street Mexican American gangs in Los Angeles. Two experienced Los Angeles gang investigators commented on this issue on the National Gang Center listserv. Sergeant Wesley McBride (2005), agreeing with fellow investigator Tony Moreno, said,

> There is no evidence of a Crip or Blood nation in California. They do not understand that concept of gang nations. Each gang is totally independent of other gangs. In Los Angeles there are over 200 Crips sets and maybe a 100 Blood sets, there is no common leader among any and they war on one another. [Most gang violence involves] Crips on Crips, Bloods on Bloods.

Thus, these warring sets in each megagang could hardly be considered tight-knit gang "alliances." Felson (2006) suggests that "the colossal gang is largely an illusion—many independent gangs just using the same name over wider space and longer time" (p. 308).

Myth 3: Gangs of the Same Name Are Connected

This myth—that big city gangs spawn small local gangs of the same name—is a key premise of the big gang theory and broadcast media presentations. Local gangs that call themselves Crips and Bloods, for example, are assumed affiliated with parent gangs of the same names in distant cities.

Reality. The common notion that local gangs are affiliated with big city gangs persists because of the similarity of their names and symbols, which mimicry or imitation explain. An analogy helps reveal the reality of the situation. Local Little League baseball teams may appear to be affiliated with major league baseball teams because of similar names and uniforms, but there is no connection between local youth teams and professional baseball clubs. So it is with gangs; there rarely is any connection whatsoever between local gangs and big city gangs known by the same names. The reality is that local gangs often "cut and paste" bits of Hollywood images of gangs and big-city gang lore into their local versions of gangs (Starbuck et al., 2001). And, they often do a poor job of this copying—perhaps using the wrong colors, distorting the original gang's symbols, and so on. To illustrate the point, Fleisher (1998) documents a gang of youth in Kansas City who said they were affiliated with the Chicago Folks gang, but when asked about the nature of their affiliation, they couldn't explain it. They said that they just liked to draw the Folks' pitchfork symbol.

Local gangs also like to create the impression that they are comprised of numerous "sets" or cliques, and as Felson (2006) suggests, to promote a nastier image. Rather than one big gang with many branches, most communities have several small gangs (discussed in Chapter 3), and even though some of them may use a common name there rarely is any connection between them. Notable exceptions are the Chicago-based Gangster Disciples (FBI, 2009) and both 18th Street and Mara Salvatrucha, Los Angeles–based gangs with connections to *maras* (gangs) they spawned in Central America (Cruz, 2010), as discussed in Chapter 7.

Myth 4: The 18th Street (M-18) and Mara Salvatrucha (MS-13) Gangs Are Spreading Across the United States

These two gangs developed in Los Angeles in the 1980s and 1990s, respectively. Since then, the FBI (2009) along with Franco (2008a) report membership in these gangs spread from the Los Angeles area and Central America to other communities across the United States, particularly Atlanta, Dallas, Washington, DC, and New York metropolitan areas. In 2008, according to Federal Bureau of Investigation (FBI) estimates, the MS-13 gang is active in 42 states and the M-18 gang is active in 36 states (Franco, 2008a), especially in Washington, DC, Virginia, and surrounding areas (Franco, 2008b). The FBI (2009) estimates membership in 18th Street "at 30,000 to 50,000 [in the U.S. and Central America and] in California approximately 80% of the gang's members are illegal aliens from Mexico and Central America," while the MS-13 gang "is estimated to have 30,000 to 50,000 members and associate members worldwide, 8,000 to 10,000 of whom reside in the United States" (p. 23).

Reality. The large estimates of membership in these two gangs cannot be verified. For one thing, typical immigration patterns are not separated from gang-related activity in these numbers. The Washington Office on Latin America (WOLA) conducted a brief evaluation and analysis of the characteristics, both local and transnational, of Central American gangs in the Washington, DC, area (McGuire, 2007). This study revealed that 18th Street, at that time, did not "have a strong presence in the DC area," but that Mara

Salvatrucha "does have a presence in the Washington area" (pp. 1–2). Although the WOLA research found that "the evidence supports the argument that [these] gangs are not a major public security issue in the Washington D.C. area," the study concluded that "Central American gangs do affect specific communities in a serious way, however, and they need to be addressed" (p. 40). Similar research to determine the actual presence of these gangs has not been conducted in other regions of the United States. Chapter 7 addresses the extent to which these gangs are transnational in scope.

Myth 5: Our Gangs Came From Somewhere Else

Gang migration refers to the movement of gang members from one geographic area to another (Maxson, 1998), and the gang-migration myth presumes that street gangs migrate across the country to establish satellite sets. Readers may have seen arrows superimposed on national maps to illustrate the supposed movement of gangs across the country. The most predominant myth is that they likely came to the local city or town to set up a drug trafficking operation. Another part of this myth is that urban gangs in the 1980s—particularly in large metropolitan areas—may have become less concerned with territorial concerns and turned to drug trafficking and other economic crimes (Bourgois, 2003; Fagan 1996; C. Taylor, 1990b).

Reality. Klein (1995, 2004) asserts that most youth gang problems are homegrown and gang *members* rather than *gangs* themselves tend to migrate. When families move, their gang-involved offspring usually move with them. This reality explains most so-called gang migration.

Mapped gang movement routes describe exceptions rather than the rule. The first recorded exception of such gang migration occurred when members of the Dirty Dozen, a Chicago gang of the 1920s, routinely migrated to Detroit whenever they needed to earn money. There they would rent a house, and earn good wages in the city's automobile factories. They frittered away money they earned, however—partying, drinking, gambling, and courting women. Adamson (1998) explains, "Even though they were making fabulous wages," a former gang member reported, "they did not save a cent, and finally came back to Chicago, broke" (p. 69).

More consistent with the reality, Maxson (1998) notes that gang networks and connections generally extend not more than 100 miles from the city of origin, and rarely further. Of course, there are exceptions. Fleisher (1995) found that some Compton and Hover Crips from Los Angeles moved north to Seattle, Washington, and set up drug trafficking gangs there. Other instances are noted elsewhere in this book. The National Alliance of Gang Investigators (2005) contends that a few gangs do have the capacity to expand into other regions, but McGuire (2007) along with Van Gemert, Peterson, and Lien (2008) debunk the notion of international migration of gangs. However, W. Miller (2001) claims some gang member migration occurred in conjunction with the enormous U.S. population shift during the 1980s and 1990s from metropolitan to suburban and rural areas. Maxson's (1998) research shows that the most common reason—in more than half the instances—behind the migration of gang members is social considerations, including family moves to improve the quality of life and to be near relatives and friends. Egley and Ritz (2006) substantiate that drug market expansion applies in not more than 2 out of 10 cases according to law enforcement sources.

Myth 6: Gangs, Drugs, and Violence Are Inexorably Linked

This myth is another product of the big gang theory that imaginative gang inmates told to researchers in the late 1980s. Their tales and the subsequent media accounts vividly described violent money-making gangs that intended to wipe out local drug dealers as they presumably marched across the country. The gangs-drugs-violence myth soon was revived again in the broadcast media (K. Johnson, 2006). In sum, the gangs-drugs-violence myth ties together three big gang theory components—(1) migrating gangs, (2) gang drug trafficking, and (3) the inevitable violence—wherever migrating gangs take their drug operations, either locally or to other cities.

Reality. The gangs-drugs-violence myth is a complex one that must be dissected in parts. The migrating gang notion is a key to the first part; the second one is gang control of drug trafficking; the third part is the related violence.

As explained previously, in clarifying the fourth myth, the migrating-gang myth has been refuted in two independent national surveys of law enforcement: in Maxson's (1998) work and by respondents in the National Youth Gang Survey (detailed in Chapter 8). Law enforcement officers do not view migrating gangs as the predominant factor contributing to gang violence. They claim that drug involvement and intergang conflicts are far more important factors.

Coughlin and Venkatesh (2003) state, "The consensus appears to be that drug trafficking is usually a secondary interest compared to identity construction, protecting neighborhood territory, and recreation" (p. 44). It may come as a surprise to many readers to learn that, although street gang members often are actively involved in drug sales, gang research confirms that *few street gangs control drug distribution operations.* While research does indicate that some drug distribution operations are managed by former youth gangs that transformed themselves into drug gangs or by drug gangs initially formed as such, further studies show most drug trafficking operations are managed by adult drug cartels or syndicates (Eddy, Sabogal, & Walden, 1988). Fagan and Chin (1989) note active street level groups include drug "crews" and "posses," while Klein and Maxson (1994) and J. W. Moore (1993) add traditional narcotics operatives. The groups may also include new adult criminal organizations formed in some cities to service the growing drug market.

Important distinctions between street gangs and drug gangs are shown in Table 2.1. Drug gangs are very common. For example, Braga, Kennedy, and Tita's (2002) assessment of homicide incidents in Baltimore identified some 325 drug trafficking groups, but few of them were youth gangs. However, gang member involvement at the level of street sales brings gangs into the mix, because their members very often use drugs and need to procure them. As Valdez and Sifaneck (2004) assert, the gang collectively encourages this and sometimes provides protection for its drug-selling members even though the gang itself may not benefit from the sales.

Table 2.1   Common Differences Between Street Gangs and Drug Gangs

Street Gangs	*Drug Gangs*
Versatile ("cafeteria-style" crime)	Crime focused on drug business
Larger structures	Smaller structures
Less cohesive	More cohesive
Looser leadership	More centralized leadership
Ill-defined roles	Market-defined roles
Code of loyalty	Requirement of loyalty
Residential territories	Sales market territories
Members may sell drugs	Members do sell drugs
Intergang rivalries	Competition controlled
Younger on average, but wider age range	Older on average, but narrower age range

Source: Klein, 1995, p. 132. Reprinted with permission from Oxford University Press.

Another important question surrounds the violence connection, whether street gang involvement in the drug business typically leads to comparable violence attended by strictly drug gangs and drug wars among cartels (Eddy et al., 1998; Gugliotta & Leen 1989; Leinwald, 2007). The reality is that it sometimes does; but, as discussed in Chapter 8, youth gang-related violence mainly emanates from other conflicts. As for homicides, studies in eight cities have shown that gangs account for a large share but the correlation between gang-related homicides and drug trafficking is actually very weak in the following cities:

- Baltimore, Maryland (Braga, Kennedy, & Tita, 2002)
- Boston, Massachusetts (Kennedy, Piehl, & Braga, 1996)
- Chicago, Illinois (Block & Block, 1993; Block et al., 1996; Curry & Spergel, 1988)
- Indianapolis, Indiana (McGarrell & Chermak, 2003)
- Los Angeles, California (Hutson, Anglin, Kyriacou et al., 1995; Klein, Maxson, & Cunningham, 1991)
- Minneapolis, Minnesota (Kennedy & Braga, 1998)
- St. Louis, Missouri (Decker & Van Winkle, 1996; Rosenfeld, Bray, & Egley, 1999)
- Stockton, California (Braga et al., 2002)

Street gang wars over market control sometimes produce a large number of homicides. Block and fellow researchers (1996) report one set of ongoing gang drug wars in Chicago that involved two "brother" gangs, the Black Gangster Disciples and the Black

Disciples; in another case, the Black P. Stones committed a substantial number of homicides in the course of their push to reestablish themselves in the drug market. But an astute gang violence researcher, George Tita, makes a revealing observation about the gangs-drugs-homicides intersection based on his Los Angeles study, that "even in situations where gangs, drugs, and homicides coincided, the motivation for those homicides was much more likely to stem from an argument over quantity/quality of the drugs, payment, or robbery of a drug dealer or customer than from two groups fighting for market control" (Tita, Riley, et al., 2003, pp. 5, 36).

Myth 7: All Gangs Are Alike

"A gang is a gang is a gang. Thus when the media discover that a gang indulges in a new horror (Satanism, for instance, or crack dealing), *all* gangs are assumed to indulge in the same behavior" (J. W. Moore, 1993, p. 28). Perceptions change when media characterize gangs in a more menacing fashion, as a group. S. Cohen coined the term *moral panic* in 1980, referring to circumstances in which the perceived threat from some group or situation is greatly exaggerated compared with the actual threat. For example, Welch, Price, and Yankey (2002) point out that media characterizations of "wilding" groups in New York City as "gangs" fueled panic. Similarly, Zatz (1987) shows how the blanket categorization of Chicano gangs as violent fighting gangs in Phoenix created a moral panic.

Reality. After studying more than 1,000 Chicago gangs, the first gang researcher, Fredrick Thrasher, observed in 1927 that "no two gangs are just alike; [there is] an endless variety of forms" (p. 5). This conclusion has not changed. Weisel (2002b) reports police respondents describe the typical gang in their jurisdiction as a loose-knit organization (45%) with no formal structure (47%). In another study, Starbuck and colleagues (2001) found many current gangs are described as having a "hybrid gang culture." As a general rule, the more structured gangs in larger cities with more longstanding gang problems are far more dangerous than others in less populated areas.

In many cases, both Fleisher (2002, 2006b) and McGloin (2005) assert that it makes more sense to view gangs as embedded in social networks. Several studies of social networks show the intertwining of gang members' social circles, often with members of other gangs and nongang youths alike. Miethe and McCorkle (1997) note that gang membership is by no means a "master status." Furthermore, McGloin suggests cliques of more or less cohesive subgroupings may help hold larger groups together. However, Decker (1996) and Klein (1995) tag their cohesive states as episodic.

Myth 8: A "Wanna-Be" Is a "Gonna-Be"

This myth speaks to the inevitability of becoming an actual gang member once a youth begins to display some affinity to gang culture. If a youth associates with gang members and toys with gang lifestyles, then joining is virtually presumed in accordance with this myth.

Reality. This is not so. Youth who associate with a gang do not necessarily become members. In a St. Louis study of middle school students across the city, Curry,

Decker, and Egley (2002) found more than half of the surveyed youngsters who reported never having been in a gang said they had engaged in at least one kind of gang involvement. More than a third of them had gang members as friends, nearly one-third had worn gang colors, nearly one-quarter had hung out with gang members, and one-fifth had flashed gang signs. In another study of a Florida sample of nearly 10,000 middle-school students, Eitle, Gunkel, and Gundy (2004) reported only 5% of the sample self-reported having joined a gang, but half of the nongang youths engaged in one or more behaviors that suggested "gang orientations": they had flashed gang signs, worn gang colors on purpose, drank alcohol or gotten high with gang members, or hung out with gang members. Al Valdez's (2007) multiyear San Antonio study focused on girls who continuously associated with gangs and gang members but never joined.

Myth 9: Children Are Joining Gangs at Younger and Younger Ages

No other gang myth is repeated in broadcast media more often than this one. The youngest reported gang member is said to be 4 years of age.[2]

Reality. Thrasher (1927/2000) first classified 18 children as young as age 6, although they were associated with what he called "child gangs" or play groups. Technically speaking, then, for a child to join a "gang" is nothing new. In the modern era, quite likely only children who are born into gangs (referred to as *blessed in*) by virtue of intergenerational traditions are actually bona fide gang members below age 10. Below this age, few children are sufficiently exposed to gangs (although Kotlowitz [1992] notes exceptions in Chicago public housing); and adolescents prefer not to hang out with children. As discussed in Chapter 3, gang joining typically begins during the transition from elementary school to middle school. It is at this point that Vigil (1993) deems children first experience some freedom from adult supervision, experience exposure to gangs, and are sufficiently alienated from parents and school to find them inviting.

Myth 10: Gang Members Spend Most of Their Time Planning or Committing Crimes

A popular notion about gangs is that they constantly and indiscriminately perpetrate violence. Bjerregaard (2003) explains how the media frequently "use narratives to help convey the danger associated with gang activities frequently relying on stories of drive-by shootings that killed innocent victims" (p. 175).

Reality. Sanders (1994) observed that, most of the time, the San Diego gangs were like any other group: going to parties, interacting with friends in social situations, and spending time at home with their families. Sanders concluded that "overall gangs did not spend most of their time being violent" (p. 63). Other researchers have made the same observation, on the basis of studies using varying methodologies. Klein (1995) summarized gang life as being "a very dull life. For the most part, gang members do very little—sleep, get up late, hang around, brag a lot, eat again, drink, hang around some more. It's a boring life; the only thing that is equally boring is being a researcher watching gang members" (p. 11).

Esbensen (2000) concurs: "For the majority of the time, gang youth engage in the same activities as other youth—sleeping, attending school, hanging out, and working odd jobs. Only a fraction of their time is dedicated to gang activity" (p. 2). W. B. Miller's (1966) well-trained street workers intensively observed seven gangs over a 2-year period and recorded some 54,000 behavioral sequences in 60 categories. Among these, only 3% related to assaultive behavior.

In addition, Maxson and Klein (1990) show that failure to distinguish *gang-motivated* crime from *gang-related* crime greatly exaggerates the extent of planned gang crime. The former term applies to crimes committed on behalf of the gang or in furtherance of a gang function; the latter term—the more general measure—requires only that a gang member was involved regardless of the type of crime or circumstances surrounding it.

In Focus 2.2
Gang-Motivated Versus Gang-Related Crimes

Gang-motivated crimes can be further divided into two categories: *self-directed* crimes that are initiated and organized by individuals or small groups of rank-and-file gang members, and *gang-directed* crimes commissioned or orchestrated by gang leaders or the gang as a whole. Finally, gang-motivated crimes can be understood in terms of *instrumental* actions that are intended to advance the material interests of the gang or its leaders, and *expressive* actions that show gang pride and demonstrate that the group is more fearless than its rivals by defending turf, avenging past injuries, and so on.

Consider the following four cases:

1. A gang member gets in a fight with another man who makes a pass at his girlfriend at a party.

2. A gang member assaults a young man affiliated with a rival gang who has ventured onto territory claimed by the subject's gang.

3. A gang member is asked by older gang members to go on a "mission" into enemy territory to find and attack rival gang members.

4. A gang member is asked by gang leaders to punish a witness who testified against another member.

All four cases would fall under the member-based definition of gang crime, but the first would not meet the motive-based definition because the fight had nothing to do with the gang. The third and fourth cases could be considered gang-directed incidents, but not the second, which was initiated spontaneously by the individual in question.

The fourth case could represent an instrumental effort to advance gang members' material interests by deterring witnesses from testifying against the gang. The third case, by contrast, depicts what is probably an expressive use of gang violence that is more likely to harm than to help gang members' material interests by generating further violence and drawing unwanted attention from law enforcement.

Tales of sophisticated criminal conspiracies and calculated use of violence dominate the public discussion of gang crime. But gang-directed, instrumental activities are the exception, not the rule. Descriptions of gang activity drawn from ethnography and survey research provide little support for the view that gangs are a form of organized crime.

Source: Greene & Pranis, 2007, p. 52. Reprinted with permission from Justice Policy Institute.

Myth 11: All Gang Members Are Minorities

This myth is mainly a product of broadcast media. Youth gang members are typically pictured as inner-city males, whether African American, Hispanic, Mexican, El Salvadoran, Vietnamese, Chinese, or another immigrant group (Esbensen & Tusinski, 2007).

Reality. The racial and ethnic composition of gangs varies considerably by locality— for example, gang members are predominantly White in poor primarily White communities and mainly African American in predominantly poor African American communities. Overall, Esbensen and Lynskey (2001) report 25% of school-aged adolescents in gangs are White and relatively similar proportions are African American (31%), Hispanic (25%), and of other racial and ethnic groups (20%). A subsequent survey by this research team found slightly higher participation among multiracial groups. Howell, Egley, and Gleason (2002) found the newest gang-problem areas (i.e., emerging within the past decade) had, on average, a larger proportion of Caucasian/White gang members than any other racial/ethnic group.

Myth 12: Racial Considerations Are Not Important

This view holds that gang formation and experiences are the same for minorities and nonminorities alike. "In other words," Perkins (1987) explains, "all street gangs are influenced by universal factors which transcend race" (p. 64).

Reality. "Although there are some common characteristics among gangs of all racial groups, Black street gangs have a unique and distinct character . . . Black street gangs are part of the Black experience and, therefore, must be examined within the context of slavery, racism, colonialism, and economic repression" (Perkins, 1987, p. 66). This principle also applies to Mexican American gangs. The emergence and growth of their gangs also is intertwined with their social and cultural history, as seen in Chapter 1.

Myth 13: The Gangs Overwhelm Youths

One supposed gang technique for controlling members is to overpower youngsters with initiation rituals—to indoctrinate them. This popular myth holds that to become a full-fledged member, without exception, youths who join a gang must participate in an initiation ritual, and perhaps commit a serious violent act against a stranger, chosen at random (Best & Hutchinson, 1996).

Reality. Much like legitimate adolescent and adult organizations, most gangs do require some ceremonial type of induction to demonstrate membership, courage, and loyalty to the gang, and Vigil (2004) asserts that gang initiations often require initiates to endure a character test in what are called "beat-downs" or "jump-ins." But requiring inductees to victimize innocent members of the public is extremely rare. Best and Hutchinson state, in sum, "The accounts of gang initiation rites promoted by contemporary legends can be regarded as melodramatic versions of press reports that routinely attribute violence to gang initiations" (p. 395). There are several versions of ritual-associated myths, which periodically circulate on the Internet in the form of "urban legends." For example, Fernandez (1998) recounts the flickered-headlights myth, which refers to a legend that gang members must drive after dark with their headlights turned off in order to choose victims. According to this myth, if an approaching motorist flashes his or her headlights at the gang members' car (presumably in a friendly attempt to alert the driver that the lights are off), the gangsters must chase down and kill the motorist. Saunders (2011) identifies another media-induced hysteria that gangs from Mexico are robbing women in Walmart parking lots. Each of these urban legends and many others are nothing but hoaxes that are quickly debunked by law enforcement and skeptical observers, and also at www.snopes.com.

Myth 14: Most Youths Are Pressured to Join Gangs

A commonly held notion about gang involvement is that youths are surely pressured to join gangs. Otherwise, why would youngsters become involved in these terrible groups?

Reality. Almost without exception, other youngsters recruit gang members. Although youths commonly report ordinary peer pressure (from friends who are in the gang and siblings), the reality is that the adult gang members' recruitment of youth is extremely rare. Just as children and adolescents are recruited into cliques, friendship groups, and gangs by peers who are members or interested in joining themselves, it typically is similar aged peers who exercise the most influence. As unlikely as it may seem, many youths who join very much *want* to belong to gangs, because gangs often are at the center of appealing social action—parties, hanging out, music, dancing, drugs, and opportunities to participate in social activities with members of the opposite sex. Other adolescents often look up to gang members because of their rebellious and defiant demeanor. For example, Wiist, Jackson, and Jackson's (1996) survey of Houston middle-school students revealed that the classmates whom they looked up to as peer leaders did not have the qualities one might expect: 1 in 4 had beaten or punched another person and nearly 2 in 10 had been in a gang fight.

Social interaction and a need for protection are main reasons that youths give when asked why they joined a gang. They want to feel safe and secure, and they want to be an integral part of the social scene. They may seek support that their own parents and family do not provide. The pressures they may feel to join the gang are usually associated with family relations and normal peer influences, or come from gang members who warn them that they may be without protection if they do not join—particularly in correctional institutions. Most youths can manage these circumstances without reprisal from other gang members (Decker & Kempf-Leonard, 1991).

The gang-joining process is generally similar to the manner in which most of us would go about joining an organization. It is a gradual process that may consume multiple years. A youngster typically begins hanging out with gang members at age 12 or 13 (even younger in some instances), and joins the gang between age 13 and 15—typically taking from six months to a year or two from the time of initial associations (Decker & Van Winkle, 1996; Esbensen & Huizinga, 1993; Huff, 1996, 1998; Vigil, 1993). But many associates never join.

Myth 15: Adults Recruit Many Adolescents to Join Gangs

It is widely believed that adult gang members apply pressure to children and adolescents to join gangs. Legislators and media reports often presume that sinister adult gang operatives are using their stealth to draw younger and younger victims into their clutches, often around schools, much like pedophiles. The illogic of this presumption is never explored in such broadcasts. A corollary view is that adult gangs recruit youngsters to act as runners in their lucrative drug trade. Because of these concerns, some state anti-gang laws include enhanced penalties for adults who recruit children into gangs.

Reality. Gangs sometimes apply peer pressure on recruits in the course of gang expansion (Moore et al., 1983). To be sure, and as Fleisher and Decker (2001) confirm, prison gangs actively recruit new members through threats, force, and protection offers; while Sheley and Wright (1995) believe this likely occurs more often than reported in juvenile correctional facilities, Petersen (2000) asserts natural formation of friendships in the latter settings is likely more common. Moreover, very few gang studies have documented the use of juveniles in drug running; the best example is likely Bynum and Varano's (2003) Detroit study. Interestingly, Hagedorn (1994) found Milwaukee's older gang members actually refused to allow juveniles to get involved in the drug trade, because of the dangers involved. More common, it seems, is the use of coercion by older gang members, "pressing younger dudes into taking the fall" when arrested (personal communication, Deborah Weisel, January 21, 2009).

Myth 16: Once Kids Join a Gang, They're Pretty Much Lost for Good

Gang involvement is seen as a permanent condition; once youths join a gang, there is no turning back. The grip of the gang is said to be permanent. This myth has its origins in the mystique of dominating gangs, first promulgated in the romantic movie *West Side Story,* with the claim that "Once you're a Jet, you're a Jet all the way, From your first cigarette to your last dying day" and also in the credo: "blood in; blood out."

Reality. Gang involvement is usually not a permanent condition (Decker & Lauritsen, 1996). Field studies, community youth interviews, and surveys of students find that "for many youth, actual membership in the gang is a short-term fling" (J. W. Moore, 2007b, p. x). Excluding cities with a large number of intergenerational gangs, multiple studies conclude approximately half of the youngsters who claim membership in a gang typically leave it within a year. In Rochester, New York, Krohn and Thornberry (2008) found half of the boys (50%) and two-thirds of the girls report being a gang member for 1 year or less; and only 22% of the boys and 5% of the girls remained members for 3 or 4 years. In Pittsburgh, Gordon and colleagues (2004) determined almost half (48%) of the boys were gang members for only 1 year, and just 25% for up to 2 years. Seattle researchers Hill, Liu, and Hawkins (2001) reported that 31% belonged for longer than 1 year, but only 1% belonged for 5 years. In Denver, Esbensen and Huizinga (1993) indicate 67% were a member for just 1 year and only 3% belonged for all

4 years. Interestingly, most (60%) of the active Denver gang members "indicated that they would like not to be a gang member and expected to leave the gang in the future" (p. 582). Nationwide data collected by Bjerregaard (2010) also show that gang involvement is "a transient phenomenon" among teenagers.

Adolescence is a time of changing peer relations and fleeting allegiances to both friends (Warr, 2002) and gangs (Decker & Curry, 2000; Fleisher, 1998). Involvement in a variety of peer groups is common during the adolescent period. However, multiyear and intergenerational gang membership is far more common in cities with long-standing gang problems, such as Chicago (Horowitz, 1983) and Los Angeles (J. W. Moore, 1978). Other studies in traditional gang cities also found that gang membership was a relatively temporary experience for the majority of gang-involved youths (Hagedorn, 1998; Klein, 1971; Short & Strodtbeck, 1965/1974; Vigil, 1988; Yablonsky, 1967). Prison gangs are a different matter, to be sure. Fleisher (1995) notes the "blood in; blood out" credo is shared among prison gangs, including the Mexican Mafia, La Nuestra Familia, Texas Syndicate, and Mexicanemi, for example. Other gangs have death penalty offenses which "include but are not limited to: stealing drugs or drug money for personal use, testifying in open court against another member, failing to kill someone after being directed to do it, or betraying gang loyalty" (p. 141).

Myth 17: The Gang's Here for Good

It is commonly believed that once gangs appear, they become a permanent fixture in communities. As J. W. Moore suggests (1993), this notion seems to be based on the view that gangs thrive only "in inner-city neighborhoods where they dominate, intimidate, and prey upon" innocent citizens (p. 28).

Reality. Howell and Egley (2005a) report national survey data showing that in cities with populations under 50,000, gang problems regularly wax and wane. In smaller areas with populations under 25,000, only 10% of the localities reported persistent gang problems. Having a gang problem is certainly not a permanent condition in sparsely populated areas. Moreover, in these smaller areas, gang problems are, comparatively speaking, relatively minor in terms of size (e.g., number of gangs and gang members) and impact on the community. Hence, Howell (2006) asserts the probability of permanent gang problems is far greater in the nation's large cities than in the smaller ones and in rural counties; although experiencing gangs forever once they appear is not by any means a certainty—even in large cities (discussed further in Chapter 7).

Myth 18: If Females Are Allowed to Join Gangs, They Are Not Violent

This myth also has its origins in sexist views of female gangs and female involvement dating from the beginning of gang research. In 1927, Thrasher suggested that "they lack the gang instinct, while boys have it" (p. 80). Later studies confined girls and their gangs to an auxiliary relationship to dominant male gangs. While it appears certain that the earliest gangs in America had few female participants, subsequent studies were slow to notice their increasing involvement. However, these accounts have been questioned for lack of validity (Chesney-Lind & Hagedorn, 1999).

Reality. Moore (1991) found that numerous independent female cliques were reported as early as the 1970s. Studies conducted in the past decade or so show that female gang members commit similar crimes to those committed by male gang members, although a smaller proportion of girls participate in serious, violent offenses, and may commit these offenses less frequently on average (discussed further in Chapter 6).

Myth 19: Sole Reliance on Law Enforcement Will Wipe Out Gangs

Because gangs are commonly believed to have come from somewhere else, it is presumed that law enforcement agencies can turn them away at the city or county borders, or remove all of them from the area by arrest, prosecution, and confinement. This deterrence strategy is called *gang suppression.*

Reality. When used as a single strategy, gang suppression tactics do not have a history of success. The Los Angeles Police Department (LAPD) has long been a leader in the use of these tactics. The most notorious gang sweep, Operation Hammer, was an LAPD Community Resources Against Street Hoodlums (CRASH) unit operation (Klein, 1995). It started in South Central Los Angeles in 1988, when a force of a thousand police officers swept through the area on a Friday night and again on Saturday, arresting presumed gang members on a wide variety of offenses, including existing warrants, new traffic citations, curfew violations, illegal gang-related behaviors, and observed criminal activities. All of the 1,453 people arrested were taken to a mobile booking operation adjacent to the Los Angeles Memorial Coliseum.

Most of the arrested youths were released without charges. Slightly more than half were gang members. There were only 60 felony arrests, and charges were filed on only 32 of them. As Klein (1995) describes it, "This remarkably inefficient process was repeated many times, although with smaller forces—more typically one hundred or two hundred officers" (p. 162). Incredibly, the Rampart CRASH officers, who were fiercely involved in fighting gangs, came to act like gang members themselves (Deutsch, 2000; Leinwand, 2000). Deutsch (2000) relays how the line between right and wrong became fuzzy for these officers as an "us against them" ethos apparently overcame them. Furthermore, Leinwand (2000) describes how the CRASH officers wore special tattoos and pledged their loyalty to the anti-gang unit with a code of silence. They protected their turf by intimidating Rampart-area gang members with unprovoked beatings and threats. Bandes (2000) depicts arrests of street gang members "by the carload" (p. M6). Rafael Perez, an officer in the Rampart Division who was arrested in 1998 for stealing cocaine from a police warehouse, provided testimony for CRASH officers' arrests when he implicated 70 officers in a variety of illegal activities: planting evidence, intimidating witnesses, beating suspects, giving false testimony, selling drugs, and covering up unjustified shootings.

The Operation Hammer incident illustrates how it is unfair and unrealistic to expect that law enforcement can succeed in extinguishing street gangs by the use of gang suppression tactics. Street gangs are a product of U.S. history and are homegrown.

Myth 20: Gang Members Are a New Wave of Super Predators

John DiIulio (1995a, 1995b) coined the term *super predator* to call public attention to what he characterized as a "new breed" of offenders, "kids that have absolutely no respect for human life and no sense of the future. . . . These are stone-cold predators!" (p. 23). Elsewhere, DiIulio and coauthors have described these young people as "radically impulsive, brutally remorseless youngsters . . . who murder, assault, rob, burglarize, deal deadly drugs, join gun-toting gangs, and create serious [linked] disorders" (Bennett, DiIulio, & Walters, 1996, p. 27). DiIulio (1995b) warned that juvenile super predators would be "flooding the nation's streets," coming "at us in waves over the next 20 years. . . . Time is running out" (p. 25). DiIulio (1995b, 1997) and J. Q. Wilson (1995) predicted a new "wave" of juvenile violence to occur between about 1995 and 2010, which they based in part on a projected increase in the under-18 population. The sharp increase in adolescent and young adult homicides in the late 1980s and early 1990s (Blumstein, 1995a, 1995b) was tied to the presumed new wave of juvenile super predators, and attributable partly to drug-trafficking gangs that presumably grew to profit from the so-called crack cocaine epidemic. Several popular magazines featured stories on the predicted crime wave, and many depicted on their covers young Black thugs—often gang members—holding handguns. Stories that played to readers' fears were common. The dire warnings of a coming-generation of super predators supported what Esbensen and Tusinski (2007) assert was a helpless feeling that the young minority gang-involved offenders were beyond redemption.

Reality. None of these assumptions proved to be correct. The new wave of super predators never arrived. Several researchers have debunked this myth and doomsday projections, and Howell (2009) concludes that the anticipated increase in juvenile violence was exaggerated. A new wave of minority super predators did not develop, nor did a general wave of juvenile violence occur. Rather than super predators, public policy analysts and the research communities attribute the dramatic growth in homicide largely to the availability of firearms—primarily handguns, the involvement of young people in illicit drug markets, and an increase in gang homicide (Block & Block, 1993; Cook & Laub, 1998; Howell, 1999).

M. Fishman (1978) first discovered *crime reporting waves*, which begin as crime themes that journalists develop, often from police sources, in the process of gathering information, organizing it, and selecting news to be presented to the public. Journalists routinely rely on one another for newsworthy crime trends. Fishman noted how additional media outlets ran a story after seeing the initial attention that it garnered and how the original story was embellished as it was repeated in another locality. In this manner, a crime theme spreads throughout a community of news organizations, as one media outlet after another repeats the story. Hence, a crime reporting wave may develop as the story is recounted, added upon, and often embellished.

An experienced gang researcher tested the veracity of a broadcast warning of a coming wave of gang activity in New York City. The media stories concentrated on large gangs that presumably were present in New York City, Los Angeles, and Chicago: Bloods, Crips, Latin Kings, and Netas. J. P. Sullivan (2006) explains, "The nationally famous gangs finally came to New York City in 1997, at least in name" (p. 22). Furthermore, "Many stories were told of the rituals supposedly associated with Blood membership" (p. 30). In the most widely circulated version, induction into the Bloods gang required recruits to commit random violent acts, such as slashing the face of a total stranger with a razor. In an embellished claim, the slashing victim had to be a family member, including one's mother. The crescendo of this hysteria was reached on Halloween, 1997: "As the day approached, rumors circulated throughout New York City that Halloween would be a day of a massive Blood initiation" (p. 30). Alas, "the mass slayings never occurred. The hysteria subsided, and the media lost interest" (p. 31). Data that Sullivan collected in the three neighborhood areas proved that nothing happened in the way of a surge in gang activity, only a spike in media reports.

Myth 21: Zero Tolerance of Gang Behaviors Will Eliminate Gangs From Schools

The term *zero tolerance* (ZT) means that certain behaviors will not be tolerated. ZT policies grew out of state and federal drug enforcement policies in the late 1980s (Skiba & Noam, 2001; Skiba & Peterson, 1999). By 1993, school boards across the country adopted ZT policies, partly in response to a spate of mass school shootings in the early 1990s. Originally intended to restrict drug use, gang involvement, and gun possession, ZT had evolved into an instrument to punish minor student misconduct (Skiba & Peterson, 1999). By the late 1990s, at least three-fourths of all schools reporting to the National Center for Education Statistics said that they had ZT policies in place for various student offenses, including bringing firearms or other weapons to school; gang activity; alcohol, drug, and tobacco offenses; and physical attacks or fighting.

These policies specify predetermined mandatory consequences or punishments for specific offenses. In the school setting, ZT is a disciplinary policy that sends this message by punishing all offenses severely, no matter how minor, and suspension from school is the most common punishment. There is no room for discretion. ZT policies are also called "one strike and you're out" policies. These policies are based on deterrence philosophy, and they originally targeted drug use, gang involvement, and gun possession.

Reality. Skiba and colleagues (2006) explain that zero tolerance policies administered in schools may serve to move many "disruptive" children and adolescents into the "highly rebellious pupil" category. Unfortunately, zero tolerance policies currently are the favored method of dealing with disciplinary issues, and an epidemic level of school suspensions prevails across the United States. According to the most recent federal data, 7% of all students are suspended each year (Snyder & Dillow, 2009), but these data seriously underestimate the actual number (Howell, 2009). For example, the North Carolina Department of Public Instruction (2011) reports 1 out of 10 North Carolina students are suspended short term at least once annually, with an average of 3 days per suspension. The high school suspension rate is even more astounding, with 1 in 6 students annually across North Carolina.

To enforce school-based zero tolerance policies, armed police are increasingly placed in schools; the end result of which appears mainly to send more children to juvenile courts for minor forms of misbehavior that should have been addressed as disciplinary matters (Figure 2.1). For example, the North Carolina Department of Juvenile Justice and Delinquency Prevention (M. Q. Howell, Lassiter, & Anderson 2011) reported more than 4 out of 10 NC court referrals came from schools in 2010. Yet "there has been no serious, sustained public consideration about whether the North Carolina taxpayer dollars annually expended to pay police officers to patrol middle and high schools create any educational or public safety benefits whatsoever" (Langberg, Fedders, & Kukorowski, 2011, p. 1).

Figure 2.1 Schoolhouse Zero Tolerance

Illustrator: Joe Lee. Reprinted by permission.

Zero tolerance policies can have a cumulative effect, as follows:

- The "difficult schools" with zero tolerance policies can increase future delinquency by imposing more severe sanctions (Kaplan & Damphouse, 1997).
- Suspension and expulsions from school often means that students are removed from adult supervision, and in turn, experience more exposure to delinquent peers, which can lead to delinquency onset (Hemphill et al., 2006).

- Delinquency involvement can increase gang membership and court referral (Esbensen & Huizinga, 1993; Hill et al., 1999; Thornberry et al., 2003).
- Teenagers who experience juvenile justice system intervention are substantially more likely than their peers to become members of a gang (Bernburg, Krohn, & Rivera, 2006).

Myth 22: Nothing Works With Gangs

There is a tendency to believe that the gang problem is too complex to be solved, or that prior attempts have been misguided. This myth was recently promoted in a book by two prominent gang researchers, Malcolm Klein and Cheryl Maxson (2006).

Reality. Klein and Maxson (2006) overlooked several programs that have demonstrated scientific evidence of effectiveness with gangs or gang members (which are reviewed in Chapters 9 and 10). One of the effective gang programs (the Comprehensive Gang Prevention, Intervention, and Suppression Program) demonstrated crime reductions in controlled studies in five cities (Chapter 10), specifically contradicting Klein and Maxson's critique of this particular program.

Concluding Observations

Gang problems are often difficult to assess, and gangs are often shrouded in myths, which can lead to ineffective community responses. For example, if it is believed that local gangs migrated from distant cities such as Los Angeles or Chicago, officials may assume that the newly arrived gang members can be driven out. If they and their families are established residents of the city, however, this approach is unlikely to work.

In the 1980s and 1990s, myths and stereotypes about gangs and gang members contributed to moral panic in America. In this state of moral panic, political and social leaders suddenly define a specific group of people as a major threat to our values and behavioral standards. A "war on gangs" was declared. The LAPD's Operation Hammer is a reminder of the futility of singular suppression strategies.

The myths and stereotypes, coupled with a lack of research to address their validity, contribute to our lack of ability to address the gang problem effectively. The first responsible step in every community that suspects it has a gang problem is an objective, interagency, and communitywide assessment to determine if in fact a gang problem exists, and if so, to identify the dimensions of the problem. Every effort must be made to discard preconceived notions in this assessment because many of these are based on gang myths. The results of an objective assessment of gang activity can be used profitably in the development of a balanced approach incorporating multiple strategies of prevention, intervention, and suppression.

DISCUSSION TOPICS

1. From your perspective, what is the most surprising gang myth?

2. Review several newspaper or magazine articles on gangs and see how many myths you can identify.

3. Why are gang myths so popular, and who benefits from gang myths besides broadcast media?

4. How can gang myths be countered?

5. Dissect the so-called crack cocaine epidemic. Did it actually happen? Why was it promoted? Why is it still considered to have happened by many criminologists?

RECOMMENDATIONS FOR FURTHER READING

Street Gang Involvement in Drug Distribution Operations

Decker, S. H. (2007). Youth gangs and violent behavior. In D. J. Flannery, A. T. Vazsonyi, & I. D. Waldman (Eds.), *The Cambridge handbook of violent behavior and aggression* (pp. 388–402). Cambridge, MA: Cambridge University Press.

Esbensen, F., Peterson, D., Freng, A., & Taylor, T. J. (2002). Initiation of drug use, drug sales, and violent offending among a sample of gang and nongang youth. In C. R. Huff (Ed.), *Gangs in America III* (pp. 37–50). Thousand Oaks, CA: Sage.

Gugliotta, G., & Leen, J. (1989). *Kings of cocaine.* New York: Simon & Schuster.

Howell, J. C., & Decker, S. H. (1999). The youth gangs, drugs, and violence connection. *Juvenile Justice Bulletin. Youth Gang Series.* Washington, DC: Office of Juvenile Justice and Delinquency Prevention.

Howell, J. C., & Gleason, D. K. (1999). Youth gang drug trafficking. *Juvenile Justice Bulletin. Youth Gang Series.* Washington, DC: U.S. Department of Justice, Office of Juvenile Justice and Delinquency Prevention.

Huff, C. R. (1989). Youth gangs and public policy. *Crime and Delinquency, 35,* 524–537.

Klein, M. W., Maxson, C. L., & Cunningham, L. C. (1991). Crack, street gangs, and violence. *Criminology, 29,* 623–650.

Leinwald, D. (2007). DEA busts ring accused of sending tons of drugs to US. *USA Today,* (March 1), A11.

Levitt, S. D., & Venkatesh, S. A. (2001). Growing up in the projects: The economic lives of a cohort of men who came of age in Chicago public housing. *American Economic Review, 91,* 79–84.

Papachristos, A. V. (2005a). Gang world. *Foreign Policy, 147*(March-April), 48–55.

Sanchez-Jankowski, M. S. (1991). *Islands in the street: Gangs and American urban society.* Berkeley: University of California Press.

Taylor, C. S. (1990a). *Dangerous society.* East Lansing: Michigan State University Press.

Taylor, C. S. (1990b). Gang imperialism. In C. R. Huff (Ed.), *Gangs in America* (pp. 103–115). Newbury Park, CA: Sage.

Venkatesh, S. A. (1996). The gang and the community. In C. R. Huff (Ed.), *Gangs in America* (2nd ed., pp. 241–256). Thousand Oaks, CA: Sage.

Venkatesh, S. A. (2008). *Gang leader for a day: A rogue sociologist takes to the streets.* New York: Penguin Press.

How Gang Members Spend Their Time

Hughes, L. A., & Short, J. F. (2005). Disputes involving gang members: Micro-social contexts. *Criminology, 43,* 43–76.

Klein, M. W. (2004). *Gang cop: The words and ways of Officer Paco Domingo.* Walnut Creek, CA: AltaMira Press.

Miller, W. B. (1958). Lower class culture as a generating milieu of gang delinquency. *Journal of Social Issues, 14,* 5–19.

Miller, W. B. (1966). Violent crimes in city gangs. *The Annals of the American Academy of Political and Social Science, 364,* 96–112.

Short, J. F., Jr., & Strodtbeck, F. L. (1965/1974). *Group process and gang delinquency.* Chicago: University of Chicago.

Spergel, I. A. (1995). *The youth gang problem.* New York: Oxford University Press.

Length of Gang Membership Period

Bendixen, M., Endresen, I. M., & Olweus, D. (2006). Joining and leaving gangs: Selection and facilitation effects on self-reported antisocial behaviour in early adolescence. *European Journal of Criminology, 3,* 85–114.

Esbensen, F., & Huizinga, D. (1993). Gangs, drugs, and delinquency in a survey of urban youth. *Criminology, 31,* 565–589.

Gatti, U., Tremblay, R. E., Vitaro, F., & McDuff, P. (2005). Youth gangs, delinquency and drug use: A test of selection, facilitation, and enhancement hypotheses. *Journal of Child Psychology and Psychiatry, 46,* 1178–1190.

Gordon, R. A., Lahey, B. B., Kawai, E., Loeber, R., Stouthamer-Loeber, M., & Farrington, D. P. (2004). Antisocial behavior and youth gang membership: Selection and socialization. *Criminology, 42,* 55–88.

Hill, K. G., Lui, C., & Hawkins, J. D. (2001). Early precursors of gang membership: A study of Seattle youth. *Juvenile Justice Bulletin. Youth Gang Series.* Washington, DC: U.S. Department of Justice, Office of Juvenile Justice and Delinquency Prevention.

Peterson, D., Taylor, T. J., & Esbensen, F. (2004). Gang membership and violent victimization. *Justice Quarterly, 21,* 793–815.

Thornberry, T. P., Krohn, M. D., Lizotte, A. J., Smith, C. A., & Tobin, K. (2003). *Gangs and delinquency in developmental perspective.* New York: Cambridge University Press.

Moral Panic

Jackson, P. G., & Rudman, C. (1993). Moral panic and the response to gangs in California. In S. Cummings & D. Monti (Eds.), *Gangs* (pp. 257-275). Albany: State University of New York Press.

McCorkle, R. C., & Miethe, T. D. (2002). *Panic: The social construction of the street gang problem.* Upper Saddle River, NJ: Prentice Hall.

St. Cyr, J. L. (2003). The folk devil reacts: Gangs and moral panic. *Criminal Justice Review, 28,* 26–45.

Notes

1. The crack cocaine epidemic is also a myth when generalized nationally, because no one has presented convincing empirical evidence that a nationwide crack cocaine epidemic in fact occurred (Hartman & Golub, 1999; Reeves & Campbell, 1994; Webb, 1999). Only a few cities can claim to have experienced such an "epidemic." In fact, a list cannot be generated with certainty (for the best effort, see Cork, 1999; Grogger & Willis, 1998; Golub & Johnson, 1997, but each of these studies was laden with methodological and measurement problems). Thus we are left with claims of a "crack cocaine epidemic" that could not be substantiated and was very short-lived if widespread elevated use occurred at all.

2. In thousands of print and broadcast reports of very young gang members over the past decade (A. Egley, personal communication, February 24, 2011).

Defining Gangs and Gang Members

Introduction

There is no single, universally accepted definition of a *gang* or a *gang member*. In fact, neither gang researchers nor law enforcement agencies can agree on a common definition, and as Spergel and Bobrowski (1989) report, a concerted national effort in the 1980s failed to reach a consensus between researchers and practitioners on what constitutes a gang, a gang member, or a gang incident. Thus, federal, state, and local jurisdictions in the United States tend to develop their own definitions. W. B. Miller (1982/1992) explains: "In general, police departments in large cities apply the term quite narrowly, police in smaller cities and towns less narrowly, most media writers more broadly, and most distressed local citizens very broadly" (p. 17). Nevertheless, rather explicit definitions of both gangs and gang members are essential for effective communitywide strategic planning. Reaching agreement on these matters is no small feat. As seen in Chapter 2, no other deviant group is shrouded in more mythic and misleading attributes than gangs.

Europeans are sometimes critical of American criminologists for the difficulties associated with defining gangs universally, and in turn, for the discordant findings that sometimes result (Hallsworth & Young, 2008; Pitts, 2008). However, researchers should not be criticized when gang members themselves have difficulty defining their gangs (Fleisher, 1998). When Petersen (2004) asked bona fide female gang members confined in a youth correctional facility to explain the difference between a gang and a peer group, "the majority of the young women in this study did not know how to respond . . . and most did not believe that major differences existed" (p. 30). In addition, observers may often fail to appreciate the diversity of gangs in the United States, and in other parts of the world as well.

This chapter first addresses the matter of defining gangs, and next explores ways to define gang members. The last section examines the demographics of gangs and gang members in the United States.

Defining Gangs

Early Gang Definitions

In 1898, educator Henry D. Sheldon conducted the first student survey that collected information on gang involvement. In Sheldon's survey of 2,508 students in five cities in different U.S. regions, children and adolescents themselves voluntarily provided written descriptions about organizations in which they participated. Sheldon noted that some of the students' organizations, which he categorized as "predatory," included organized *fighting gangs* (a term the students freely used to describe small groups that fought between schools or sections of a town or city) along with bands of robbers, hunting clubs, play armies, and other amusements.[1] Several of the predatory sorts of gangs in which the young students participated dabbled in property crimes and some fighting. For the most part, these were child-dominated play groups—although a few of them persisted throughout the teenage years, with the typical boys who claimed involvement in predatory groups aged 10 to 13. Sheldon (1898) explained the "predatory organization" as "the typical association of small boys. After twelve years of age boys transfer their interest from these loose predatory bands to more definitely constructed athletic clubs" (p. 428).

In the early 1900s, the interests of professional groups—including psychologists, sociologists, and youth workers—turned to the grouping behavior of very young boys. An intense interest in this matter developed and remained for almost half a century due to the emerging notion that child and adolescent grouping might well be instinctive. Several reports described the juvenile and amateur gangs. In fact, so many reports were published on "boy life" that a bibliography, written by R. T. Veal in 1919, contained as many as 2,000 references to works that appeared in print during the last decade of the 19th century and up to 1918.

Early research investigated organized gaming among children, and such young groups were often considered to be "gangs of boys," which often were loosely defined. Scott (1905) considered groups containing less than five boys to be too small to constitute a gang (these he called *chums*). His Springfield, Massachusetts, survey included boys ages 10 to 18 and revealed that 68% belonged to a group—which Scott referenced as a *gang*. Among these youngsters, two-thirds said that their gang had secrets, although this often was nothing more than a distinctive whistle or call-out. One-half of the group members said that their gang had some sort of unspecified initiation.

Other early 20th-century groups of children and juveniles evidenced a considerable degree of organization, many of which rightfully could be considered gangs. In 1912, at least 66 gangs were revealed in Puffer's extensive interviews and observations of boys under his direction as principal of a Massachusetts industrial school for wayward youth. These gangs usually had names—"the Hicks Street Fellows, The Bleachery Gang, Morse Hollow Athletic Club, Warf Rats, Crooks, Liners, Egmen, Dowser Glums" (1912, p. 30). Typically ages 10 to 16, the members tended to gather daily, often on a street corner, and they expressed a sense of ownership of that area along with their clubroom tents or camps. Puffer attributed a social purpose to the gatherings: "Boys from bad, broken, or inefficient homes are forced to provide their own social life, and the gang is their one instinctive reaction to their social environment" (p. 28). Two-thirds of the young groups Puffer studied had a leader, and most gangs had some form of initiation, "to test the new fellow's grit and strengthen his spirit of loyalty" (p. 35). These gangs also were typically governed by rules, particularly governing snitching and lying on one another, and standing by each other in trouble. Although many of the gangs to which Puffer's juvenile reformatory inmates belonged were comparable to play groups, a great deal of the participating youths were involved in mischievous deeds, fighting, stealing, and some criminal activities. "Apparently, then, some gangs at least were pretty thoroughly bad. On the other hand, some gangs proved to be almost as thoroughly good. Their members were real boys, but on the whole the gang was helping them to become worthy citizens and upright men" (p. 12). In the course of his research, Puffer is presumed to have provided the earliest gang definition:

> The gang is, in short, a little social organism . . . with a life of its own which is beyond the lives of its several members. It is the earliest manifestation . . . of that strange group-forming instinct, without which a beehive, an ant hill and human society would be alike impossible. (p. 38)

Puffer also characterized gangs as "for the boy one of three primary social groups [which include] the family, the neighborhood, and the play group; but for the normal boy the play group is the gang" (p. 7). While Puffer attributed gangs' existence to lack of parental supervision, it is nevertheless apparent that the prevailing view of that era—that boys and adolescents have an instinctive "grouping" tendency, sometimes leading to prolonged ganging together (Spaulding, 1948)—influenced Puffer's views. In particular, Puffer believed that boys begin to develop "the gang-forming instinct" about the age of 10 (p. 72).

Similarly, Thrasher (1927/2000) defined the gang as

> an interstitial group originally formed spontaneously, and then integrated through conflict. It is characterized by the following types of behavior: meeting face to face, milling, movement through space as a unit, conflict and planning. The result of this collective behavior is the development of tradition, unreflective internal structure, esprit de corps, solidarity, morale, group awareness, and attachment to a local territory. (pp. 18–19)

Thrasher's study was so comprehensive and widely respected that half a century would pass before more specific gang definitions were developed. It also is important to note that several recent studies confirm the importance—if not the universality—of conflict for gang formation and identity.

Various scholars have developed detailed criteria for the term *gang*, particularly the following:

> Gangs are groups whose members meet together with some regularity, over time, on the basis of group-defined criteria of membership and group-defined organizational characteristics; that is, gangs are non-adult-sponsored, self-determining groups that demonstrate continuity over time. (Short, 1996, p. 3)

[A gang is] any denotable adolescent group of youngsters who a) are generally perceived as a distinct aggregation by others in the neighborhood, b) recognize themselves as a denotable group (almost invariably with a group name), and c) have been involved in a sufficient number of delinquent incidents to call forth a consistently negative response from neighborhood residents and/or law enforcement agencies. (Klein, 1971, p. 13)

A youth gang is a self-formed association of peers united by mutual interests with identifiable leadership and internal organization who act collectively or as individuals to achieve specific purposes, including the conduct of illegal activity and control of a particular territory, facility, or enterprise. (W. B. Miller, 1982/1992, p. 21)

A collective of youths—most likely young adults (16 and over)—that has a discernible organizational structure, whose members recurrently interact and congregate in particular areas or neighborhoods, use collective and individual symbols for identification purposes, and engage predominantly in acts of violence (including threats and intimidation) and drug-related crimes. (Oehme, 1997, p. 67)

The two latter definitions were generated empirically, in a multicity U.S. survey and in a statewide North Carolina study, respectively. Oehme's definition reflects specific offenses in which local gangs were involved. Indeed, for law enforcement, gang definitions are usually described in the context of criminal activity. For example, the Chicago Police Department (Chicago Crime Commission, 1995) asserts that gang characteristics consist of a gang name and recognizable symbols, a geographic territory, a regular meeting pattern, and an organized, continuous course of criminality.

Currently, federal law defines the term *gang* as

an ongoing group, club, organization, or association of five or more persons: (A) that has as one of its primary purposes the commission of one or more of the criminal offenses [...]; (B) the members of which engage, or have engaged within the past five years, in a continuing series of offenses [...]; and (C) the activities of which affect interstate or foreign commerce. (18 USC § 521[a])

Concurrent to the definition, current federal law describes the term *gang crime* as

(1) A federal felony involving a controlled substance [...] for which the maximum penalty is not less than five years. (2) A federal felony crime of violence that has as an element the use or attempted use of physical force against the person of another. (3) A conspiracy to commit an offense described in paragraph (1) or (2). (18 USC § 521[c])

Also, current federal law describes the term *gang member* as

a person who: (1) Participates in a criminal street gang with knowledge that its members engage in or have engaged in a continuing series of offenses [...] (2) Intends to promote or further the felonious activities of the criminal street gang or maintain or increase his or her position in the gang. (3) Has been convicted within the past five years for: (A) An offense described in subsection (c). (B) A state offense [...] (C) Any federal or state felony offense that by its nature involves a substantial risk that physical force against the person of another may be used in the course of committing the offense. (D) A conspiracy to commit an offense described in subparagraph (A), (B), or (C). (18 USC § 521[d])

State law works alongside the federal definitions. The National Gang Center (2009) reveals 14 states having legislation that defines a *gang member*. Only 6 states have a list of criteria, some of which a person must meet to be considered a gang member. Of those, 5 states require that a person must meet at least two criteria to be considered a gang member. Kansas requires an admission of gang membership or three or more of its criteria.

State designations of *gang* are more widespread. The National Gang Center (2009) identifies 39 states and Washington, DC, with legislation that defines *gang*. Of these, 30 states and Washington, DC, define a gang as consisting of three or more persons; 23 states include a common name, identifying sign, or symbol as identifiers of gangs in their definitions; and 24 states refer to a gang as an "organization, association, or group." The term *criminal street gang* describes a gang in 22 states and Washington, DC, and 22 states define *gang crime/activity*, with 12 referring to gang crime as a "pattern of criminal gang activity," while 18 states enumerate the exact crimes that are to be considered criminal gang activity.

Three Common Characteristics of Early Gangs

Despite the varying scope of characteristics employed in legal and academic communities, and as discussed in Chapter 1, gang historians and researchers consistently observe three common traits of serious street gangs, regardless of their location. First, urban street gangs typically claim a turf or territory, and they defend it resolutely. Second, they fight and even kill others over the personal and collective honor of the group. Third, tough, aggressive, fighting postures provide the individual status that is often craved by gang members.

Claiming and defense of turf or territory. From the time of their origin in the United States, "control over turf has been the basis of street gangs' social honor (Adamson, 1998, p. 60). As early as 1902, while the earliest gangs in the United States emerged in the criminal and slum districts of the city, urban researcher J. A. Riis documented staking claim to a particular territory in the inner city of New York as a necessary means of survival. Under the worst slum conditions, territorial claims are inexorably linked with financial, human, and social capital. The generation of fear in competitors for these scarce resources, to prevent invasions by others, is a fundamental strategy. In the slum areas, the gang's very existence sometimes depended upon its capacity to stand its ground in areas of great population density and ward off incursions from hostile groups.

Although modern-day gangs are more mobile than those in times past, and some do not claim physical turf, there is evidence that the more established street gangs naturally persist in maintaining a *set space:* the well-defined gathering place where gang members come together as a social group (Tita, Cohen, & Endberg, 2005; Tita & Ridgeway, 2007). According to Tita and Ridgeway (2007), once established, this "set space serves as a niche within the greater community that allows the gang to survive and even flourish" (p. 215). Furthermore, as Tita (1999) asserts, the set space promotes fear in the community.

Honor-based conflict. "Gang ideology is simple. The members of these groups believe that they must respond to insults in kind. In theory, derogation of one of their members affects the collective honor" (Horowitz & Schwartz, 1974, p. 239). In addition, Papachristos (2009) tags "collective honor" as "a function of a group's cohesion and ability to fend off perceived threats" (p. 82).

At the individual level, for males in the Chicago Mexican American gangs Horowitz (1982) studied, honor is founded on three fundamental principles. First, it is humiliating for a man to be placed in a situation of inferiority, particularly in which he cannot deal with a situation alone. Second, a respected man protects the sexual purity of the women in his family and must be the sole source of support. Third, "a man of honor is particularly sensitive to actions that indicate a lack of respect, and he must be ready to redress any discerned slight on his claim to precedence. Honor is an expressive value requiring direct action" (p. 11).

Tough, aggressive, fighting postures. This characteristic provides gang members individual status among peers. According to Papachristos (2009), "One of the street code's most pervasive norms is that of *retribution,* a perversion of the 'golden rule' stipulating that personal attacks (verbal or physical) should be avenged" (p. 79). Showing "toughness" is a particularly important personal attribute, for two reasons. First, as Vigil (2010) explains, "physical appearance counts in a street world where protective posturing often deters potential aggressors" (163). Second, those who succeed in attaining and maintaining an image of toughness are rewarded with the admiration of fellow gang members and the street community in general. Ethnographic studies in several cities reveal that notions of personal honor are generally associated with acts of hypermasculinity—extreme aggressiveness and the use of violence and fear to protect one's reputation. Recent studies demonstrate that young women similarly experience the threat of violence and, in turn, work "the code" to mediate those threats (Jones, 2004, 2008).

Typologies of Gang Members and Gang Structures

In 1927, Thrasher first characterized the members of the typical gang as resembling three concentric circles, with the core members (the power structure consisting of the leader and his lieutenants) forming the inner circle, the full-time members or regulars in the second ring, and an outside ring composed of occasional members or hangers-on.

Later, Klein (1971) and Spergel (1990) described gangs or alliances among them as either "vertically" or "horizontally" structured. Gangs structured vertically implies a hierarchy that differentiates between leaders and followers or age groups who move from one level to another as they become older or otherwise earn higher status in the gang. Gangs structured horizontally have no such hierarchy as vertical gangs. Rather, these gangs consist of social networks that might be allied with other gangs as *nations* or *supergangs.*

Subsequent researchers introduced variations on this basic structure. For example, Vigil (1988) classified Hispanic gang members as regular members, peripheral members, temporary members (who typically join the gang later and remain a part of it for a shorter length of time), and situational members (who join in gang party activities but avoid the more violent activities whenever possible).

In Chicago, Hutchison and Kyle (1993) described Hispanic gangs in the city as having chapters (with each gang containing subgroups differentiated by age, typically between pee-wees, juniors, and seniors), and evidencing an intergenerational pattern. Tita and colleagues (2003) explain further: "Most gangs have some sort of hierarchy that includes 'shot callers,' or leaders who tend to be older and more isolated from day-to-day activity of the gang; 'shooters,' or those most likely to commit an attack against another gang; and 'active soldiers,' or those most likely to associate with a gang but not necessarily involved in attacking rivals. Most gang members are in the latter group" (p. 12).

In the mid-1960s, New York City Youth Board's street workers classified more than 200 New York City street gangs (containing an estimated 3,100 boys) as corner groups, social groups, conflict groups and "the thoroughly delinquent and pathological group" (Gannon, 1967, p. 121). Nearly 9 out of 10, the street workers agreed that distinctions between "core" and "peripheral" membership was valid, and that approximately half of them indicated that the gangs typically had "a significant relationship to some older or other group," mainly an older group (p. 122).

Many gangs have no clear age-differentiated structure. For example, Spergel and associates (2006) found that in Chicago, the Latin Kings had an estimated 1,200 members and the Two Six had some 800 members, and each gang was comprised of 15 subunits, or sections, that operated on the streets in their respective territories. Larger gangs in many cities are composed of age-differentiated segments, each bearing a name W. B. Miller (1975) asserts to denote its status (e.g., Pee-Wees, Juniors, Old Heads). Monti (1993) concurs: "The fundamental building block of gangs remains the age-graded set or clique of local youngsters" (p. 10).

In Focus 3.1
Ways of Classifying or Typing Youth Gangs

1. Age
2. Race/ethnicity
3. Gender (all male, all female, mixed)
4. Setting (street, school, prison, motorcycle)
5. Type of activity (social, delinquent, violent)
6. Purpose of gang activity (defensive, aggressive)
7. Degree of criminality (serious, minor, mixed)
8. Level of organization (scavenger, territorial, corporate; vertical, hierarchical, horizontal)
9. Stage of group formation or development (early, marginal, well-established)
10. Degree of activity (active, sporadic, inactive)
11. Level of member involvement (regular, peripheral, temporary, situational)
12. Group function (expressive, instrumental)
13. Drug involvement (use, selling, organized distribution)
14. Cultural development (traditional, nontraditional, transitional)
15. Gang purpose (social gangs, party gangs, serious delinquents, incipient organizations)

Source: Modified from Spergel, 1995, p. 79

Because of this diversity, several widely respected gang researchers have cautioned the field about defining and categorizing gangs in a potentially misleading way (Bookin-Weiner & Horowitz, 1983; Curry & Spergel, 1992; Horowitz, 1990; Short, 2006; Sarnecki, 2001; Sullivan, 2006), particularly if a given definition happens to reify certain gangs by attributing real features to their essential fictive existence. A few gang researchers "have shown how network analysis can transform perceptions of a 'gang problem' from a fuzzy inkblot into a comprehensive blueprint by producing a highly detailed and visual way of looking at a specific gang situation" (Papachristos, 2005b, p. 645). This perspective obviates the arduous task of settling on an academically pleasing definition for *gang*. Papachristos (2005b) offers an alternative:

> Rather, [one] can begin with the definition of a problem: Who are the most violent gang offenders? What does the current gang war look like? How are the various members connected with each other? What does the gang's structure look like, and how stable is it? Who is the core and peripheral members or groups? And where should effective strategies be directed? (pp. 645–646)

The main premise of the network approach to understanding gangs is this: "If gangs are important institutions or actors in specific contexts, patterns of interaction among groups or members should produce a structure that affects social behaviors of gang members, the corporate behavior of the group, and/or the communities they inhabit" (Papachristos, 2006, p. 103).

Shifting membership and an intermittent existence characterize many gangs, especially those with younger members. Because involvement in a variety of peer groups is common during adolescence, in many situations, gangs should be viewed as social networks rather than as bounded "organizations" (Fleisher, 2006b; Papachristos, 2005b). Youth drift in and out of these groups, and as Fleisher (1998) reminds us, even members may be unable to name all current members. The degree of gang involvement is thus difficult to specify.

Children who also are alienated from both school and parents are susceptible to delinquent group formation and participation. After school officials sanction them, these children begin to identify with each other and they may give themselves a name. The next step toward transformation of these delinquent groups into gangs is the adoption of elements of the gang culture, including distinctive clothing, colors, rituals, regular gatherings, and exclusion of other children (Fleisher, 1995).

In Columbus and Cleveland, Ohio, Huff (1989) observed neighborhood gangs formed in three ways, two of which were conflict situations. First, break dancing or "rappin" groups sometimes experienced conflict with other groups as a by-product of dancing, rappin, or skating competitions. "This competition would sometimes spill over into the parking lots of skating rinks, where members frequently had concealed weapons in their cars" (p. 526). Second, street corner groups evolved into gangs as a result of ongoing conflicts with other groups that gathered on the streets (Whyte, 1941). "These groups were more typical of distinctive neighborhoods, such as housing projects" (p. 526). Third, experienced gang members moved with their families to

Ohio from Chicago or Los Angeles. "These more sophisticated leaders often were charismatic figures who were able to quickly recruit a following from among local youths" (p. 527). In New York City, more than three-fourths (76%) of street workers observed clique formation first, typically around a number of delinquent activities (Gannon, 1967). In Milwaukee, gangs emerged in the 1980s, where economic restructuring was seriously affecting Black communities. In modern-day Great Britain, Pitts (2008) suggests that "the term 'gang' has had its widest currency in primary school playgrounds, where younger children have used it to describe their friendship groups" (p. 15).

There is considerable evidence of very young gang members in the United States, and this is by no means a new development. Two large-scale studies produce data that permit an assessment of the involvement of children in the late–childhood age block in gangs. According to Gottfredson and Gottfredson (2001), in the late 1990s, almost 6% of boys and 4% of girls in the last year of *elementary school* said they were gang members. In a nine-city sample of students in grades 6 through 9, Esbensen's (2008) research team revealed almost 2% of children under age 12 and 5% of 12-year-olds self-identified as gang members. However, the youngest gangs are small in size and likely don't survive long. When the entire group of surveyed students claiming to be active gang members was asked to describe features of their gang, 13% said their gang had five or fewer members, and 25% said it had been in existence for 1 year or less. The brief existence of gangs in less populated areas is also evident. Nationwide, Howell and Egley (2005a) found only 4% to 10% of gangs survive in small cities, towns, and rural areas.

A Spectrum of Gangs and Other Groups

Developing a gang definition that captures the younger gangs, yet excludes adult criminal organizations that are not considered youthful street groups, is challenging. To complicate this matter, multiple terms are used interchangeably in defining gangs— *youth gang, street gang, criminal street gang,* and *gang*—and whether or not each of these terms refers to a common problem in practical applications is not clear. Because the focus of this book encompasses both gangs comprised of children and adolescents as well as older gangs populated by young adults, the two terms most commonly used in the gang research literature are employed: *youth gangs* or *street gangs.* Youth gang is a useful term for drawing attention to the younger gangs—and an important target for prevention and early intervention initiatives. The street gang term makes a point of emphasis, denoting older gangs that have a presence on city streets and commit characteristically urban crimes, especially assault, robbery, gun crimes, and murder. This term also applies to some "crews" and "posses" that are specialized groups engaged in predatory crimes or drug trafficking (W. B. Miller, 1982/1992). Many of these appear to be bona fide street gangs who are given an alternative name in the interest of denying a "gang problem."

In the interest of inclusiveness with respect to street presence, it is important, however, to avoid overidentification of street gangs with adult organized crime groups. A number of state legislatures have modeled their gang definitions after California's gang law that defines a *criminal street gang* as "any ongoing organization, association, or group of three or more people, whether formal or informal, having as one of its primary activities the commission of criminal acts" (Street Terrorism Enforcement and Prevention Act, 1988, California Penal Code sec. 186.22[f]). This definition encompasses adult criminal enterprises (organized crime) that typically are not considered to be street gangs. Interestingly, there is no widely accepted definition of organized crime.

W. B. Miller (1982/1992), an anthropologist, remains the only researcher to have attempted to empirically categorize an identifiable law-violating youth group across multiple cities. Based on his 26-city study, he identified 20 such groups, of which 3 were gangs: turf gangs, gain-oriented gangs (typically "extended networks"), and fighting gangs. The 17 remaining law-violating youth groups included burglary rings, established predatory cliques, robbery bands, extortion cliques, drug-dealing cliques and networks, assaultive affiliation cliques, and others.

At the other end of the spectrum, it is important to distinguish between actual young gangs and small groups of youngsters who commit delinquent acts in concert. Unsupervised peer groups are small groups (typically three to four members) of adolescents that are highly transitory and not well organized (Warr, 1996, 2002). Short (1996) points out that many of these groups are involved in occasional delinquent behavior but lack a commitment to a criminal orientation. Hence, they may be considered as only "troublesome" and not of great concern as a threat to society. These adolescent groups lack the size, formal organization, and permanence of youth gangs, and their delinquency typically is not as frequent, serious, or violent.

The most important point to keep in mind in any attempt to define youth gangs is that such groups are not necessarily an integral feature of the experiences of young people during adolescence. R. M. Gordon (1994) offers an age-graded continuum of social and criminal groups to view gangs, anchored at one end by childhood play groups and at the other by adult criminal organizations. The following groups are represented along this continuum:

- *Childhood play groups*—harmless groups of children that exist in every neighborhood.

- *Troublesome youth groups*—youths who hang out together in shopping malls and other places and may be involved in minor forms of delinquency.

- *Youth subculture groups*—groups with special interests, such as "goths," "straight edgers," and "anarchists," that are not gangs. (Goths are not known for criminal involvement, but some members of other youth subcultures have histories of criminal activity [Arciaga, 2001].)

- *Delinquent groups*—small clusters of friends who band together to commit delinquent acts such as burglaries.

- *Taggers*—graffiti vandals. (Taggers are often called gang members, but they typically do nothing more than engage in graffiti contests.)

- *School-based youth gangs*—groups of adolescents that may function as gangs only at school and may not be involved in delinquent activity, although most members are involved.

- *Street-based youth gangs*—semistructured groups of adolescents and young adults who engage in delinquent and criminal behavior.

- *Adult criminal organizations*—groups of adults that engage in criminal activity primarily for economic reasons.

Two general gang types are noted in Gordon's list: school-based and street-based youth gangs. Finding these in a community seems simple enough, but distinguishing between them is not. Moreover, the distinction between youth gangs and adult criminal organizations is not always clear. Figure 3.1 illustrates the overlap of delinquent groups, criminal groups, street gangs, adult criminal organizations, and juvenile gangs.

Gang Subculture

A distinctive gang subculture often distinguishes street gangs. The *sub*culture forms when two national cultures are brought into close physical contact. Incomplete adaptation or assimilation of the two cultures can produce a subculture. This process may take several generations.

Dual cultural identity. The formation of a Mexican American subculture in the western United States is an excellent example of the cultural merge. From the mid-19th century, Mexican immigrants had been culturally marginalized between their society of origin and the dominant American culture to which they had migrated—or within which

Figure 3.1   Overlapping Gangs With Other Groups

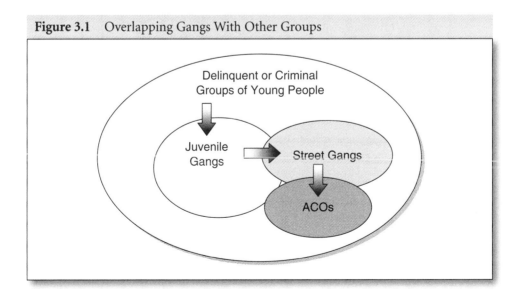

native Mexicans had become enveloped following U.S. annexation of their land. As the result, a *cholo* (a derivative of the Spanish *solo,* meaning "alone") subculture emerged, an American product. Cholo youth, the poorest of the poor marginalized immigrants, could not fully assimilate into Anglo culture or develop a unique identity incorporating aspects of both Anglo and Latino culture (Horowitz, 1983; Lopez & Brummett, 2003; Lopez, Wishard, Gallimore, & Rivera, 2006; Vigil, 1988, 1990). Vigil (1988) documents the cholo style as further developed in the 1930s and 1940s when the second generation was under intense conflict: It grew out of "the effects of racism, the persistent cycle of underclass involvement with little chance of social mobility, and the general malaise engendered by problems with getting jobs and staying in school" (p. 39).

Synonymous with development of the cholo subculture, Vigil (1988) shows how a street gang subculture emerged, made up of elements or fragments, of past Mexican culture. This youth-gang subculture "is a response to the pressures of street life and serves to give barrio youth a source of familial support, goals and directives, and sanctions and guides" (p. 2). It provides a social structure and a cultural values system—norms, goals, and behavioral expectations, age-graded cohorts, and together, these undergird formidable street gangs. In this way, the youth-gang subculture functions to socialize and acculturate barrio youth. "The pressures and anxiety of urban poverty, of the struggle toward a better life, and of overcoming feelings of ethnic and racial inferiority made immigrant cultural adaptation problematic. Such an experience often resulted in gangs" (p. 4). After decades of continuous migration, social isolation, prejudice, and discrimination, the Mexican American youth-gang subculture became so pervasive that all Mexican immigrant youth are compelled to come to terms with it—even though most of them never join. The children in families that suffer the most acute poverty and limited social mobility opportunities are the youths who come to be "most intensely involved" in gangs (p. 5). Vigil (2002) elaborates:

> What established gangs in the neighborhood have to offer is nurture, protection, friendship, emotional support, and other ministrations for unattended, un-chaperoned resident youths. In other words, street socialization fills the voids left by inadequate parenting and schooling, especially inadequate familial care and supervision. This street-based process molds the youth to conform to the ways of the street. (p. 10)

Being a cholo allows such youth to assert a Latino identity, take pride in it, and deny "being *enbacheado* (Anglicized)" (Vigil, 1988, p. 42). In other words, these street youth shaped their own cholo subculture. Thus, Vigil concludes that the cholo subculture "has become an institutionalized entity that provides many poor, barrio youths with human support networks and a source of personal ego identity that are unavailable to them elsewhere" (p. 39).

Ritual and symbolic representation. In the Southwest, Mexican American gangs have a unique subculture that is rich in ceremonial ritual and symbolic representations expressed in Mexican language, specific clothing styles, hair styles, tattoos, and graffiti. With the passage of time, each successive generation contributed features to the gang culture, including a distinctive speech form and a style of dress. Sanders (1994) details the clothing: "A uniform consisting of a Pendleton shirt, buttoned at the top over a white T-shirt, a hair net over short hair combed straight back, a bandanna tied around the head pulled down just above the eyes, dark sun glasses, a hat, and baggy khaki pants makes up the outward *cholo* style" (p. 129). Arciaga, Sakamoto, and Jones (2010) claim a simpler style, the *pachuco* look—khaki pants, white T-shirt, and plaid shirt—came about through economic necessity. This style of dress remains common with many gangs around the United States, and because of its youth subculture foundation, Mexican American gang culture is remarkably uniform across the United States.

Baseball caps and clothing of a professional sports team are also commonly worn to signify identification with that team's reputation, not actual endorsement. Baggy clothing (and sometimes particular brands) is another gang trademark—once worn only by gang members—but this also is a popular style in general adolescent subculture. "Gang members wear gang colors and symbols on many items: shirts, bandanas, jewelry, sports clothing, hats (including altering or writing graffiti on the insides of hats), hairbands, shoelaces, belts, belt buckles, and specific shoe brands whose names or logos have a specific meaning to the gang" (p. 11). Walker and Schmidt (1996) note how students would carry their clothes in bags so that they could change into conforming colors as they cross gang boundaries. Of course, as Al Valdez (2007) reminds us, tattoos are widely used in all gang cultures across the United States, typically to show one's level of commitment to a gang.

There are multiple forms of graffiti, only one of which is the gang type, as shown in Table 3.1. Gangs of all origins employ graffiti for various motives—to mark turf, threaten violence, boast achievements, honor the slain, and to insult or taunt other gangs. Klein (1995) reports that graffiti is not only useful in establishing gang identity, but it also may be used as a symbolic form of gang conflict. Hutchison and Kyle (1993) add that some individual "taggers" and "tagger groups" have become conflict-oriented gangs as a result of conflict over graffiti.

Nicknames, tattoos, and places are uniformly used in the Mexican gang culture. Fellow gang members create and assign nicknames to signify acceptance, which tattoos containing nickname and barrio confirm. Vigil (1988) asserts that public tattoos in the form of *placas*—literally meaning plaques, public tattoos, and also graffiti—on walls and fences publicize a person's affiliation and commitment. This graffiti consists of the gang member's nickname, barrio, and *klika* signature or territorial coordinates. Furthermore, gang graffiti is informative in educating the viewer about gangs that are active in the community, denoting their boundaries, their symbols, and names or monikers of members, other messages, and also a way to claim territorial dominance in communications intended for members of other gangs to see. Hutchison (1993) explains that traditional Mexican American gangs did not adopt particular colors in the early 1990s, except, according to Al Valdez (2007), in California where northern Mexican American gangs identify with red and southern ones favor blue.

According to Al Valdez (2007), gang graffiti is often called the "gangland's newspaper." Thus, "the analysis of gang graffiti may also be used to learn about the structure of the gangs, conflicts and alliances between particular street gangs, and possible linkages among street gangs. . . . Regardless of the particular region of the country, the content of gang graffiti is generally consistent; it is designed to convey specific messages about

Table 3.1 Types of Graffiti and Associated Motives

Type of Graffiti	Features	Motives
Gang[a]	Gang name or symbol, including hand signs Gang member names, nicknames, or sometimes a roll-call list of members Numbers: code in gang graffiti, with a number representing the corresponding letter in the alphabet (e.g., 13 = M, for the Mexican Mafia) or a penal or police radio code Distinctive, stylized alphabets: bubble letters, block letters, backwards letters, and Old English script Key visible locations Enemy names and symbols or allies' names	To mark turf To threaten violence To boast of achievements To honor the slain To insult or taunt other gangs
Common Tagger[b]	High-volume, accessible locations High-visibility, hard-to-reach locations May be stylized but simple name, nickname tag, or symbols: single-line writing of a name usually known as a *tag*, whereas slightly more complex tags, including those with two colors or bubble letters, known as *throw-ups* Tenacious (keeps retagging)	Notoriety or prestige Defiance of authority
Artistic Tagger	Colorful and complex pictures known as masterpieces or pieces	Artistic Prestige or recognition
Conventional Graffiti	Spontaneous Sporadic episodes or isolated incidents Malicious or vindictive	Play Rite of Passage Excitement Impulse Anger Boredom Resentment Failure Despair
Ideological	Offensive content or symbols Racial, ethnic, or religious slurs Specific targets, such as synagogues Highly legible Slogans	Anger Hate Political Hostility Defiance

[a]Copycat grafts looks like gang graffiti and may be the work of gang wannabes or youths seeking excitement.

[b]*Tagbangers*, a derivative of tagging crews and gangs, are characterized by competition with other crews. Therefore crossed-out tags are features of their graffiti.

Source: Weisel, 2004, p. 9

group identity, gang allegiances, and individual membership" (Hutchison, 1993, pp. 138, 163). Another consistency Al Valdez identifies is the practice of insulting or challenging another gang by using a "puto-out," that is, crossing out or writing over the name or symbol of that gang. But, there are also some interesting differences in gang graffiti between Mexican American gangs in Chicago and Los Angeles. This, Hutchison (1993) points out, is proof that LA gangs did not migrate to Chicago. Each Chicago gang has a specific symbol; LA gangs do not. LA gangs have specific lettering styles; Chicago gangs do not. LA gang graffiti identifies the gang's barrio; this is not so in Chicago. Certain elements of gang culture have been incorporated into the general adolescent subculture in LA, but not in Chicago. Hutchison contends that this is because the Chicago barrios were settled much later; he asks, does this imply that the Mexican American gang culture in Chicago will someday become integrated?

In addition to identity signified via clothing, Arciaga and colleagues (2010) discuss how gang members use hand signs to show affiliation to their set, show disrespect to rivals, challenge rivals, and provoke confrontations. To a gang member, a hand sign is a nonverbal form of communication that can lead to violence. It follows that many gangs have their own hand signs. To throw a hand sign at a rival gang member is considered an insult or a challenge, called *set tripping*, and is often the catalyst for gang-related violence.

Gang graffiti may also reveal an incident of importance and who did it. Fallen comrades are often memorialized with the RIP (Rest in Peace) message. "Unfortunately, graffiti can be the prelude to gang violence. Monitoring graffiti in the schools, neighborhoods, and homes can be a good indicator to show rising tensions between individual gang members and rival gangs" (Valdez, 2007, p. 77).

Tobin (2008) coined the term *groupthink* to describe the adolescent subculture's view of the world, which is important for understanding gang patterns and cohesion. "This groupthink is amplified in street gangs" (p. 66). The group mentality or beliefs and symbolic views of manhood and brotherhood that the gang provides for its members include social codes and other ways of integrating identity in everyday behavior (Garot, 2007, 2010; Garot & Katz, 2003). Although these codes are unwritten and, according to Decker and Van Winkle (1996), largely have "evolved out of practice, lore, or common sense" (p. 100), Tobin (2008) asserts that they "include rules regarding expressions of conformity and rules of prohibited behavior" (p. 66).

Expressions of conformity, again, include symbolic ways of dressing and other displays of gang affiliation. Thus, prohibited behaviors and consequences include disrespecting one's gang colors, fighting members of one's own gang, ratting out a member of one's own gang, running from a fight, and pretending to be a member of a rival gang (Decker & Van Winkle, 1996). The consequences can include demotion in gang rank and violent beatings.

Key Elements of a Gang Definition

Bjerregaard (2002b) catalogued various gang definitions in the literature according to the key definitional criteria of gangs that are found in them. These criteria are intended "to verify or confirm that an individual is a member of a traditional street 'gang' and not just a group of friends or individuals that hang out together" (p. 37). Bjerregaard's review is updated in Table 3.2 to add several recent

Table 3.2   Key Elements of Gang Definitions

Researchers	Name/ Identity	Leadership	Colors/ Dress/ Symbols	Territory/ Turf	Meetings/ Continuous Association	Organized	Criminal Activity
Thrasher, 1927/2000				X	X	X	
B. Cohen, 1969	X	X		X		X	
Klein, 1971	X						X
Cartwright et al., 1975					X	X	X
W. B. Miller, 1982/1992		X		X	X	X	X
Johnstone, 1981		X	X	X		X	X
Spergel, 1984	X	X		X		X	X
Hagedorn, 1988	X			X			
Spergel, 1995			X		X	X	X
Short, 1996					X	X	
Howell & Lynch, 2000	X		X	X	X		X
J. W. Moore, 1991				X		X	X
Huff, 1993	X		X	X	X		X
Chicago Crime Commission, 1995		X		X		X	X
Nat'l Drug Intelligence Center, 2008	X		X				X
Decker & Van Winkle, 1996			X	X			
Oehme, 1997			X	X	X	X	X
Curry & Decker, 2003					X		X

Researchers	Name/ Identity	Leadership	Colors/ Dress/ Symbols	Territory/ Turf	Meetings/ Continuous Association	Organized	Criminal Activity
CA Council on Criminal Justice, 1989	X			X	X		X
CA STEP Act, 1998	X		X			X	X
Esbensen, Winfree, et al., 2001		X	X			X	X
Esbensen et al., 2008	X						X
Federal Law 18 USC § 521(a)					X	X	X
Total	10	6	9	13	11	14	18

Source: Modified from Bjerregaard, 2002b, p. 34

definitions, in accordance with Bjerregaard's original framework. The most common features of gangs in the 23 cited gang definitions follow.

Criminal activity. The bottom row of Table 3.2 shows the total number of gang definitions that incorporate each of the seven key characteristics. The most common element for defining gangs is involvement in criminal activity or commitment to a criminal orientation—cited in three-fourths (18) of the 23 listed definitions. Law enforcement and legislative definitions emphatically emphasize this criterion for classifying gangs. The major source of state and law enforcement definitions is the previously cited California Street Terrorism Enforcement and Prevention Act. However, consideration of this factor separates gang researchers into two camps. In one camp are those who contend that gang definitions should not incorporate this element. These advocates argue that the inclusion of criminal behavior in the gang definition presents a tautology for researchers who wish to investigate and explain delinquent and criminal behavior by looking at the individual's gang membership (Bjerregaard, 2002). In the opposing camp are those who contend that, in the absence of this criterion, gang definitions could envelop nongang groups such as boy scouts, girl scouts, and private school students. Advocates in both camps make valid points. However, when a gang definition is used to help communities assess gang activity that represents a threat to their safety, it certainly should include the criminal behavior criterion, because it is precisely criminally active gangs that must be identified in this practical application.

Organizational structure. The organizational structure of the gang is the second most frequently specified aspect of gangs, in nearly two-thirds (14) of the 23 definitions. Unfortunately, as Sanders (1994) points out, "Sociologists tend to promote an 'over-structured' view of gangs. The various roles that make up legitimate public and private organizations are often sought out in gangs, and somehow are found by sociologists" (p. 134). In fact, organizational characteristics of gangs are among the common elements in virtually all gang studies. Others place an emphasis on the frequency of association in specifying the gang's organizational structure (Huff, 1993; W. B. Miller, 1982/1992; Oehme, 1997). Having identifiable leadership is another indicator of organizational structure of the gang. In addition, a hierarchy of membership is sometimes present. The key issue here, and as Weisel (2002a) contends, is whether a "continuous organizational structure is present" (p. 185). Available evidence of gang structure suggests that the answer is a qualified "yes"; that is, a minimum set of structural features is typically present. As Weisel explains, "For day-to-day activities of the gang, members indicate a reliance on subdivisions—the groups, cliques, sets and chapters which are the building blocks of the gang—and routinely emphasize the trust relationship between members" (p. 145), rather than between themselves and a leader above them in the gang. To the extent that a "continuous organizational structure" exists, gangs "appear to feature flat age-graded layers rather than a vertical organizational structure" (pp. 185–186). In day-to-day associations, "everyone appears to take some responsibility for the success of the organization or the small group in which the member is predominantly involved" (p. 186).

Esbensen, Winfree, and colleagues (2001) discovered that members of gangs that were somewhat "organized" (i.e., had initiation rites, established leaders, and symbols or colors) self-reported higher rates of delinquency and involvement in more serious delinquent acts than other youths. In another study, Decker and fellow researchers (2008) found that individuals who were members of well-organized gangs (i.e., had leaders, rules, meetings, and symbols) reported higher victimization counts, more gang sales of different kinds of drugs, and more violent gang offenses than did members of less organized gangs. Having a sufficient number of members to be considered a "group" is also a common indicator of an organizational structure. While sociologists consider three persons to form a group, only a few gang definitions make this criterion explicit.

Territory or turf. The third most common defining characteristic of gangs is claiming a turf or specific location in the neighborhood (13 of 23 definitions). Identification with a particular territory is another form of symbolism that helps identify groups as gangs, although this indicator has not been widely used in the past 30 years because of the growing mobility of gangs—owing mainly to suburban sprawl, more widespread use of automobiles, and expanding highway networks. Many gangs do not claim a geographic territory, or only episodically, if they do practice this tradition. However, a cautionary note should be issued here. Territory or turf claiming may not have diminished appreciably in large cities where most bona fide street gang members reside and continue to control the streets. In fact, Egley and associates (2006) report that approximately 85% of all gang members are estimated to be located in cities with populations of 50,000 or more and suburban counties.

It is a curious thing that gang definitions do not place more emphasis on named gangs as an identification criterion, particularly given the intertwined status-providing function of gangs and individual identity needs of alienated youth (Vigil, 2004). This is an important matter in the case of Hispanic gangs, because the gang name is often linked with neighborhood or barrio of residence, and what Vigil (1993) calls the youth's "'psycho-spatial identity' with their territory or 'turf'" (p. 98), with *mi barrio* (my neighborhood) becoming synonymous with *my gang*.

A new gang definition. The Eurogang Network definition was developed by consensus among participating scholars for the purpose of researching European gangs: "A youth gang, or troublesome youth group, is any durable, street-oriented youth group whose involvement in illegal activity is part of its group identity" (Esbensen et al., 2008, p. 117). The absence of a gang name criterion in the Eurogang definition may explain why it has not applied well to Hispanic/Latino gangs in the U.S. (Esbensen et al., 2008, p. 129).

This definition incorporates five indicators. Klein and Maxson (2006) explain:

1. Durable is a bit ambiguous, but at least an existence over several months can be used as a guideline. . . . The durability refers to the group.

2. Street-oriented implies spending a lot of group time outside home, work, and school—often on streets, in malls, in parks, in cars and so on.

3. Youth can be ambiguous. Most street gangs are more adolescent than adult, but some include members in their 20s and even 30s.

4. Illegal activity generally means delinquent or criminal, not just bothersome.

5. Identity refers to the group, not the individual self-image. (p. 4)

In Klein and Maxson's view, these "are the minimal necessary and sufficient elements to recognize a street gang" (p. 4).

This broad gang definition—currently applied in multiple Eurogang studies, in collaboration with the Eurogang Research Network—has caused some consternation among European scholars. Deuchar (2009) explains why: "The word gang is a highly contested term, and the criteria for classifying someone as a gang member are debatable (Bradshaw, 2005), since some people in deprived communities may regard their peer networks simply as 'friendship groups' [or other similar terms]" (p. 10). Hallsworth and Young (2008) and White (2008) challenge such attempts to interpret urban violence in the United Kingdom as a problem of gangs or as a burgeoning gang culture. These scholars and Pitts (2008) object to the characterization of youth groups in that region as "gangs" in any event because of the bad example that American media have set in overuse of this arguably pejorative term, and also owing to the reservations of other European scholars. Yet Bennett and Holloway's (2004) study illustrated that 15% of subjects in the New English and Welsh Arrestee Drug Abuse Monitoring research program had experience of gang life, and that 4% were current gang members. Notwithstanding these findings, the argument has been made that Great Britain has only a ganglike subculture, rather than what Hallsworth and Young (2008) and

Muncie (2004) identify as street gangs that resemble the American forms of these groups. This point may well be valid because the Eurogang Research Network's gang definition seriously overclassified American gangs in a nine-city U.S. study in which only 38% of the youth classified as gang members considered their group to be a gang (Esbensen et al., 2008). Future studies surely will resolve this measurement issue (see Weerman & Esbensen, 2005).

A Recommended Gang Definition for Practical Purposes

To guide local assessments of gang problems (see Chapter 10 for details on this activity), communities need a more practical definition. Practitioners would have difficulty applying the Eurogang definition—particularly in making a determination as to whether or not involvement in illegal activity *is part of its group identity* because delinquent activity is normative among adolescents. It also is unfortunate that the Eurogang Network definition encompasses troublesome youth groups because, as Thrasher (1927/2000) asserted, this definitional element may erroneously mark groups as "gangs" that have not reached the tipping point of becoming a bona fide gang.

As an alternative, the following criteria for classifying groups as youth gangs take into account practical application of gang definitions in community assessments:

The group has five or more members. Thirty states and Washington, DC, define a gang as consisting of three or more persons (National Gang Center, 2009). Federal law (18 USC § 521[a]) specifies that a *criminal street gang* must have five or more members to qualify as a gang. The federal standard of five or more members is preferred because this helps identify bona fide gangs that are not small delinquent cliques. (Howell [2010a] previously recommended three or more participants, but this higher standard of five members is consistent with Warr's [1996] research on group offending and will help exclude small friendship groups that happen to be involved in delinquency.)

Members share an identity, typically linked to a name and often other symbols. These features are covered in two key characteristics of gangs (Table 3.1): name/identity and colors/dress/symbols. Having a gang name helps distinguish gangs from the many other law–violating youth groups. Bjerregaard (2002b) believes this to be the most potent criterion of the symbolic aspects of gangs. The inclusion of a name requirement also should help community stakeholders distinguish gangs from other delinquent groups. Indeed, the two main criteria students use nationwide to identify gangs are having a name and time spent with other gang members (Howell & Lynch, 2000).

Members view themselves as a gang and are recognized by others as a gang. This definitional element recognizes the importance of establishing affiliation with bona fide gangs, which have crossed the tipping point from a delinquent group. Hence, the use of this criterion ensures that groups designated as gangs will not be merely troublesome youth groups, because approximately 9 out of 10 youngsters who self-identify themselves as a gang member are members of gangs that are involved in delinquent activity (Esbensen et al., 2010).

The group associates continuously, evidences some organization, and has some permanence. This criterion contains two common characteristics of gangs (continuous association and having some degree of organization, Table 3.2), and the addition of the "permanence" element serves to distinguish gangs that exhibit persistence over time. There is no empirical basis for a specified period. Several months seems reasonable, as the Eurogang Network suggests. Furthermore, Howell and Egley (2005a) contend that transitory or emerging gangs are far less likely to present a public safety threat to the community.

The group is involved in an elevated level of criminal activity. Two-thirds of the definitions in Table 3.2 include this element, because it addresses public safety. The inclusion of this definitional element is based on Krohn and Thornberry's (2008) research showing delinquency rates are higher among gang members than other youth.

These criteria can be used in communitywide assessments of gang activity (see Chapter 9 for the assessment process). Communities that develop or otherwise adopt their own definitions will find that this leads to better understanding of local gangs, more targeted programming, and greater success in reducing gang crime.

Defining Gang Members

Defining gang members is surprisingly complex. Most important, several levels of gang membership are conceivable. Many youth who associate with gang members on occasion never join. This group of associates, or affiliates, may be quite large (Curry et al., 2002; Eitle et al., 2004). Other studies in traditional gang cities also report that youth typically are members of gangs for relatively short periods of time (Krohn & Thornberry, 2008; Hagedorn, 1988; Klein, 1971; Short & Strodtbeck, 1965/1974; Vigil, 1988; Yablonsky, 1967). In their Denver gang research, Esbensen and Huizinga (1993) concluded: "It appears that the majority of gang members are peripheral or transitory members who drift in and out of the gang" (p. 582).

Chapter 9 describes how very young gangs form, referred to as *starter gangs* (which can be distinguished from a pregang in Chapter 1, or Hoyt's [1920] *embryo gang*). The youngest gangs—with members approximately ages 10 to 13—are formed in small groups of rejected, alienated, and aggressive children (Craig, Vitaro, Gagnon, & Tremblay, 2002). Debarbieux and Baya (2008) delve even deeper, showing how the groups of children that form gangs are more likely to attend "difficult" schools characterized by high student victimization rates, self-reported violence, poor student–teacher relations, and harsh punishments for minor behavior.

A common approach for reducing the complexities associated with determining gang involvement, and to reduce errors of false identification, is to ask respondents to self-identify gang membership (Esbensen, Winfree, et al., 2001; Winfree, Fuller, Vigil, & Mays, 1992). However, Rennison and Melde (2009) caution that "this method relies on the respondent's interpretation or perception of what constitutes a gang" (p. 495). As Howell and Lynch (2000) assert and noted previously, students most often employ three criteria: (1) having a name, (2) spending time with other members of the gang, and (3) wearing clothing or other items to identify their gang membership.

In an ingenious study, Esbensen, Winfree, and colleagues (2001) examined the impact of varying definitions of gang membership on the prevalence of gang members:

> Five types of gang members were created. The first two types were identified by use of single items: "Have you ever been a gang member?" and "Are you now in a gang?" Three increasingly restrictive definitions of gang membership were then created. The third type, "delinquent gang" member, included respondents who indicated that their gang was involved in at least one of the following illegal activities: getting in fights with other gangs; stealing things; robbing other people; stealing cars; selling marijuana; selling other illegal drugs; or damaging property. The fourth type, "organized gang" member, included delinquent gang members who also indicated that their gang had some level of organization. Specifically, the survey respondents were asked whether the following described their gang: "There are initiation rites; the gang has established leaders; the gang has symbols or colors." The last characteristic used to determine gang membership was an indicator of whether individuals considered themselves to be "core" or "peripheral" gang members. The impact of definitional criteria on the prevalence of youth gang membership is quite pronounced. Depending on which of the five different definitions of gang member is used, anywhere from 2% to 17% of the sample would be regarded as involved in a gang. Almost 17% of the respondents indicated that they had ever belonged to a gang; . . . [but] as the definition of "gang" became more restrictive, the number of youths reporting involvement decreased: 7.9% reported belonging to a "delinquent gang"; 4.6% were in "organized delinquent gangs"; and only 2.3% were "core" members of an "organized delinquent gang." (Esbensen et al., 2010, p. 77)

Based on this study, Esbensen and colleagues (2010) conclude that a single-item self-identification definition of gang membership ("Are you now in a gang?") is a highly reliable indicator; the *gang* term "connotes something unique and distinguishable" (p. 78). "The vast majority (89%) of youths who indicated that they were currently in a gang also indicated that their gang was involved in delinquent activity" (p. 78).

Esbensen and colleagues (2008) applied a similar funneling approach with the Eurogang Network's gang definition in a nine-city American survey of almost 1,500 youths. They first restricted their gang sample to those respondents who indicated the following: "They had a group of friends with whom they spent time and who were aged 12–25; the group had been around for more than three months and hung out in public places; and, importantly, members of the group believed it was okay to do illegal things, and they reported committing illegal acts together" (p. 123). Only 7.6% of the sample met all four of the criteria and thus were classified as bona fide gang members. Esbensen and colleagues (2008) also reported that exploratory analyses on this group's involvement in juvenile delinquency showed substantive differences in rates of involvement in violent crimes among active gang members of different racial/ethnic backgrounds.

In an earlier survey of nearly 6,000 eighth graders in 11 cities with known gang problems, 9% were currently gang members and 17% percent said they had belonged to a gang at some point in their lives (Esbensen & Deschenes, 1998; Esbensen et al., 2010). However, this percentage varied from 4% to 15% depending on location (see Table 3.3). Most of this variation is undoubtedly linked to city-size and seriousness of gang problems. Gang membership is even greater among representative samples of youth in high-risk areas of large cities, according to studies in Seattle (15%); Denver (17%); Pittsburgh (24%); and Rochester, NY (32%) (Hill et al., 1999; Lahey et al., 1999; Thornberry et al., 2003).

Although the prevalence of gangs and gang members vary from one geographical area to another, the degree of restrictiveness in gang definition also accounts for variations in membership calculations. Klein and Maxson's (2006) review of most of the high-quality studies provides this range for "ever" gang involvement in risk-selected populations of youngsters: 6% to 30% using an unrestricted gang definition, and 13% to 18% using a restricted gang definition; and 6% to 8% for "current" gang involvement in a general study population.

Nationwide, the most current estimate comes from the National Longitudinal Survey of Youth (a nationally representative sample of 9,000 adolescents). Using the "ever" membership metric, Snyder and Sickmund (2006) found 8% of the youth surveyed had belonged to a gang at some point between the ages of 12 and 17.

Other studies have validated self-reported gang involvement, with self-reported delinquency (Bjerregaard, 2002a; Krohn & Thornberry, 2008), arrest data (Curry, 2000), and violent crime drug sales, along with degree of gang organization (Decker, Katz, & Webb, 2008). In sum, research has demonstrated that self-identification is a robust indicator of gang involvement.

Table 3.3   Prevalence of Gang Membership in 11 Locations

Location	Percentage of Youth Who Are Gang Members
Milwaukee, Wisconsin	15.4%
Phoenix, Arizona	12.6%
Omaha, Nebraska	11.4%
Las Cruces, New Mexico	11.0%
Kansas City, Missouri	10.1%
Orlando, Florida	9.6%
Philadelphia, Pennsylvania	7.7%
Torrance, California	6.3%
Providence, Rhode Island	6.0%
Pocatello, Idaho	5.6%
Will County, Oregon	3.8%

Source: Esbensen et al., 2010, p. 74

Al Valdez's Hierarchy of Gang Membership

Al Valdez (2007) has described four membership levels of the typical street gang. The first level is the *target group* for recruiting future members. These potential associates may be as young as 8- to 11-years-old, and they typically are intrigued by the gang lifestyle. They frequently imitate actual gang members by wearing stylized clothing, throwing gang signs, or using gang slang. As Tiny, a 9-year-old Hispanic child, said to a gang investigator, "We just hang around with the guys, we just watch them . . . we try to be like them" (p. 42).

The second level of gang membership is the *associate gang member*, also called "gang dabbler" and "wannabe," although gang-involved youths at this stage are actually more like apprentices or part-time members. They frequently are photographed with other gang members, may even have a tattoo, and effectively simulate the mannerisms of regular gang members' ways of talking, dressing, and acting. In addition, they often visit the areas where the gang hangs out and even participate in criminal activity with other gang members. In short, the associates are often hard to distinguish from active gang members.

Regular gang member is the third level of gang membership. This level is the total group of full-time members, typically ranging in age from 10 to 25 years, the newest of which often are referred to as "youngsters." They have specific responsibilities, and "blind loyalty," an unconditional commitment, is often required:

> Turf gangs are responsible for helping to protect the area claimed by the gangs, or assist in gang operations. . . . Gang members assist in retaliatory actions, drug sales, transport and usage. At this level, some gang members may even have some major leadership responsibilities. The gang lifestyle may demand dedicating life and all resources to the gang. (p. 43)

The fourth level of gang membership is *hard core gang member*, the nucleus of the gang. These members typically are the oldest participants, veterans or OGs (original gangsters), to which a great deal of notoriety is attached, although very young members sometimes qualify. Most of the OGs have been incarcerated in local or state facilities. They often have numerous arrests and police contacts: "Hard core gang members know gang violence intimately. Many have been arrested for attempted homicide, murders and assaults with deadly weapons. Many have been injured in gang-related violence" (p. 43).

Demographic Characteristics of U.S. Gangs and Gang Members

We now turn to an examination of demographic characteristics of gangs and gang members, beginning with the number of gangs, then gang member characteristics: age, race and ethnicity, and gender.

National Data on the Number of Gangs and Gang Members

The number and size of gang memberships vary directly with population sizes of gang problem areas.

• The rural counties and small cities, towns, and villages (with 50,000 or fewer populations) typically report 6 or fewer gangs and 50 or fewer members.

• In the second category, larger cities (with populations greater than 50,000) typically report up to 15 gangs and 54% estimate more than 50 members.

• Cities with populations between 100,000 and 250,000 typically report up to 30 gangs and 32% estimate more than 500 members.

• Cities with populations greater than 250,000 typically report more than 30 gangs and more than 1,000 members. (Egley, 2005)

Across these population groups, localities that report a persistent gang problem have far more gangs and gang members. For example, in Egley, O'Donnell, and Howell's (2009) National Youth Gang Survey (NYGS), respondents who consistently reported gang problems from 2002 to 2008 estimated 15 gangs and 250 members. In contrast, the respondents who inconsistently reported gang problems in this period averaged only 6 gangs and 74 members.

Age of Gang Members

Youth gangs have long been an adolescent phenomenon. A large majority of youths who join gangs do so at very early ages, typically between ages 11 and 15, with ages 14 to 16 the peak for gang involvement. Esbensen and coresearchers' (2008) nine-city sample of students drawn from known gang-problem areas illustrates this progression, finding 7% joined by age 12, 9% were members at age 13, and 16% were members at age 14 and older.

The ages of typical street gang members are older. NYGS respondents are asked in alternate years to provide information on the ages of members of all gangs in their respective jurisdictions. Figure 3.2 shows the relative proportion of juvenile and adult gang members on which respondents are able to provide information. Law enforcement representatives have always said that half or more of all members of gangs with a street presence are adults, from 50% in 1996 to 59% in 2008. However, as Barrows and Huff (2009) and also Howell, Moore, and Egley (2002) assert, this increasing proportion of adults may reflect the parallel increase in gang intelligence systems. Absent regular purging of these databases—which Barrows and Huff contend is typically the case —young gang members age in the database with the passage of time.

Race and Ethnicity of Gang Members

As discussed in Chapter 1, early in U.S. history, gang members were largely Caucasians of various European origins. Thus, Riis's (1902/1969) and Thrasher's (1927/2000) early

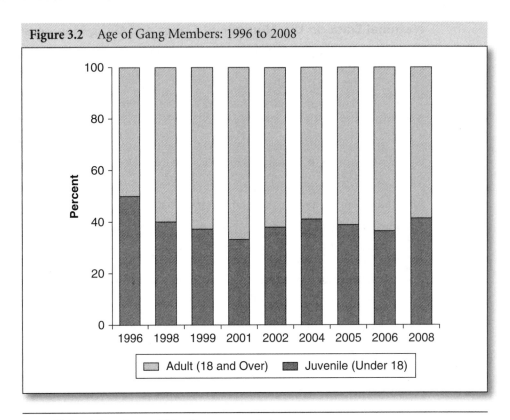

Figure 3.2 Age of Gang Members: 1996 to 2008

Source: National Gang Center, 2010 online survey analysis

gang studies focused on the nationality of the members, and it was not until the 1950s that scholars began to note the changed racial/ethnic composition of youth gangs. But Mexican American and African American youth became predominant members with successive periods of immigration and internal migration, respectively. Still, it must be underscored that minority youngsters have no special predisposition to gang involvement. As Vigil (2002) observes, they happen to be overrepresented in inner-city areas in which concentrated disadvantage and social disorganization disproportionately give rise to highly visible gang activity. From the beginning of gang activity in the United States to the present time, Esbensen and Lynskey (2001) suggest that gangs reflect the racial/ethnic makeup of the locality, and this mixture can vary enormously. For example, Vigil reports that in Los Angeles, Mexicans, Blacks, Vietnamese, and Salvadorians are the predominant membership groups. Kontos and colleagues (2003) and M. L. Sullivan (1993) find a different mixture in New York City, where Puerto Ricans, non–Puerto Rican Latinos (especially Dominicans), followed by Central and South Americans, Mexican Americans, and African Americans predominate.

It must be emphasized that studies in which youth are asked to self-report gang membership show more equal proportions of racial/ethnic groups. In a 1995 survey of middle-aged students in 11 diverse cities in the United States, Esbensen and Lynskey (2001) found that overall 25% of the gang members were Hispanic, 31% African

American, 25% were Caucasian, and 20% were of other race/ethnicities. This is the first multisite study to document the growing involvement of Caucasian youth in gangs. In a more recent nine-city student survey, Esbensen's (2008) research team calculated the proportion of the main racial/ethnic groups that self-reported gang membership. The racial/ethnic proportions were relatively even for Caucasians (7.3%), African American (8%), and Hispanics (9%), but larger (13%) for multiracial groups (reflecting the increased immigration from outside the United States after 1995, discussed in Chapter 1). Studies in other countries also show more similarities than differences among racial/ethnic categories—in Europe (Van Gemert, Peterson, & Lien, 2008), Toronto, Canada (Wortley & Tanner, 2006, 2008), and in a study comparing five countries (Winfree et al., 2007).

Almost half (46%) of all gangs in the United States were reported to be multiethnic/multiracial in the 1996 NYGS. To obtain more specific information on this matter, the 1998 survey asked recipients to estimate the percentage of youth gangs in their jurisdiction with "a significant mixture of two or more racial/ethnic groups." Respondents estimated that more than one-third of their youth gangs met this criterion. Interestingly, Starbuck, Howell, and Lindquist (2001) found the largest proportion of this type of *hybrid gang* was reported in small cities, towns, and villages (54% of all gangs), and the smallest proportion was in large cities (32%).

There are, however, some important racial/ethnic differences related to gang activities and structure. Self-reported prevalence of violence involvement for White gang youth is almost one-third that of Black and Hispanic gang youth in the Esbensen team's (2008) survey of students in 15 schools in nine U.S cities. "Black gang members reported the lowest rates of property offending but the highest rates of drug sales and weapon carrying, 22%, indicating that they had carried a gun for protection" (p. 126). Gang characteristics also varied with racial/ethnic makeup. "Among the gang youth, whites tended to report smaller gangs than did the other racial/ethnic groups; only 36% of whites reported more than ten members in their group compared to 58% of black, 68% of Hispanic and 58% of multiracial respondents" (p. 128).

In Focus 3.2
A Research Challenge: Racial/Ethnic Differences

Gang research has identified gangs of a wide variety of racial/ethnic origins.

- Thrasher's (1927/2000) first gang study described varieties of gangs of European origins.
- Spergel (1964) observed differences in the structure of youth gangs and delinquent groups in the three ethnic communities that he studied: Puerto Rican, Italian, and mixed European ethnic gang involvement.
- Suttles (1968), another early gang researcher, distinguished Italian, Puerto Rican, Mexican American, and African American gangs in Chicago.

(Continued)

(Continued)

- In describing female gangs, Campbell (1984/1991) contrasted differences in family organization between African Americans and Puerto Ricans.
- Curry and Spergel (1988) analyzed differences in patterns of gang crime and delinquency across racial/ethnic communities in Chicago.
- J. W. Moore (1988) and Hagedorn (1988) contrasted distinguishing features of African American gangs studied by Hagedorn in Milwaukee with Mexican American gangs that Moore (1978, 1985) studied in Los Angeles.
- Shortly thereafter, Huff's (1990) book of gang works by noted authors featured participant observation studies of Mexican American, African American, Chinese, and Vietnamese gangs. These studies reported an array of ethnically distinct cultural and organizational features of gangs.
- Next, Curry and Spergel (1992) contrasted features of gangs in which Chicago Mexican American and African American youth participated.
- In the current century, Vigil (2002) describes four varieties of Los Angeles gangs in some detail: (1) Mexican American gangs in the barrios of the city, (2) Blacks in various Los Angeles communities, (3) Salvadorans in the city, and (4) Vietnamese in Los Angeles and other Southern California areas.
- Al Valdez (2007) describes an even greater variety of gangs of different racial/ethnic origins.

Homogeneity of membership is high among all racial/ethnic groups. More than two-thirds (72%) of White gang members and Black gang members (73%) indicated that most or all of their gang members were of the same race/ethnicity. In contrast, 90% of Hispanic gang members made this claim. It is also noteworthy that only 55% of the multiracial gang members indicated that they had a meeting place of their own, versus 67% of White gang members and 75% of Black and Hispanic gang members (Esbensen et al., 2008). Hence, territoriality appears to be more characteristic of Black and Hispanic youths than of the White and multiracial gang youths, and especially for Hispanic youth, in keeping with their social and cultural history. In addition, when asked about allowing others into their territory, Esbensen and associates found 68% of White gang members and 50% of the multiracial gang members said that they did, compared with only 40% of Hispanic gang members and just 20% of Black gang youth.

According to the 2008 NYGS, half (50%) of all gang members are Hispanic/Latino, 32% are African American/Black, 11% are Caucasian/White, and only 8% are of other race/ethnicities (Figure 3.3). There are large discrepancies between these figures and the proportions of racial/ethnic groups that are represented in self-report student surveys and law enforcement surveys. The largest discrepancy is for Whites and other racial/ethnic groups versus Hispanic/Latinos and Blacks. These four groups are about equally represented in self-report samples. These large discrepancies are not easy to explain other than the widespread overrepresentation of minority groups in the juvenile (Snyder & Sickmund, 2006) and criminal justice systems (Tonry, 2009). Another plausible explanation is that more law enforcement officers are deployed in inner-city areas where gang-involved Hispanic/Latino and African American youth are more visible.

Figure 3.3 Race/Ethnicity of Gang Members: 1996 to 2008

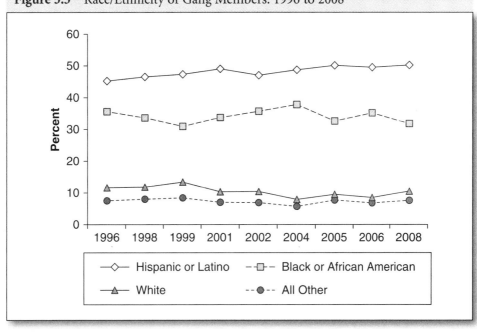

Source: National Gang Center, 2010 online survey analysis

The most noteworthy trend observed in Figure 3.3 is the greater representation of Hispanic/Latino gang members relative to the proportions of Black and White gang members in the NYGS. The proportion of Hispanic/Latino gang members only increased from 45% of all gang members in 1996 to 50% in 2008. However, this increase proved of greater significance when coupled with the drop in the proportion of Blacks from just under 36% to slightly less than 32%. Together, these changes mean that the differential between Hispanic/Latino and Black representation doubled from 9% in 1996 to 18% in 2008. This increase is partially explained by the fact that Latino immigrants make up the majority of the growth of foreign-born people who were living in the United States in 2005. And, as Vigil (2008) contends, second-generation Hispanic/Latino youth are at high risk of gang involvement.

Immigrants are a large and growing segment of the U.S. population (Vericker, Pergamit, Macomber, & Kuehn, 2009). In the past 25 years, the United States has experienced a 150% increase in the foreign-born population, such that more than 35 million foreign-born people were living in the States in 2005. Latino immigrants make up the majority of this growth, and 53% of the foreign-born persons emigrated from Latin America. As a result of this increase in the foreign-born population, the share of all U.S.–born children with at least one immigrant parent has more than tripled. Currently, about 2 in 10 children are growing up in immigrant families.

Gender

Law enforcement agencies consistently report a far greater percentage of male gang members versus female gang members. A typical finding from law enforcement surveys is about 10% female membership, but this estimation is challenged by other research methodologies, discussed in Chapter 6. Hughes (2005) underscores the enormous discrepancy between this source and self-report survey data: "Although female youth are shown to be relatively underrepresented in gangs and gang activity, they self-report gang membership at a rate up to 4½ times higher (20% to 46%) than typically indicated in surveys of law enforcement" (p. 100). Two factors may account for some of this discrepancy. First, law enforcement officers mainly observe older gang members (girls drop out earlier than boys), and, second, a smaller proportion of female gang members may be actively involved in violent activity.

In sum, research on demographic characteristics, including NYGC results reported here, underscores the diversity of youth gang members. Presumptions can be very misleading. Two discoveries in recent studies seem especially important. First, most communities can be expected to have gang members that represent a wide variety of racial/ethnic backgrounds. Second, expect to find far more female gang members than in the past. In addition, they are more likely to be active in male-dominated gangs—or somewhat gender-balanced gangs—than in independent female gangs.

Concluding Observations

Debates surrounding definitions of gangs and identification of gang members will continue indefinitely. There are numerous dimensions along which gangs can be classified. Gangs constitute a complex group phenomenon and there is considerable variation in them from one locality to another:

> No two gangs are alike, and they change constantly in membership, structure, and behavior; new gangs are formed and old ones fade away or merge with others. They do not sit still to be examined under a microscope, or by any other method devised by researchers. When we study them, whether by observation, interviews, administering questionnaires and other research instruments, or by arrest rates, the best we can hope for are snapshots at a given time or over a period of time. (Short & Hughes, 2009, p. 406)

Knowledgeable law enforcement officials are particularly adept at enumerating older gang members and more established gangs, and no doubt, undercount young gangs and gang members. Everyone has difficulty recognizing and counting these, and prison gangs as well. But the reason for law enforcement's limitation is that the youngest gangs and gang members may not yet be actively involved in criminal activity. In addition, much of the criminal acts of young gang members are not yet carried out in concert and in the streets. "Commits crimes together" is ranked the most important criterion for defining gangs among national samples of law enforcement—in large cities, suburban counties, small cities, towns, and villages, and also in rural counties (Howell et al., 2002). Hence, as discussed in Chapter 8, NYGS data show that law enforcement representatives' estimates of the number of gang members and gangs correlate with homicides.

That gang researchers disagree on the key elements of gang and gang members is not inconsequential. However, it is of utmost importance that stakeholders in neighborhoods, communities, and cities agree upon a practical definition for the assessment of their own gang problem. If not, the danger is present that stakeholders may characterize their local gang problem in terms of images of big city gangs that may not apply. Most statutory gang definitions are too broad to be of practical value in gang prevention and intervention initiatives. A practical definition is offered here that local stakeholders can modify for the purposes of their assessment.

DISCUSSION TOPICS

1. Why are gangs so difficult to define and classify?

2. Why do police departments in large cities have a tendency to apply the term very narrowly, police in smaller cities and towns more broadly, most broadcast media very broadly, and most distressed local citizens perhaps most broadly of all?

3. How do myths about gangs factor into these tendencies?

4. Class exercise: Assume that your neighborhood or community is concerned about a potential gang problem. Construct a gang definition that you would take to a community meeting to discuss the problem.

5. After reading the Decker and Kempf-Leonard (1991) article, test the acceptability of your definition from Exercise 4 in discussions with local stakeholders such as police, school officials, parents, court staff, social workers, middle and high school students, and others.

RECOMMENDATIONS FOR FURTHER READING

Peak Ages for Gang Involvement

Battin, S. R., Hill, K. G., Abbott, R. D., Catalano, R. F., & Hawkins, J. D. (1998). The contribution of gang membership to delinquency beyond delinquent friends. *Criminology, 36,* 93–115.

Esbensen, F., & Winfree, L. T. (1998). Race and gender differences between gang and non-gang youths: Results from a multi-site survey. *Justice Quarterly, 15,* 505–526.

Gottfredson, G. D., & Gottfredson, D. C. (2001). *Gang problems and gang programs in a national sample of schools.* Ellicott City, MD: Gottfredson Associates.

Hill, K. G., Lui, C., & Hawkins, J. D. (2001). Early precursors of gang membership: A study of Seattle youth. *Juvenile Justice Bulletin. Youth Gang Series.* Washington, DC: U.S. Department of Justice, Office of Juvenile Justice and Delinquency Prevention.

Lahey, B. B., Gordon, R. A., Loeber, R., Stouthamer-Loeber, M., & Farrington, D. P. (1999). Boys who join gangs: A prospective study of predictors of first gang entry. *Journal of Abnormal Child Psychology, 27,* 261–276.

Defining Gangs

Ball, R. A., & Curry, G. D. (1995). The logic of definition in criminology: Purposes and methods for defining "gangs." *Criminology, 33,* 225–245.

Bookin-Weiner, H., & Horowitz, R. (1983). The end of the gang: Fact or fiction? *Criminology, 21,* 585–602.
Horowitz, R. (1990). Sociological perspectives on gangs: Conflicting definitions and concepts. In R. Huff (Ed.), *Gangs in America* (pp. 37–54). Newbury Park, CA: Sage.
Klein, M. W. (1995). *The American street gang.* New York: Oxford University Press.

Common Characteristics of Gangs

Anderson, E. (1999). *Code of the street: Decency, violence, and the moral life of the inner city.* New York: W. W. Norton.
Horowitz, R. (1983). *Honor and the American dream: Culture and identity in a Chicano community.* New Brunswick, NJ: Rutgers University Press.
Miller, W. B., Geertz, H., & Cutter, H. S. G. (1962). Aggression in a boys' street-corner group. *Psychiatry, 24,* 283–298.
Papachristos, A. V. (2009). Murder by structure: Dominance relations and the social structure of gang homicide. *American Journal of Sociology, 115,* 74–128.
Polk, K. (1999). Males and honor contest violence. *Homicide Studies, 3,* 6–29.
Stewart, F. H. (1994). *Honor.* Chicago: University of Chicago Press.
Tita, G. E. (1999). *An ecological study of violent urban street gangs and their crime.* Unpublished dissertation. Pittsburgh, PA: Carnegie Mellon University.
Wilkinson, D. L., & Fagan, J. (1996). The role of firearms and violence "scripts": The dynamics of gun events among adolescent males. *Law and Contemporary Problems, 59* (Special Issue), 55–89.

Diversity Among Gangs

Cohen, B. (1969). The delinquency of gangs and spontaneous groups. In T. Sellin & M. E. Wolfgang (Eds.), *Delinquency: Selected Studies* (pp. 61–111). New York: John Wiley & Sons.
Covey, H. C. (2010). *Street gangs throughout the world.* Springfield, IL: Charles C Thomas.
Cummings, L. L. (1994). Fighting by the rules: Women street fighters in Chihuahua, Chihuahua, Mexico. *Sex Roles, 30,* 189–198.
Cummings, S. (1993). Anatomy of a wilding gang. In S. Cummings & D. J. Monti (Eds.), *Gangs* (pp. 49–74). Albany: State University of New York Press.
Cummings, S., & Monti, D. J. (Eds.). (1993). *Gangs: The origins and impact of contemporary youth gangs in the United States.* Albany: State University of New York Press.
Decker, S. H., & Pyrooz, D. C. (2010a). Gang violence worldwide: Context, culture, and country. *Small Arms Survey 2010.* Geneva, Switzerland: Small Arms Survey.
Decker, S. H., & Weerman, F. E. (Eds.). (2005). *European street gangs and troublesome youth groups.* Lanham, MD: AltaMira Press.
Gannon, T. M. (1967). Dimensions of current gang delinquency. *Journal of Research in Crime and Delinquency, 4,* 119–131.
Howell, J. C., Moore, J. P., & Egley, A., Jr. (2002). The changing boundaries of youth gangs. In C. R. Huff (Ed.), *Gangs in America* (3rd ed., pp. 3–18). Thousand Oaks, CA: Sage.
Klein, M. W. (1995). *The American street gang.* New York: Oxford University Press.
Miller, W. B. (1974a). American youth gangs: Fact and fantasy. In L. Rainwater (Ed.), *Deviance and liberty: A survey of modern perspectives on deviant behavior* (pp. 262–273). Chicago: Aldine.
Miller, W. B. (1974b). American youth gangs: Past and present. In A. Blumberg (Ed.), *Current perspectives on criminal behavior* (pp. 410–420). New York: Knopf.
Morash, M. (1983). Gangs, groups, and delinquency. *British Journal of Criminology, 23,* 309–335.
Short, J. F. (2006). Why study gangs? An intellectual journey. In J. F. Short & L. A. Hughes (Eds.), *Studying youth gangs* (pp. 1–14). Lanham, MD: AltaMira Press.

Spergel, I. A. (1964). *Racketville, Slumtown and Haulberg: An exploratory study of delinquent subcultures.* Chicago: University of Chicago Press.

Spergel, I. A. (1990). Youth gangs: Continuity and change. In M. Tonry & N. Morris (Eds.), *Crime and justice: A review of research* (Vol. 12, pp. 171–275). Chicago: University of Chicago.

Starbuck, D., Howell, J. C., & Lindquist, D. J. (2001). Into the millennium: Hybrids and other modern gangs. *Juvenile Justice Bulletin. Youth Gang Series.* Washington, DC: U.S. Department of Justice, Office of Juvenile Justice and Delinquency Prevention.

Sullivan, M. L. (2005). Maybe we shouldn't study "gangs": Does reification obscure youth violence? *Journal of Contemporary Criminal Justice, 21,* 170–190.

Valdez, Al. (2007). *Gangs: A guide to understanding street gangs* (5th ed.). San Clemente, CA: LawTech.

Van Gemert, F., Peterson, D., & Lien, I.-L. (Eds.). (2008). *Street gangs, migration and ethnicity.* Portland, OR: Willan Publishing.

Vigil, J. D. (2002). *A rainbow of gangs: Street cultures in the mega-city.* Austin: University of Texas Press.

Walker-Barnes, C. J., & Mason, C. A. (2001). Ethnic differences in the effect of parenting on gang involvement and gang delinquency: A longitudinal, hierarchical linear modeling perspective. *Child Development, 72,* 1814 -1831.

Notes

1. Based on the students' descriptions, Sheldon classified the organizations in which they participated into seven categories: (1) secret clubs; (2) predatory organizations; (3) social clubs; (4) personal development associations; (5) philanthropic associations; (6) athletic clubs; and (7) organizations for the promotion of literacy, artistic, and music training.

Early Gang Theories and Modern-Day Applications

Introduction

This chapter reviews theories of gangs themselves. Why do these theories exist? First, given that the existence of gangs is rooted in community and city conditions, it seems reasonable that the most relevant theories should tie gang existence to those historical and environmental conditions, or at least place a premium on those economic, social, and political contexts. Second, valid theories should provide explanations of gang activity, that is, gang-related crime. These are the two lenses through which gang theories are examined in this chapter, beginning with the earliest theories and then advancing to more recent ones. This arrangement is deliberate, to give readers a sense of the generation of gang theories that matched to some degree the historical evolution of the gangs that theories purport to explain (Table 4.1).

Macrolevel, or *ecological*, theories concern themselves with characteristics of delimited geographical areas—neighborhoods, communities, census tracts, cities, counties, states, and nations—that are related to crime (Pratt & Cullen, 2005). These theories "seek to explain why certain characteristics of ecological areas, but not others, account for the distribution of crime" (p. 373). Priority attention is given to social disorganization theory because it is the backbone of gang theorizing. Prior to the new millennium, gang theory development had been at a standstill for 35 years—since Cloward and Ohlin's (1960) strain theory/differential opportunity theory. The next explicit gang theory is Vigil's (2002) multiple marginalization theory. It is discussed subsequently along with Papachristos and Kirk's (2006) seminal contribution to social disorganization theory. Interactional theory, which is used to account for why some youths join gangs, is considered in Chapter 5.

The Chicago School and Development of the Social Disorganization Perspective

Given the rapid immigration and widespread concentration of poverty in the first large U.S. cities, it is no surprise that social scientists first focused their attention on urban social problems. A group of scholars in the University of Chicago's Department of Sociology is universally recognized for having ushered into the field of sociology a strong emphasis on the study of urban areas, using both quantitative and qualitative methodologies in studying society. The term *Chicago school* referred to that collection of faculty and graduate students during the period from around 1915 to 1930–1935. Bursik and Grasmick report the sociology discipline was first established as a department at the University of Chicago in 1892, and except for strain theory, early delinquency theories originated at the University of Chicago—the Chicago school of criminology grew out of this tradition.

Table 4.1   Main Gang-Related Theories

Theory	*Primary Scholar(s)*
Social disorganization	Burgess, 1925; Morenoff, Sampson, & Raudenbush, 2001; Park, 1936a, 1936b; Park & Burgess, 1921, 1925; Papachristos & Kirk, 2006; Sampson & Laub, 1993, 2005; Shaw & McKay, 1942, 1969; Shaw, Zorbaugh, McKay, & Cottrell, 1929; Thrasher, 1927/2000
Cultural norms of the lower-class	W. B. Miller, 1958
Subculture of violence	Wolfgang & Ferracuti, 1967
Routine activities/opportunity	Cohen & Felson, 1979; Felson & Cohen, 1980
Multiple marginalization	Vigil, 2002, 2008
Interactional	Thornberry, 1987, 2005; Thornberry & Krohn, 2001; Thornberry, Krohn, et al., 2003; Thornberry, Lizotte, et al., 2003
Reaction formation	Cohen, 1955
Strain/differential opportunity	Cloward & Ohlin, 1960
Underclass	Wilson, 1987, 1999
Conflict	Alonso, 2004; Anderson, 1998; Cureton, 2009; Currie, 1998; Pastor, Sadd, & Hipp, 2001

Under the direction of Robert Park, the Chicago school researchers accomplished what Dimitriadis (2006) says "is to this day the most systematic and comprehensive study of a single urban center" (p. 345). This was no easy task: documenting, describing, and explaining urban life during a time of rapid growth and transition in the period between World War I and the depression. Like New York City and other large urban centers, Chicago was undergoing "an unprecedented economic and cultural revolution" that presented both dangers and opportunities—a point underscored by Robert Park time and again" (p. 338). Park brought a conception of the city and communities not simply as a geographical phenomenon but as a kind of social organism. In 1921, Park explained: "Sociology, speaking strictly, is a point of view and a method for investigating the processes by which individuals are inducted into and induced to co-operate in some sort of permanent corporate existence which we call society" (p. 20).

Burgess's (1925) concentric zones theory describes the manner in which a city grows outward from the central business district to the suburbs. Of particular interest to Park and Burgess (1921) was the area immediately surrounding the central business district, known as the zone of transition, consisting of "ghettos" and "slums" located between the downtown financial and shopping center and the working-class homes. Hence Park and Burgess constructed maps as a primary means of analyzing social life in the city. Their work drew colleagues' attention to the disorganized transition zone.

As Burgess (1925) explains, this zone typically attracted real estate developers who bought large parcels of land in hopes of later growth outward from the central city. In the meantime, these speculators were not inclined to spend much money on property upkeep. It was natural for new immigrant groups to settle in these less expensive housing areas. The zone was also considered transitional in the sense that Bursik and Grasmick (1993) suggest is "a tendency for the centralization of populations into a geographic area to be followed by a period of decentralization during which these groups attempt to move into adjacent neighborhoods" (p. 8). Park (1936a, 1936b) referred to this normal process as invasion, dominance, and succession and he argued that a city's development and growth could best be understood according to these processes. Hence, Park and Burgess observed that this zone was characterized by high rates of residential mobility, deteriorated housing, abandoned buildings, and crime—characteristics that contributed to *social disorganization* within the zone.

These conditions of social disorganization were illuminated in a variety of studies: *The Polish Peasant in Europe and America* (Thomas & Znaniecki, 1920); *The Unadjusted Girl* (Thomas, 1923); *The City* (Park & Burgess, 1925); *The Ghetto* (Wirth, 1928); *The Hobo* (Anderson, 1926); *Suicide* (Cavan, 1927); *The Gang* (Thrasher, 1927/2000); *Gold Coast and Slum* (Zorbaugh, 1929); *The Taxi Dance Hall* (Cressey, 1932); *The Jack Roller* (Shaw, 1930); and *Street Corner Society* (Whyte, 1943b). Among these studies, widely applauded particularly for their richness of description and connection to human context, Park and Burgess's (1925) study proved to be a landmark one for having mapped the location of a wide variety of social problems and attendant social disorganization.

The application of social disorganization theory to crime and gangs grew specifically upon Shaw and McKay's (1942, 1969) mapping of tens of thousands of the residences of male adolescents referred to the Cook County (Chicago) Juvenile Court from 1900 through 1965. Two very important conclusions immediately emerged from this massive database. First, as Bursik and Grasmick (1993) explain, "The relative distribution of delinquency rates remained fairly stable among Chicago's neighborhoods between 1900 and 1933 despite dramatic changes in the ethnic and racial composition of these neighborhoods" (p. 31).

Second, Shaw and McKay discovered that delinquency rates correlated negatively with the distance from the central business district. This finding provided the foundation for their seminal contribution to social disorganization theory, which Bursik and Grasmick (1993) describe as follows: "that the patterns of neighborhood delinquency rates were related to the same ecological processes that gave rise to the socioeconomic structure of urban areas" (p. 33).

Next, having been influenced by Park and Burgess's discovery that economically deprived neighborhoods tended to have high population turnover, and that poor neighborhoods tended to be characterized by racial and ethnic heterogeneity, Shaw and McKay "argued that these characteristics made it very difficult for the neighborhood to achieve the common social organization goal of its residents, a situation they called social disorganization" (Bursik & Grasmick, 1993, p. 33). Thomas and Znaniecki (1920) had earlier defined social disorganization as "a decrease of the influence of existing rules of behavior upon individual members of the group" (p. 2). It appears that it was this definition that influenced Thrasher's research perspective and helped him explain the existence of street gangs.

Thrasher's Contributions to Social Disorganization Theory

In his study of gangs, Thrasher made major contributions to the social disorganization theory of the Chicago school, of which he was a vital part. Clearly, his gang research was influenced by the pioneering studies of Park, Burgess, Thomas, and Znaniecki, and other colleagues of that time and place—with whom he "was in constant dialogue" (Dimitriadis, 2006, p. 336). First, Thrasher (1927/2000) describes the gangs' physical location. This focus is, of course, a hallmark of the Chicago school studies: "There is nothing fresh or clean to greet the eye; everywhere [one sees] unpainted, ramshackle buildings, blackened and besmirched with the smoke of industry. In this sort of habitat the gang seems to flourish best" (p. 9). In addition, "Gangland represents a geographically and socially interstitial area in the city," in the slums and ghettos (p. 6). Thrasher further detailed gang activity as follows:

> The feudal warfare of youthful gangs is carried on more or less continuously. Their disorder and violence, escaping the ordinary controls of the police and other social agencies of the community, are so pronounced as to give the impression that they are almost beyond the pale of civil society. In some respects these regions of conflict are like a frontier; in other [respects], like a 'no man's land,' lawless, godless, wild. (as cited in Dimitriadis, 2006, p. 340)

Thrasher is said to have studied an astonishing total of 1,313 gangs. Short (2006), however, describes Thrasher's 1,313 gangs as "legendary." Solomon Kobrin, a long-time associate of Shaw and McKay, told colleagues that the precise (1,313) number of studied gangs was an in-joke played on Thrasher by his graduate assistants, and that it was actually the house number of a nearby brothel.

Regardless of an exact number, Thrasher (1927/2000) himself emphatically states the value of this work: "Probably the most significant concept of [his] study is the term interstitial—that is pertaining to spaces that intervene between one thing and another" (p. 6). He adds more interpretation of the meaning of the term *interstitial*:

> In nature foreign matter tends to collect and cake in every crack, crevice, and cranny—interstices. There are also fissures and breaks in the structure of social organization. The gang may be regarded as an interstitial element in the framework of society, and gangland as an interstitial region in the layout of the city where unfettered ganging occurs. (p. 6)

Therefore, the gang is "an interstitial group [i.e., between childhood and adulthood] providing interstitial activities for its members" (p. 12). Thrasher even superimposed his "gangland" map over Park's concentric zone model to demonstrate the location of gangland as virtually confined within the parallel circular boundaries of Burgess's (1925) zone in transition.

In his further contextualization of gangs, Thrasher (1927/2000) connected ganging in relation to five community conditions: (1) community disorganization, (2) ineffective families, (3) poor-quality schooling, (4) association with undesirable peers, and (5) lack of leisure-time guidance. First, with respect to community disorganization, he contended that gangs form and flourish in communities in which normally directing and controlling customs and institutions fail to function efficiently and effectively. Contributing conditions included poverty, deteriorating neighborhoods, disintegration of family life, ineffective religion, lack of opportunity for wholesome recreation, inefficient schools, corruption and indifference in local politics, and unemployment. "All of these factors enter into the picture of the moral and economic frontier, and, coupled with deterioration in housing, sanitation, and other conditions of life in the slum, give the impression of general disorganization and decay" (p. 12). These conditions, Thrasher argued, resulted in the breakdown of social control because of the weakening of social institutions, particularly family life, education, and religion. Youths exposed to these conditions enjoyed "an unusual freedom from restrictions of the type imposed by the normal controlling agencies in the better residential areas of the city" (p. 170). In Thrasher's view, this freedom from normal social controls allows ganging to occur.

In addition, Thrasher observed that immigration dynamics contribute significantly to community disorganization. Rapid immigration, he argued, created a "chaotic milieu" that lessened parental controls on their children. "Stated in another way, the problem is one of too quick and too superficial Americanization of the children of the immigrant. The more rapid this type of assimilation, the rapid will be the disintegration of family control" (p. 170). It became impossible to maintain family discipline in this context "because of the lack of tradition in the American community to support it, and the constant coming and going out (immigrant succession).... In this sense, then, the gang becomes a problem of community organization" (p. 170).

Thrasher's second contextual factor that explains ganging (or prepares youths for gang life) is ineffective families. By this, he refers to "any condition of family life which promotes the gang by stimulating the boy to find the satisfaction of his wishes outside the plan and organization of family activities" (p. 170). As noted, Thrasher found that immigration dynamics undermined parental control and nourishment of children. He also called attention to other sources of family deficiencies—including poverty and antisocial parents—contributing to neglect, and ignorance of proper childrearing practices or immorality, "but their general effect is the same: the family fails to hold the boy's interest, neglects him, or actually forces him into the street" by alienating him from family life (p. 170).

The quality of schooling that youngsters received in Chicago was the third key factor in preparing youths for gang life in Thrasher's view. In particular, the educational process during school hours did not interest future gang members; thus they resorted to truancy. In addition, the compulsory education law permitted youth to drop out at age 14, but they were prohibited from taking employment before age 16. Lastly, the lack of school extracurricular activities that supplemented the work of other agencies (during and after school hours) freed youths from adult supervision. As Thrasher noted, "it often happens that boys expelled either as individuals or as a group from some formal organization are drawn together to form a gang" (p. 11).

Association with delinquent peers is Thrasher's fourth contextual factor. Intimate "grouping" had long been viewed as a normal social tendency, especially among children and adolescents, beginning about the age of 10, according to early observers such as Puffer (1912) and Spaulding (1948). Although this tendency is no longer viewed as an innate trait, gang scholars still put plenty of currency into Thrasher's notion that, at the individual level, boys—gang and nongang alike—from an early age are influenced enormously by relations with their peers, and problem associations are often formed when the above conditions "give the boy an opportunity to roam about and choose his associates and his amusements for himself" (Thrasher, 1927/2000, p. 171).

Thrasher's fifth negative factor that contributes to conditions favorable to ganging is the lack of proper guidance for boys during leisure-time activities: "The common assumption that the problem of boy delinquency will be solved by the multiplication of playgrounds and social areas is entirely erroneous" (p. 171). Rather, it is guidance on the part of trained parents, teachers, and recreational leaders who can organize and direct wholesome activities that was largely absent in gangland. Freeing the boy from ordinary controls allows gangs to form.

It is readily apparent that Thrasher not only saw multiple factors that account for gang involvement, but also that the interaction of these factors increases the potency of them. Dimitriadis (2006) calls this perspective Thrasher's "situation complex," his recognition that the collection of community, family, school, and peer factors form "a web of influences that could not be understood in isolation form, but only in relation to each other" (p. 338). This is the strength of Thrasher's *contextualization* of ganging, which he ties directly to immigration in a "matrix of gang development" to wit: "It seems impossible to control one factor without dealing with the others, so closely are they interwoven, and in most cases they are inseparable from the general problem of immigrant adjustment" (Thrasher, 1927/2000, p. 491). Bourgois (2003) notes that Thrasher devoted the last 15 years of his life to applying his theory to juvenile delinquency in Italian Harlem with great academic success, through his institute at New York University.

In sum, Thrasher's gang social disorganization theory developed from the premise that youth gangs were rooted in the struggles of ethnic groups to establish themselves or maintain their standing in urban social systems—if only what J. W. Moore (1998) calls "a toehold." Thrasher (1927/2000) contended that his theory applied not only in Chicago but also in New York City, Boston, Minneapolis, Cleveland, Los Angeles, New Orleans, Denver, and San Francisco. The key element of his theory—and the Chicago school in general—was the weakening of social controls in the fabric of key institutions, social structural weaknesses, particularly in families, schools, communities, and religious institutions. It can be said that the lasting legacy of Thrasher's gang theory is that, in the tradition of the Chicago school, it cast a spotlight on "the processes by which gangs influence and are influenced by their physical environment . . . the human element" (Hughes, 2007, p. 44).

A key limitation of Thrasher's gang social disorganization theory is its focus on the development of gangs in lower-class areas while ignoring the presence of gangs in suburban and working- and middle-class areas. But gangs had not yet been reported in these areas; these groups were far more prevalent in inner-city areas. Both Hughes (2007) and Short (1998) assert that critiques of Thrasher's theory also have noted the near absence of a focus on group processes and their relationship to individual and collective behavior. Other critics have been less charitable, including Becker (1999) who characterizes the Chicago school as promoting an "eclectic" mix of sociological approaches rather than an actual "unified school of thought." Subsequent development and testing of the theory would dispel this notion.

Application of Social Disorganization Theory to Gangs

For nearly a century, social disorganization theory has endured as an explanation of crime and the existence of gangs, despite many academic challengers (J. W. Moore, 1998; Pizarro & McGloin, 2006). Numerous studies have found that gang behaviors and gangs themselves are more likely to be concentrated in poor and disorganized cities and communities (Curry & Spergel, 1988; Short & Strodtbeck, 1965/1974; Vigil, 1988, 2002).

Interestingly, the very survival of social disorganization theory may owe to an early Chicago school study, which found that gangs could provide at least a semblance of order in some "disorganized" communities. According to Papachristos (2004), Whyte's (1943b) *Street Corner Society* research demonstrated that gang turfs "were not *dis*organized, but rather *differently* organized and that 'street corner groups' provided a clear and *organized* response to the disruption of formal social institutions" (pp. 7–8). Geis (1965) adds, "The Whyte book is jammed with shrewd observations and telling insights into the wellspring of gang action" (p. 20). In his discovery of *The Social Order of the Slum,* Suttles (1968) also observed mechanisms by which street gangs contributed to a "skeletal" frame of order in the world of street corner gangs. These included social interactions on the streets and relational patterns such as modes of dress, dating rituals, public interaction, and familial patterns. Along with others' contributions, Whyte's and Suttles's studies appear to have prompted a significant modification of social disorganization theory, the identification of social networks that can bring order to otherwise disorganized communities. In this context, gangs organized youths' behavior, a point made emphatically by Whyte (1943a): "In other words, the boy's behavior is not un-organized; it is organized—by the gang." Whyte's research revealing the social organization of slums convinced him that "it is fruitless to study the area simply in terms of the breakdown of old groupings and old standards; new groupings and new standards have arisen" (p. 38).

The term *collective efficacy* (informal social control) was created to describe this dimension of community disorganization (Sampson and Laub, 1997; Sampson, Raudenbush, & Earls, 1997). According to Sampson and colleagues (1997), "Collective efficacy refers to the process of activating or converting social ties to achieve any number of collective goals, such as public order or the control of crime" (p. 919). More specifically, "It is the linkage of mutual trust and the shared willingness to intervene for the public good that captures the neighborhood context of what [these scholars] term *collective efficacy*" (Sampson, Morenoff, & Gannon-Rowley, 2002, p. 457). Coleman (1988, 1990) notes that the collective efficacy variable has also been operationalized in reference to specific dimensions of social capital, pinpointed specifically by Short and Hughes (2010) at the neighborhood level of social organization. Sampson and colleagues (1997) suggest this form of community or neighborhood organization refers to cooperative efforts in defining, monitoring, and condemning undesirable behaviors that occur within a community. Examples include a willingness to intervene in such everyday situations as graffiti spray painting, truant children hanging out on a street corner, a fight in front of one's house, and a disrespectful child. These types of interference make it possible for people to walk the streets at night or to permit their children to play outside the home without fear. In community-level social disorganization studies, Sampson and Graif (2009) identify rapid population turnover, concentrated disadvantage, and racial or ethnic heterogeneity as the key components of an explicit definition for the term *social disorganization*.

Modern-Day Application of Social Disorganization Theory to Gangs

The central thesis of social disorganization theory, according to Shaw and McKay (1942, 1969), is that ethnic heterogeneity, low socioeconomic status, and residential mobility reduce the capacity of community residents to control crime. Prevailing thought, for which there is growing research support, is that these structural disadvantages in communities make it difficult for residents to develop informal social control and collective efficacy, both of which can reduce crime and violence in communities. Several studies, most notably Bursik and Grasmick (1993), suggest that where collective efficacy is lacking, the diminished private, parochial, and public forms of social control permits gangs to flourish. Furthermore, leading researchers such as Papachristos and Kirk (2006) find the principles of collective efficacy also apply to the control of gang crime. Short and Hughes (2010) suggest that "the extent to which detached workers—and other therapeutic or change agents and programs develop social capital and collective efficacy among gang members and communities is, we believe, an important missing ingredient in arguments concerning their effectiveness" (p. 133).

Shaw and McKay (1931, 1942) specified three mechanisms by which social disorganization theory accounts for gang existence: neighborhood population mobility, ethnic heterogeneity, and poverty. They viewed mobility and population heterogeneity as disruptive influences on community institutions and networks, but these effects would vary according to, first, the rate of immigration, and second, the similarities between the newly arriving immigrants and the established residents of a neighborhood. As Papachristos and Kirk (2006) explain, these factors disrupt the normative foundation that supports effective social control, thereby contributing to a state of social disorganization. "The extension of social disorganization theory to gang behaviors is straightforward: Gangs arise either to take the place of weak social institutions in socially disorganized areas, or because weak social institutions fail to thwart the advent of unconventional value systems that often characterize street gangs" (p. 64). In this statement, Papachristos and Kirk offer two interpretations of social disorganization theory. The first is that gangs fill a void created by weak social institutions, as Thrasher suggested. Second, Papachristos and Kirk posit that the weakened social institutions allow destructive social organisms to flourish, including gangs. Each of these interpretations nicely fits the theory.

Key variables were specified in the next iteration of social disorganization theory, and the addition of neighborhood social control would remedy its blind spot to ganging outside of inner-city areas. These additions were explicitly codified and tested by Papachristos and Kirk (2006) in an apropos Chicago study. These scholars identified key crime-producing structural features of neighborhoods that had previously received substantial research support: concentrated disadvantage, residential instability, and immigrant concentration (Figure 4.1). Their explication of social disorganization theory also incorporated collective efficacy and social control as key social processes in neighborhoods. Their test of whether or not social disorganization theory predicted gang homicide encompassed the entire city of Chicago (847 census tracts combined into 343 neighborhood clusters).

Papachristos and Kirk (2006) found that the social disorganization theory predicted gang homicide, but not nongang homicide: "The neighborhoods with a high level of general violence are not necessarily the same neighborhoods that have high levels of gang violence" (p. 80). However, the researchers found that "collective efficacy operates similarly on violent gang behavior as it does on other forms of violent behavior" (p. 75) in mediating the effects of concentrated disadvantage. In addition, Papachristos and Kirk found that greater immigrant concentration was related to Hispanic homicides but not Black homicides.

Other gang studies have produced findings that also support social disorganization theory although not without exceptions. In an Arizona study, Katz and Schnebly (2011) found that some disadvantaged areas produce gangs while others do not. This

Figure 4.1 Causal Model of Social Disorganization Theory Applied to Gang Behavior (Homicide)

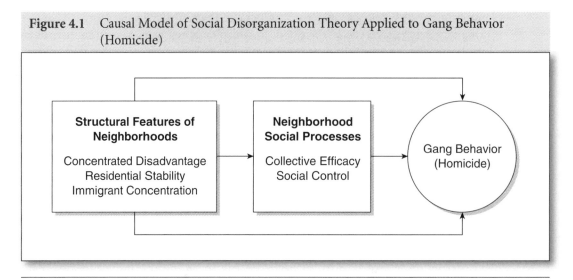

Source: Papachristos & Kirk, 2006, p. 68

study examined the relationship between neighborhood structure, violent crime, and concentrations of gang members at the neighborhood level. The researchers used official police gang lists, police crime data, and two waves of decennial census data (that characterized the socioeconomic and demographic conditions) on 93 neighborhoods in Mesa, Arizona. Two research questions were addressed: (1) How well do neighborhood-level structural conditions account for variability in gang member concentrations (gang member rates based on the police gang lists)? (2) What impact, if any, do levels of crime or community conditions have on the concentration of gang members in a given neighborhood? The study revealed that neighborhoods characterized by relatively high levels of economic deprivation and social or familial disadvantage indeed tend to have higher rates of gang membership. However, this study produced two surprising findings. First, "neighborhoods with high concentrations of gang members were not necessarily those with comparatively high levels of officially recorded violence" (p. 23). This anomaly could be a result of underreporting of gang crime. Second, "concentrations of gang members within neighborhoods was not significantly related to population density" (p. 23). This finding, according to W. Miller (2001), could be attributed to the long-term trend of outward movement of inner-city residents to suburban areas.

Independent St. Louis studies produced contrasting outcomes. First, in a study of gang formation in five geographic regions in the city, Monti (1993) observed that "gangs can be found in a variety of communities [and they] may be more prevalent in areas that are characterized as 'unsettled' [but] gangs can arise in communities with substantial strength and a varied organizational base" (p. 249). For example, Monti found that gangs formed in north St. Louis and adjacent suburbs in conjunction with migration of Black residents across the metropolitan area. "The emergence of youth gangs in this case . . . is better explained as an expression of community reorganization than disorganization" (p. 250).

In the second St. Louis study, Rosenfeld and fellow researchers (1999) examined gang homicides between 1985 and 1995 in St. Louis (in 588 census block groups comprising the entire city), and found few differences between neighborhoods that had high levels of gang and nongang homicide. Both gang and nongang homicide tended "to cluster in areas with high levels of disadvantage and large concentrations of African-American residents" (p. 510). All three homicide types examined in the study—gang-affiliated, gang-motivated, and nongang youth homicides—also were concentrated in racially isolated, disadvantaged neighborhoods with moderate levels of instability, but gang-motivated events exhibited a somewhat distinctive spatial patterning, related to turf battles or retaliatory motives.

In the only test to date of social disorganization theory's ability to account for gang activity at the national level, Pyrooz, Fox, and Decker (2010) examined the effects of economic disadvantage and racial and ethnic heterogeneity on gang membership in the 100 largest U.S. cities using National Youth Gang Survey (NYGS) data. In contrast with macrolevel gang studies Jackson (1991) and Wells and Weisheit (2001) conducted, asking only whether a city had a gang problem, this study analyzed gang membership rates (changes in the number of gang members). The key finding from this research is that both economic disadvantage and racial and ethnic heterogeneity—two key cornerstones of social disorganization theory—exhibit independent, additive, and multiplicative effects on gang membership rates as reported in the NYGS. However, the results indicated that the effect of heterogeneity was twice as great as that of economic disadvantage. In cities with high heterogeneity, disadvantage had a larger effect. Conversely, the effect of disadvantage was modest in cities with low heterogeneity. Pyrooz and colleagues (2010) emphasize the main implication for gang research, that heterogeneity and disadvantage work together to produce higher levels of gang membership—in the 100 largest cities.

It is important to keep in mind that social disorganization theory was based on Chicago populations, and thus applies mainly to cities with large urban populations. The discordant findings in several studies suggest that even when communities are characterized by concentrated disadvantage, residential instability, and immigrant concentration, one can find different patterns of criminal gang activity and nongang crime. "The diverse findings . . . may be the result of a common problem in empirically oriented gang research: data are often aggregated in units that are too large to capture localized gang phenomena" (Tita et al., 2005, p. 275). Long-term intergang wars can also explain these variations. Examples of the former case are plentiful in gang literature (W. B. Miller, 1974b, 1982/1992), with the Crips versus Bloods and MS-13 versus 18th Street in Los Angeles serving as a good example of long-term gang battles. In Chicago, an ongoing conflict involved two "brother" gangs, the Black Gangster Disciples and the Black Disciples, which Block and colleagues (1996) report resulted in 45 homicides over an 8-year period. Each of these feuds likely was driven simultaneously by macrolevel (e.g., social disorganization) and microlevel (e.g., drug wars) factors.

Moreover, "it is unknown whether neighborhood social processes operate in a similar way across different types of disadvantaged neighborhoods. It is possible that some social processes are unique to economically depressed areas" (Kingston, Huizinga, & Elliott, 2009, p. 53). This could be attributable to the condition that most studies lack a sufficiently large sample of disadvantaged neighborhoods to test these kinds of neighborhood-specific hypotheses. This situation is particularly evident in studies that have tested gang phenomena. Findings that are consistent at the national level often are not at the neighborhood or community level. Cities and communities have their own social, economic, and gang histories, which appears to account for disparate findings seen in several of these studies. Nevertheless, Papachristos (2004) notes social disorganization theories encounter two main problems in their application to gangs. First, as seen in the work of Whyte (1943b) and Suttles (1968), social disorganization theories do not account for the development of gangs in more stable, lower-class and nonghetto neighborhoods. Second, social disorganization theories do not account for social and demographic changes that do not fit within the original social disorganization paradigm. The key condition is the "underclass," which has not experienced the same ethnic succession as seen among European White ethnics. A related limitation is that factors other than the variables that comprise social disorganization theories may be important in accounting for gang presence in communities. Other prominent theoretical frameworks are examined next.

Other Pioneering Gang Theorizing

In addition to Thrasher's "unparalleled" research, Hardman (1967) draws attention to several other important contributions of the 1900 to 1930 eras, notably Puffer, Cooley, Furfey, and Bolitho. It was Puffer (1912) who mainly carried forward Hall's (1904) persisting notion of the 18th century that ganging together was a manifestation of primitive behavior. Hardman (1967) explains, "Our ancestors bagged game by throwing stones; hence all boys throw stones; they play in caves, climb trees, and play [other games] for the same reason. Savages learned that they could survive better in groups than alone" (p. 6). Hence, Puffer (1912) identified a boy's ganging tendency as the "direct result of his inheritance from some thousands of generations of savage ancestors who, willy-nilly, have been doing these things all their lives" (p. 76). Sheldon (1898) noted the tendency of children to engage in organized activity at a very young age: "American children, left [unattended will] organize" (p. 430). This principle soon lost currency, in favor of Cooley's (1909) competing primary group theory that emphasized the importance of intimate face-to-face association and cooperation for group actions. Furfey (1928) later identified some key elements of group cohesion: age; mental development; and common environment, interests, and moral standards. But it was Bolitho (1930) who saw the same influence of immigrant clashes on gang formation as Thrasher (apparently working independently of each other). Very shortly, Landesco (1932) confirmed Bolitho's thesis.

As noted earlier, following Thrasher, Shaw, and McKay (1942) found ganging to be concentrated in transitional slum areas, particularly where social conditions were deteriorating). Over the next three decades—after a period of neglect following the pioneering studies of Thrasher and Shaw and McKay—theoretically oriented scholars made numerous attempts to explain the existence of gangs and the violent behavior of their members. After reviewing pertinent studies of this period, Klein (1969) formulated a schematic framework that depicts the presumed causal relationships among key theoretical variables that were prominent in the first four decades of gang literature (Figure 4.2). The early gang studies produced several important findings.

Figure 4.2   A Causal Paradigm of Gang Violence

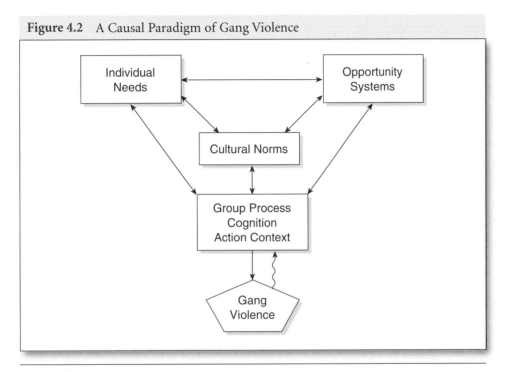

Source: Klein, 1969, p. 1436

1. Delinquency is a group phenomenon.

2. Better structured and more enduring gangs are found in the lower-class culture.

3. Racial, ethnic, and cultural conflict are conducive to ganging.

4. Parental delinquency, indifference, laxity, and neglect are correlates of both ganging and delinquency.

5. There is a wide variety of sizes, types, structures, and purposes of gangs.

(Continued)

(Continued)

6. There are many more examples of nebulous, causal, unstructured cliques than of tightly organized, highly structured gangs.

7. Primary groups similar to gangs are constantly forming and dissolving at all levels of society, from preschool play groups to formalized adult groups.

8. There is greater interest in and need for ganging during adolescence (although there is discord as to why this is so).

9. If gang organization exists, there will be a leadership elite, lay members, and hangers-on.

10. There are more similarities than differences between social groups and antisocial gangs.

11. Ganging fills strong emotional needs.

12. Gang structure and cohesiveness increase with external threat or conflict.

13. Better structured gangs live longer.

Source: Hardman, 1967, p. 26. Reprinted with permission from Sage Publications, Inc.

Individual needs. Early gang scholars concentrated their work on the characteristics of criminals in their quest to provide an answer to the burning question "why do they do it" (Short & Meier, 1981, p. 462). The most commonly identified individual needs among gang members in the early studies, particularly Klein's in 1969, were self-esteem, self-expression, and status. Whyte (1943b) found a premium placed on status management in gangs, while Hughes and Short (2005) detected violent behavior frequently initiated in response to status threats. Short and Strodtbeck (1965 /1974) concluded that a primary source of violent individual conduct resided in gang members themselves, their own social disabilities: "Caught in a cycle of limited social abilities and other skills, and experiences which further limit opportunities to acquire these skills or to exercise them if acquired. These disabilities, in turn, contribute to the status dilemmas of these youngsters and in this way contribute to involvement in delinquency" (p. 243). Gordon (1967) concurred, emphasizing members' social disabilities while providing support for Thrasher's view of conflict as a main cohesion-builder for the gang. Similarly, Yablonsky (1959, 1967) contended that violent gang members were so impaired socially that he labeled them as "sociopathic" and he suggested that the gang itself drew social misfits and defective personalities to it.

Opportunity systems. The theorizing of the 1950s and 1960s reflects a discernible shift away from the individual's status in relation to the gang setting toward status in relation to the larger class system (Hughes & Short, 2005). Social class differentials were the focus of theories in this era. Major contributors to gang theorizing with respect to social opportunities include Cohen (1955), Bloch and Niederhoffer (1958), and Cloward and Ohlin (1960). Cohen (1955) suggested that the lower-class boy's social position of having been denied status in the respectable society leads to a "reaction formation" and a collective solution, turning to the delinquent subculture, within which status criteria could be met with ease. Bloch (1963) and Bloch and Niederhoffer (1958) also emphasized the social isolation of gang members, that is, their striving for manhood, adult status, and autonomy in a society that could not bridge the cleavage between adolescence and adulthood.

Cloward and Ohlin (1960) provided a challenge to Cohen's notion that gang boys are primarily reacting to middle-class values. Instead, in their strain theory (also called a theory of differential opportunity), Cloward and Ohlin argued that special problems of adjustment (strain) are created because of the disparity between cultural goals and legitimate opportunities to achieve them. Therefore, boys turned to deviant opportunities in order to achieve goals because opportunities for conformity are limited—not out of negative, malicious, reaction to established goals, as Cohen (1955) had suggested. Bordua (1961) poked fun at Cloward and Ohlin's theory with the observation that "Cloward and Ohlin's delinquents seem suddenly to appear on the scene, sometime in adolescence, to look at the world, and to discover, 'Man, there's no opportunity in my structure'" (p. 134). On a more serious note, Bordua notes that their theory is devoid of family variables.

Importantly, Cloward and Ohlin (1960) identified three types of subcultures: (1) *conflict* (prone to violence), (2) *criminal* (prone to mainly property crimes), and (3) *retreatists* (drug addicts), all of which drew widespread attention for a number of years. Their theory, influenced by Merton's (1957) theories of anomie (normlessness) and strain, as well as Sutherland's (1947) differential association theory, explained why boys would choose one type or the other. As might be expected—given the history of criminal and conflict orientations of street gangs, evidence of these types has been reported from time to time but not with the retreatist gangs (Short & Strodtbeck, 1965/1974).

Cultural norms. W. B. Miller (1958) developed a new theoretical perspective on gangs, which differed significantly from those of his predecessors. He offered an exclusively cultural explanation, based on 3 years of extensive research on 21 corner groups in Boston. He saw gang members' behavior as influenced by the cultural system of the lower-class community itself, rather than middle-class expectations. That is, youths who engaged in gang delinquency were behaving in a manner consistent with the "lower-class way of life" (p. 6). Miller concluded that, in the case of "gang" delinquency, the cultural system that exerts the most direct influences on behavior is that of the lower-class community itself, rather than a so-called delinquent subculture, which is oriented to the deliberate violation of middle-class norms. He said, "The standards of lower-class culture cannot be seen merely as a reverse function of middle-class culture—as middle-class standards 'turned upside down'; lower-class culture is a distinctive tradition many centuries old with an integrity of its own" (p. 19). Miller identified and described in normative terms six "focal concerns of lower class-culture": trouble, toughness, smartness, excitement, fate, and autonomy—as they are defined within lower-class society. Miller saw these focal concerns as "attitudes, practices, behaviors, and values characteristic of lower class culture . . . designed to support and maintain the basic features of the lower class way of life" (p. 19).

Wolfgang and Ferracuti (1967) soon proposed the existence of a subculture of violence in certain segments of the lower class in which there exists a culturally transmitted lifestyle. These scholars suggested that code-related values and beliefs were widespread in high-crime communities. In the *subculture of violence thesis,* violence is "a part of the lifestyle," a means "of solving problems and problem situations" that Wolfgang and Ferracuti found to be widespread among lower-class individuals in Philadelphia. However, Pratt and Cullen (2005) conclude from their systematic review that research support for the general subculture of violence thesis is weak.

Three decades later, based on his Philadelphia ethnographic research, Anderson (1999) coined the phrase *code of the street* as a set of informal rules or street culture governing interpersonal public behavior, particularly violence. At the heart of Anderson's code is the idea of respect. Youngsters on dangerous streets who follow the street code attempt to build reputations for themselves as commanding respect, or *juice,* thereby warding off potential attackers. One of the street code's most pervasive norms is that of *retribution,* stipulating that personal attacks (verbal or physical) must be avenged (Anderson 1999).

Brezina and colleagues (2004) found some support for Anderson's basic thesis using national youth survey data. Stronger support has emerged from studies in Georgia and Iowa with neighborhood-level data that measured adolescents' adoption of Anderson's street code (Stewart, Schreck, & Simons, 2006; Stewart & Simons, 2010). This research found that, indeed, adolescents who follow the street culture have a higher probability of offending. Moreover, at the individual level, adoption of the street code "seems to aggravate the risk of victimization because most of the potential offenders are themselves operating by the street code" (Stewart et al., 2006, p. 448). At the neighborhood level, Stewart and Simons (2010) conclude,

> The positive effect of individual-level street code values on violent delinquency is increased when an adolescent lives in a neighborhood where the street culture is endorsed. This pattern of results adds to the growing literature that describes how neighborhood structural characteristics combined with deviant cultural and situational codes lead to the perpetuation of violence. (p. 592)

Stewart and fellow researchers' (2006) earlier study also drew attention to this cycle of violence: "As a consequence, the potential for relatively minor infractions to escalate into major violent encounters (e.g., lethal violence) increases, thereby leading to a cycle of violence in African American communities" (p. 449).

Underclass Theory

Before underclass theory was developed, there existed what W. B. Miller (1990) describes as "a virtual taboo among public figures against characterizing low-status populations as a class" (p. 278). Opposition to characterizing low-status populations as a class all but ended in the 1980s with widespread acceptance of the *underclass* concept. Auletta (1982) first used this term to promote his characterization of socially dysfunctional behaviors among extremely poor minorities in New York City. W. J. Wilson (1987) next described certain Black peoples as underclass of Chicago.

Underclass theory proposes that in inner-city Black communities, the social impact of deindustrialization have been exacerbated by the departure from cities by middle-class and working-class residents (W. J. Wilson, 1987, 1999). J. W. Moore (1998) emphasizes these people who were able to relocate "were the beneficiaries of affirmative action in jobs and anti-segregation in housing" (p. 76). A distinguishable social class (underclass) was created by a new set of demographic, technological, and economic conditions that had reduced the demand for low-skilled workers—permanently locking them out of the labor market and slicing off upward mobility routes that had been available to earlier generations. In her critique of this underclass theory, Moore (1998) argues that there are exceptions to disappearance of urban jobs for the lowest rungs of workers. "It should be noted that deindustrialization is not the only story: low-wage manufacturing—which, however is largely staffed with immigrant labor—has burgeoned in many cities, and, in addition, there has been growth of low-wage service jobs" (p. 76). Therefore, she prefers the term *economic restructuring* (the replacement of manufacturing work with services jobs) over underclass theory. As seen in Chapter 1, and as Moore and Pinderhughes (1993) assert, the underclass theory never applied in Los Angeles to the Mexican immigrants, nor in several other cities.

Routine Activities (Opportunity) Theory

Whereas the macrolevel theories attempt to explain why collectives of individuals are motivated to engage in criminal and gang activity, the routine activities theory (also called opportunities theory) (Cohen & Felson, 1979; Felson & Cohen, 1980) assumes that propensity is present and intends to account for the opportunities that motivated offenders have to victimize other people or property. These opportunities consist of attractive or suitable targets (e.g., vulnerable persons with money, and unlocked automobiles) and a lack of capable guardians. It is the convergence of motivated offenders, suitable targets, and lack of capable guardians in time and place that accounts for the nonrandom distribution of crime and violence in this theory. Although this theory is very popular and intuitively quite appealing, Pratt and Cullen (2005) contend it presently lacks the research support to elevate it on a par with other macrolevel theories. The empirical support to date is almost entirely on attenuated guardianship and very few studies have examined the independent effects of either the motivated offender or the suitable targets variables of routine activities theory.

Wikstrom and Treiber's (2009) situational theory of violence suggests that certain persistent forms of violence may be driven by habitual processes, which occur in a setting which the actor regularly takes part in and which consistently presents opportunities or frictions conducive to violence, and few deterrents. Examples include "gang violence, which occurs in specific geographic areas (territories), presents regular opportunities and frictions (via the presence of fellow gang members, rival gang members and those 'transgressing' on gang 'turf') and few controls (is regulated by the rules of gang membership)" (p. 89). Indeed, such gang settings tend to become familiar for regular members, thus automating expected responses.

Griffiths and Chavez's (2004) more recent research extends routine activities, or opportunity, theory in suggesting that certain other community characteristics may increase criminal opportunities. "For instance . . . increases in criminal facilitators (for example, guns), and demographic changes (for example, increases in young males who become involved in gangs) may explain large changes in crime and violence over a relatively short time" (p. 945). Griffiths and Chavez suggest that "neighborhoods may demonstrate differences in quantity and lethal violence, not only between one another at any given time, but also within their own boundaries over time. That is, they may have distinct trajectories" (p. 943). Following Clarke (1995), Griffiths and Chavez call attention to increases in guns and increases in young males who join gangs as "facilitators" that "may explain large changes in crime and violence over a relatively short time" (p. 945). In other words, such changes "need not, counter to the expectations of social disorganization theory, accompany structural or demographic changes in neighborhoods" (p. 945).

Although their analysis of Chicago homicides did not include gang presence, Griffiths and Chavez (2004) demonstrate in their space and time analysis that gun homicide trends from 1980 to 1995 were volatile in the most violent communities—in which social disorganization theory would predict stable high rates. Moreover, they extend routine activities/opportunity theory in finding some "high and increasing street gun communities" that surrounded the most violent ones. This finding, they suggest, provides "tentative support for the proposition that facilitators, such as guns, increase homicide victimization by bringing together victims and offenders in time and space, but that these opportunities diminish as one moves farther from the violent core" (p. 970). Without question, this observation applies to street gangs.

Multiple Marginalization Theory

Just as "young immigrant men of Irish, Italian, German, and Polish origin gathered on the street corners of their respective neighborhoods to confront together the rigors of their new life in the industrialized cities of the eastern United States," Vigil (2002) explains, "today, too, the processes young immigrants go through in dealing with life in a new country are essentially the same as in the past" (p. 3). These processes involve the *marginalization* of ethnic minorities, particularly Mexican American youth (Vigil, 1979, 1988, 2002, 2008). Having been left out of mainstream society because of language, education, cultural, and economic barriers, this situation "relegates these urban youths to the margins of society in practically every sense. This positioning leaves them with few options or resources to better their lives. Often, they seek a place where they are not marginalized—and find it in the streets" (Vigil, 2002, p. 7).

As Adler, Ovando, and Hocevar (1984) note, economic, psychological, and cultural conflicts are common among Mexican American families. Domestic violence, alcoholism, drug addiction, inadequate housing, poverty, and discrimination are widespread problems in these families. In stress-filled families, there is a tendency for primary social control networks to be loosened. Vigil (1988) explains in more detail, "As a result, an unsupervised child spends more time outside the home with street peers. These street peers . . . become agents of socialization" (p. 43). Thus, Vigil (2002) identifies a result of multiple marginalization experiences as the emergence of street gangs and the generation of gang members.

In addition to the emphasis that Vigil places on the contribution of educational deficits to gang involvement, Telles and Ortiz's (2008) multi-intergenerational study revealed just how critical educational achievement is for Mexican American youth: "The die for the social disadvantage of Mexican Americans is cast during their educational process, which ultimately determines their position in the American social structure" (p. 156). The evidence is clear that "Mexican Americans have lower levels of schooling than any other major racial-ethnic group" (p. 265). Vigil's (2008) more recent elaboration of his multiple marginalization theory emphasizes the acute period during which gang involvement is most pronounced among Mexican American youth—in the second generation of migrants. This observation is also supported in Telles and Ortiz's landmark study and by Waters (1999).

Conflict Theory

Conflict theory seeks to explain why economic disadvantage leads to higher crime rates. Among conflict theory's several theoretical variations, Turner's (1978) popular version is that the dynamics of conflict between social groups produces higher crime rates. And more contemporary conflict theorists contend that conditions of economic deprivation may generate markets for drugs, firearms, and gangs (Currie, 1998).

The history of street gangs discussed in Chapter 1 provides a plethora of conflict examples. In New York City, after 1840, gang warfare replicated ethnic conflict in the climate of economic restructuring and intense competition for jobs. By the 1940s, three-way race riots involved Italian Americans, Puerto Ricans, and African Americans in East Harlem. In the 1990s, post–World War II urban renewal, slum clearances, and ethnic migration pitted gangs of African American, Puerto Rican, and Euro-American youth against each other in battles in New York City to dominate changing neighborhoods.

Chicago's early history saw large Irish gangs terrorizing the German, Jewish, and Polish immigrants who settled there from the 1870s to the 1890s. The reigning Irish gang, Ragen's Colts, attacked both Mexican American and Black youth, in marking the racialized boundaries of "their" space. The race riot of 1919, in which Black males united to confront hostile White gang members who were terrorizing the Black community, also contributed directly to gang formation. Partly in response to growing racial and ethnic violence, Black, Puerto Rican, and Mexican American gangs alike proliferated in the late 1950s.

In Los Angeles, physical and cultural marginalization (Vigil, 1988, 2002, 2008) gave rise to street gangs of Mexican origin, and two interracial events in the 1940s proved pivotal in the growth of Mexican American gangs in the West: the Sleepy Lagoon murder and the zoot suit riots. Later, both Alonso (2004) and Cureton (2009) suggest the Black civil rights movement (1955 to 1965) produced an underclass-specific, socially disorganized, and isolated Black community. Several gang historians show that the resurgence of new emerging street groups in Chicago and Los Angeles coincided with the political, social, and civil rights movements, particularly the Black Panther Party and the U.S. Organization (both of which were racially motivated entities) (Alonso, 2004; Cureton, 2009; Diamond, 2001). According to Alonso (2004),

"Racialization and disenfranchisement of Black youth in specific geographic locales was a driving force behind initial gang formation" in Los Angeles (p. 669).

In each of these three cities, public housing isolated and ghettoized large numbers of Blacks and Mexican Americans and served to incubate more gangs with greater capacities for violence. From his study comparing Chicago with several European cities, Wacquant (2007) tells us that a form of *neo-apartheid* characterizes major American cities, Chicago in particular—the most racially segregated of all American cities, he claims—that produces the *hyper-ghetto,* in which there are no governing rules and regulations. The victims experience what Wacquant calls *advanced marginality,* which may account for a variety of violent social conflicts, gang violence included.

Racial/ethnic conflicts may well be more neighborhood-specific than generalized. Pastor, Sadd, and Hipp (2001) coined the term *ethnic churning* to describe the process of racial/ethnic transition in a neighborhood, specifically, changes in the proportions of each racial/ethnic group. Hipp, Tita, and Boggess's (2009) subsequent research finds that racial/ethnic transition in nearby tracts leads to greater levels of intergroup violence by both Black and Latino groups. Along this line of thinking, recent research suggests that the degree of immigrant concentration may be less important than proximity to other hostile immigrant groups and conflict with dominant local groups. In a study using data from the National Neighborhood Crime Study to examine the racial-spatial dynamic of violence for neighborhoods in 36 U.S. cities, R. Peterson and Krivo (2009) found that "proximity to more disadvantaged areas and especially to racially privileged (heavily white) areas is particularly critical in accounting for the large and visible inequality in violence found across neighborhoods of different colors" (p. 105). However, this effect of disadvantage in neighboring areas operates indirectly. "Our [research] approach points to particular aspects of neighboring conditions (e.g., disadvantage, residential instability, percentage white) that are relevant for generating violence" (p. 106). It appears that "proximate white concentration directly affects race-ethnic inequality in focal neighborhood violent crime" (p. 106).

Concluding Observations

This chapter has given particular attention to macrolevel theories of crime and gang-related indicators. Social disorganization theory has received substantial empirical support. The central thesis of this theory is that ethnic heterogeneity, low socioeconomic status, and residential mobility reduce the capacity of community residents to control crime. Three earlier reviews of research that tested social disorganization theory suggested a modest relationship between neighborhood structural disadvantage and delinquency, violence, and other child, adolescent, and adult behaviors (Bursik & Grasmick, 1993; Sampson, Morenoff, & Gannon-Rowley, 2002; Sampson & Groves, 1989).

A more systematic and inclusive review indicates that the strength of the evidence in favor of social disorganization theory as an explanation of crime is more robust when compared with other ecological theories of crime. This observation comes from Pratt and Cullen's (2005) systematic review (employing meta-analysis) of more than 200 studies of ecological correlates of crime. "Concentrated neighborhood disadvantage" (most commonly measured in terms of racial heterogeneity, poverty, and family disruption) proved to be the strongest and most consistent predictor of crime. "Across all studies, social disorganization and resource/economic deprivation theories receive strong empirical support; anomie/strain, social support/social altruism, and routine activity theories receive moderate support; and deterrence/rational choice and subcultural theories receive weak support" (p. 373). In examining the strength and stability of macrolevel predictors, Pratt and Cullen's systematic review revealed that just five of these scored high on *both strength of effects and stability* across studies: (1) percent non-White, (2) percent Black, (3) poverty, (4) family disruption, and (5) the effect of incarceration on crime rates. Unemployment, low firearms ownership, collective efficacy, religion, and unsupervised peer groups are other variables that scored *high on strength of effects.* Racial heterogeneity and residential mobility were among six variables that showed *moderate strength of effects and stability* across studies (along with household activity ratio, social support/altruism, inequality, and urbanism).

The second noteworthy observation is that support for social disorganization theory as an explanation of gang membership and gang violence is substantial. Several studies reviewed in this chapter suggest a relationship that is conditioned on the features of the supporting research, such as samples, research methods, and level of analysis (individual, micro, and macro). Arguably most important, Pyrooz and colleagues (2010) demonstrated the generalizable effects of economic disadvantage and racial and ethnic heterogeneity on gang membership rates, in the 100 largest U.S. cities with National Youth Gang Survey data. Other city-level studies showed strong relationships between social disorganization and gang violence in multiple-year data, in Chicago and in St. Louis. Coupling these results with Pyrooz and colleagues' (2010) national analysis provides additional support for social disorganization theory as an explanation of gang presence and associated violence.

However, even social disorganization theory is inadequate to account for the variability in cities' gang activity trajectories and distinctive gang homicide trajectory patterns (discussed in Chapter 8). As Griffiths and Chavez (2004) note, "Overall, there is little theoretical work on how [temporal changes in violence] may manifest over time and across space in urban areas" (p. 946). Griffiths and Chavez incorporate community-level mediators including aspects of the socioeconomic structure (increased unemployment), and increases in criminal facilitators (street gun usage) to help account for homicide trends. Their research suggested that neighborhoods experiencing homicide increases along with growing street gun weaponry might also be spaces that experience drug markets or gang activity. Indeed, the highest-rate homicide neighborhoods were located within the turfs of the notorious Vice Lords or the Black Gangster Disciples.

Similarly, the locations of extremely violent gang *set spaces* need explanation. Tita and colleagues (2005) explain, "Studying the places where gang members congregate as a sociological group is analogous to studies of crime-prone locations rather than of crime-prone individuals." Furthermore,

Even within high crime neighborhoods, crime exhibits non-random patterns of highly localized concentration in crime "hot spots." The current research demonstrates that gangs display a similar pattern. Gangs are spatially concentrated among disadvantaged neighborhoods, but gang set space represents a sub-neighborhood phenomenon, with gang members hanging out in relatively small, geographically defined areas within a neighborhood. What "hot spots" are to the study of crime, "set space" is to the study of gangs. (p. 273)

Research to date suggests several contexts in which gang set space and violence are more likely. Griffiths and Chavez (2004) attribute an increase in lethal violence to street gun usage, and Cohen and Tita (1999) more specifically identify the joint location of drug trafficking and guns as key factors. Of course, adjacent location of gang territories and high crime density is important. Last, concentrated disadvantage, residential instability, and immigrant concentration, and racial/ethnic heterogeneity (racial/ethnic churning) in urban areas appears to contribute both to violence and gang activity.

DISCUSSION TOPICS

1. How did the early social disorganization theories provide a foundation for later theories?

2. Are the social disorganization theories still relevant today to modern-day gangs? Why or why not?

3. How much of modern-day gang activity does each of the most relevant theories explain? Arrange them in a pie chart and justify the relative proportion.

4. Which is the more valid theoretical construct—racial/ethnic succession or racial/ethnic churning—for explaining the existence of gangs?

5. Can the most popular gang theories explain *hot spots* and *set space* influences on gang violence?

RECOMMENDATIONS FOR FURTHER READING

Chicago School Reviews

Kornhauser, R. R. (1978). *Social sources of delinquency*. Chicago: University of Chicago.

Short, J. F., Jr. (1998). The level of explanation problem revisited—The American Society of Criminology 1997 Presidential Address. *Criminology, 36*, 3–36.

Snell, P. (2010). From Durkheim to the Chicago school: Against the "variables sociology" paradigm. *Journal of Classical Sociology, 10*, 51–67.

Social Disorganization Theory

Chavez, G. M., & Griffiths, E. (2009). Neighborhood dynamics of urban violence: Understanding the immigration connection. *Homicide Studies*, 13, 261–273.

Deuchar, R. (2009). *Gangs, marginalised youth and social capital*. Sterling, VA: Trentham Books.

Hagedorn, J. M. (1988). *People and Folks: Gangs, crime and the underclass in a Rustbelt City*. Chicago: Lakeview Press.

Hughes, L. A., & Short, J. F. (2005). Disputes involving gang members: Micro-social contexts. *Criminology, 43*, 43–76.

Miller, W. B. (1974b). American youth gangs: Past and present. In A. Blumberg (Ed.), *Current perspectives on criminal behavior* (pp. 410–420). New York: Knopf.

Monti, D. J. (1993). Gangs in more- and less-settled communities. In S. Cummings & D. J. Monti (Eds.), *Gangs: The origins and impact of contemporary youth gangs in the United States* (pp. 219–253). Albany: State University of New York Press.

Moore, J. W. (1978). *Homeboys: Gangs, drugs and prison in the barrios of Los Angeles.* Philadelphia: Temple University Press.

Rosenfeld, R., Bray, T. M., & Egley, A., Jr. (1999). Facilitating violence: A comparison of gang-motivated, gang-affiliated, and nongang youth homicides. *Journal of Quantitative Criminology, 15,* 495–515.

Collective Efficacy

Bursik, R. J., Jr., & Grasmick, H. G. (1993). *Neighborhoods and crime: The dimensions of effective community control.* New York: Lexington.

Morenoff, J. D., Sampson, R. J., & Raudenbush, S. W. (2001). Neighborhood inequality, collective efficacy, and the spatial dynamics of urban violence. *Criminology, 39,* 517–559.

Papachristos, A. V., & Kirk, D. S. (2006). Neighborhood effects on street gang behavior. In J. F. Short & L. A. Hughes (Eds.), *Studying youth gangs* (pp. 63–84). Lanham, MD: AltaMira Press.

Sampson, R. J. (2002). The community. In J. Petersilia & J. Q. Wilson (Eds.), *Crime: Public policies for crime control* (pp. 225–252). Oakland, CA: Institute for Contemporary Studies Press.

Short, J. F., Jr., & Hughes, L. A. (2010). Promoting research integrity in community-based intervention research. In R. J. Chaskin (Ed.), *Youth gangs and community intervention: Research, practice, and evidence* (pp. 127–151). New York: Columbia University Press.

Tita, G. E., Riley, K. J., & Greenwood, P. (2005). *Reducing gun violence: Operation Ceasefire in Los Angeles.* Washington, DC: National Institute of Justice.

Studies of Social Disorganization and Gangs

Curry, G. D., & Spergel, I. A. (1988). Gang homicide, delinquency and community. *Criminology, 26,* 381–405.

Hipp, J. R., Tita, G. E., & Boggess, L. N. (2009). Intergroup and intragroup violence: Is violent crime an expression of group conflict or social disorganization? *Criminology, 47,* 521–564.

Kingston, B., Huizinga, D., & Elliott, D. S. (2009). A test of social disorganization theory in high-risk urban neighborhoods. *Youth & Society, 41,* 53–79.

Mares, D. (2010). Social disorganization and gang homicides in Chicago: A neighborhood level comparison of disaggregated homicides. *Youth Violence and Juvenile Justice, 8,* 38–57.

Monti, D. J. (1993). Gangs in more- and less-settled communities. In S. Cummings & D. J. Monti (Eds.), *Gangs: The origins and impact of contemporary youth gangs in the United States* (pp. 219–253). Albany: State University of New York Press.

Pyrooz, D. C., Fox, A. M., & Decker, S. H. (2010). Racial and ethnic heterogeneity, economic disadvantage, and gangs: A macro-level study of gang membership in urban America. *Justice Quarterly, 14,* 1–26.

Rosenfeld, R., Bray, T. M., & Egley, A., Jr. (1999). Facilitating violence: A comparison of gang-motivated, gang-affiliated, and nongang youth homicides. *Journal of Quantitative Criminology, 15,* 495–515.

Shaw, C. R., & McKay, H. D. (1969). *Juvenile delinquency and urban areas* (2nd ed.). Chicago: University of Chicago Press.

Suttles, G. D. (1968). *The social order of the slum.* Chicago: University of Chicago Press.

Whyte, W. F. (1943a). Social organization in the slums. *The American Sociological Review, 8,* 34–39.

Whyte, W. (1943b). *Street corner society: The social structure of an Italian slum.* Chicago: University of Chicago Press.

5

Gang Involvement as a Developmental Trajectory

Introduction

The theories reviewed in Chapter 4 help explain the existence of gangs. In this chapter, the focus is on individuals and how they become involved in gangs. Studies show that, like other adolescent problem behaviors, gang involvement develops gradually over time, and life-course development theories (Elder, 1985, 1997) best explain this process. Two technical terms are used in this chapter that need some explanation. First, a *developmental pathway* refers to a specific pattern in the sequence of behaviors, such as a pattern of delinquency from less serious problem behaviors to more serious offenses. Second, *developmental trajectories* refer to different types or groups of offenders that are distinguished by their behavioral development over time. Gang members as a trajectory group, the focus of our attention in this chapter, are developmental processes that can inform prevention and intervention strategies. These strategies are considered in the final two chapters of this book.

Developmental Theories

The basic premise of an *ecological* model is that environmental (especially social and economic) characteristics of communities and neighborhoods determine how well a child does in school and later in life. Developmental psychologist Bronfenbrenner's (1979) conceptualization of the different spheres of influence that affect a child's behavior inspire this framework, such as relations in the family, the peer group, and schools. Generally speaking, then, developmental and life-course theories aim to explain the course of development of offending, the influence of risk factors, and the effect of life events (Farrington, 2005). Loeber and colleagues' (1993) research reveals a remarkably orderly progression from less to more serious problem behaviors and delinquency from childhood to adolescence. Along with describing a segment of a delinquent or criminal career trajectory that progresses from less serious problem behaviors to more serious offenses, Loeber, Keenen, and Zhang (1997) also consider developmental pathways as "the stages of behavior that unfold over time in a predictable order" (p. 322).

Developmental Pathways

Developmental criminology has an interesting history. First, Farrington (1986) constructed a "stepping-stones" model in which he arranged risk factors that predict adult criminality at key stages of chronological development (childhood, adolescent, and young adulthood). Next, Loeber (1990) conceived of the "developmental stacking" of problem behaviors; that is, new problem behaviors are added to earlier ones. He enumerated five characteristics of a developmental progression in problem behaviors. First, some behaviors have an earlier onset than others; second, there is usually escalation in the seriousness of problem behavior over time; third, early behaviors usually are retained as new ones are added; fourth, each behavior is best predicted by the developmentally adjacent behavior; and fifth, the ordering of behaviors in a pathway is sequential.

Loeber and colleagues (1997, 1999) discovered three main but overlapping pathways in the development of delinquency from childhood to adolescence (Figure 5.1). These are the authority conflict pathway, the covert pathway, and the overt pathway. The authority conflict pathway consists of predelinquent offenses, the covert pathway consists of concealing and serious property offenses, and the overt pathway consists of violent offenses. Loeber's pathways model has four important dimensions. First, the model shows an orderly progression over time from less serious to more serious offenses and delinquent behaviors. Second, the progressively narrowing width of the triangles illustrates the decreasing proportion of youth (from *many* to *few*) involved in particular problem behavior and delinquent offenses. Third, the model shows the general age of onset (from *early* to *late*). Fourth, the pathways are hierarchical in that those who have advanced to the most serious behavior in each of the pathways usually have displayed persistent problem behaviors characteristic of the earlier stages in each pathway.

Problem behavior typically begins in the authority conflict pathway with stubborn behavior, followed by defiance or disobedience, then truancy, running away, or staying out late. Persistent offenders then typically move into either the overt pathway or the covert pathway. The first stage of the covert pathway is minor covert behavior (shoplifting, frequent lying); this is followed by property damage (vandalism, fire setting), and then moderately serious (fraud, pickpocketing) and serious delinquency (auto theft, burglary). The first stage of the overt pathway is minor aggression (bullying, annoying others); physical fighting follows (often including gang fighting) and then

Figure 5.1 Developmental Pathways to Serious and Violent Behavior

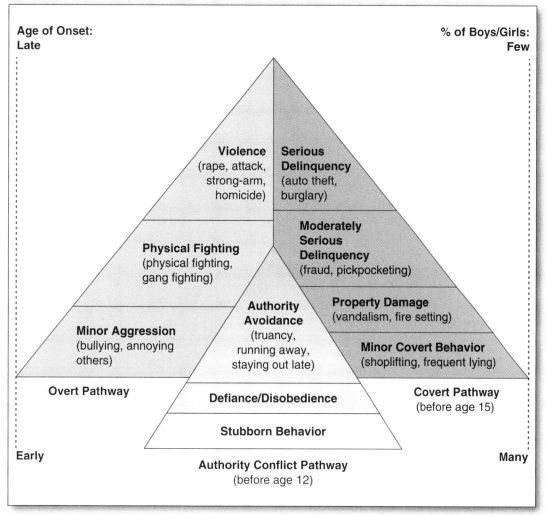

Source: Loeber et al., 1999, p. 247

more serious violence (rape, physical attacks, and strong-arm robbery). Thus, gang involvement (gang fighting) is an intermediate step in the overt pathway.

With age, Loeber, Burke, and Pardini (2009) reported a large proportion of Pittsburgh Youth Study boys progressed on two or three pathways, indicative of an increasing variety of behaviors over time. In particular, boys who were escalating in the overt pathway were more likely to progress in the covert pathway. This model also accounted for the self-reported high-rate offenders and court-reported delinquents.

Kelley's (1997) and Loeber's (1999) associates validated the model on delinquents in Denver and Rochester samples, while Tolan and Gorman-Smith (1998) confirmed in a Chicago study and the National Youth Survey. In addition, Tolan, Gorman-Smith, and Loeber (2000) replicated the model in a sample of African American and Hispanic adolescents in Chicago and in a nationally representative U.S. sample of adolescents. The pathway model has also been largely replicated for girls (Gorman-Smith & Loeber, 2005; Loeber, Slot, & Stouthamer-Loeber, 2007).

This developmental model clearly illustrates how gang involvement is intertwined with delinquent behavior, including serious and violent offenses. The elegance of this developmental theory is that it explains delinquency involvement as an outcome of the relative preponderance of risk factors over *promotive* factors. Stouthamer-Loeber and colleagues (2002) describe promotive factors as not only serving to buffer individuals from negative effects of risk factors (the common conceptualization of protective factors), but also promoting conventional social behavior and desistance. In extremely high-risk conditions, people need more than a simple majority of protective factors to overcome multiple risk factors (Stouthamer-Loeber et al., 2008). Moreover, Stouthamer-Loeber and fellow researchers (2002) note the condition of considerable exposure to risk domains paired with the relative absence of protective domains "dramatically increases the risk of later persistent serious offending" (p. 120).

Unfortunately, the youth gang field has seen less research on protective factors than other human service fields. Numerous possible protective factors have been suggested in the gang literature. However, the research base in this area is far too scant (largely consisting of cross-sectional studies) to compile a "research-supported" list of protective factors; hence, potential factors are not discussed here for fear of misleading the field. According to Stouthamer-Loeber and associates (2008), the strongest protective factors for desistance from overall serious offenses are parents' good supervision, the youth's high-perceived likelihood of getting caught when delinquent, and low parental stress.

Other Developmental Theories

Of course, there are other developmental theories of offenders that have received empirical support (see Table 5.1). While few of these address co-offending, and none do in detail, each of them holds potential for application to an individual's involvement in gangs. Two of these theories, although neither claim to account for gang membership, are widely recognized and tested by others: (1) life-course–persistent and adolescence-limited offenders and (2) age-graded informal social control. These are discussed subsequently along with Thornberry and Krohn's interactional theory.

Life-course–persistent and adolescence-limited offenders. Moffitt (1993) and others identified two main groups of offenders in the childhood and adolescent period: life-course–persistent offenders and adolescence-limited offenders. Life-course–persistent offenders begin offending in childhood, and adolescence-limited offenders begin their offending later. According to Moffitt, four characteristics distinguish life-course–persistent offenders: (1) early onset of offending, (2) active offending during adolescence, (3) escalation of offense seriousness, and (4) persistence in crime in adulthood. In

Table 5.1 Leading Life-Course Developmental Theories

Theory	*Primary Scholar(s)*
Age-graded informal social control	Sampson & Laub, 1993, 2005
Coercion/early onset model	Patterson, Capaldi, & Bank, 1991
Developmental pathways	Loeber, 1996; Loeber et al., 1993
Life-course–persistent and adolescence-limited offenders	Moffitt, 1993
Integrated cognitive antisocial potential (ICAP) theory	Farrington, 2003, 2005, 2006
Integrative multilayered control	LeBlanc, 1993
Interactional	Thornberry, 1987, 1996, 2005; Thornberry & Krohn 2001, 2005
Social development model	Catalano & Hawkins, 1996
Strain theory (modified)	Agnew, 2005

contrast, Farrington (1986) asserts adolescence-limited offenders represent a far larger proportion of the adolescent peak in the age-crime curve. These offenders do not have childhood histories of antisocial behavior; rather, they engage in antisocial behavior only during adolescence.

Moffitt's typology has stimulated a large volume of investigations into the actual existence of various delinquent subgroups with distinct offending patterns. However, most research, and notably Loeber and colleagues in 2009, has revealed four main juvenile offender groups: (1) those whose problem behavior remains high over time, (2) those whose problem behavior remains low over time, (3) those whose problem behavior increases, and (4) those whose problem behavior decreases. In addition, the Pittsburgh Youth Study did not find clear evidence for two separate life-course–persistent and adolescence-limited offenders in either of two large cohorts (Loeber et al., 2008). Sampson and Laub (2005) note that their own study and other empirical research has "firmly rejected the notion that there are only two groups of latent-class offenders" (p. 30). In the Rochester Youth Development Study, eight groups with diverging patterns of offending were observed (see Figure 5.2), prompting Thornberry (2005) to conclude that "offending can start earlier or later, but onset is not distinctly divided into neat patterns of *early starters* and *late starters*" (p. 160). Moffitt (2007) has revised her typology to feature additional types of low-level chronic offenders and adult-onset offenders.

Age-graded informal social control. The key construct in Sampson and Laub's (1993, 2005) developmental theory is age-graded informal social control, particularly the

Figure 5.2   Eight Trajectories of Offender Careers

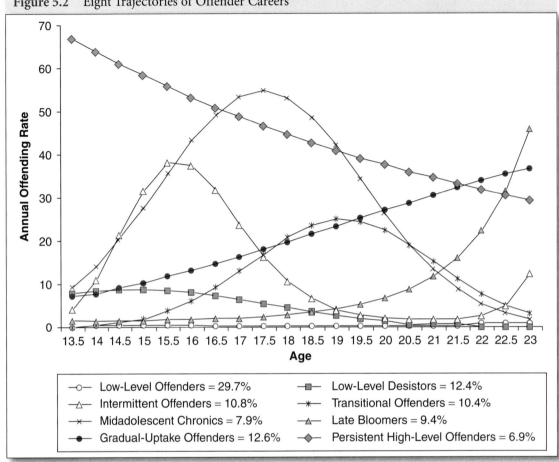

Source: Thornberry, 2005, p. 164

strength of bonding to family, peers, schools, and later adult social institutions (marriage and jobs). Like other popular developmental theories, their theory emphasizes structural background variables (e.g., social class, ethnicity, large family size, criminal parents, disrupted families) and individual factors (e.g., low intelligence, difficult temperament, early conduct disorder) that have indirect effects on offending because of their effects on informal social control (via attachment and socialization processes).

Sampson and Laub use Elder's (1985, 1997) concept of *turning points* to describe the mechanisms that operate when a change occurs in the life course, when a risk pathway is recast to a more adaptive path, such as desistance from delinquency. They contend that individual development over the life course is a dynamic process whereby the interlocking nature of trajectories, or life paths, and life-course transitions generates turning points, or changes in the life course. In childhood, according to Sampson and Laub's theory, the main causes of delinquency are found in family dynamics such as erratic, threatening, and harsh discipline; low levels of supervision; and parental rejection. In adolescence, other causal factors become important, including lack of attachment to school, attachment to delinquent peer groups, and involvement in the juvenile justice system. This theory is particularly relevant to gang involvement because of the importance placed on attachment to delinquent friends in the teenage years when offending peaks.

Interactional theory. This is the only developmental theory that portends to explain gang involvement. Various studies led by Thornberry depict the path model of the origins of gang membership, derived from the researchers' own interactional theory. This path model contains three fundamental premises. First, this theory adopts a developmental or life-course perspective, which posits that causal influences are not unidirectional:

> Indeed, a core argument is that delinquent behavior feeds back upon and produces changes in both bonding and associations. The more the individual engages in delinquency, the more that involvement is likely to increase alienation from parents, reduce commitment to school and render conventional success goals moot. (Thornberry, Lizotte, Krohn, et al., 2003, p. 13)

Second, this theory emphasizes that the casual influences vary developmentally:

> For example, during childhood, family influences are predicted to be more powerful than school or peer influences in shaping behavior. As the individual moves through adolescence, the search for and attainment of autonomy increases the impact of school and peer influences, while the impact of the family fades. (p. 13)

Third, interactional theory suggests that all of these processes vary by structural position in society. "Youth growing up in socially disadvantaged families and neighborhoods, especially if they are people of color, are apt to have more difficult life-course trajectories in which the previous processes leading to delinquent careers are exacerbated" (p. 13). Conditions and circumstances in their environment diminish the chances that strong prosocial bonds and opportunities will be available, which in turn and heighten the chances that deviant opportunities will be available as they reach adolescence. Of course, these include factors such as delinquent peers, street gangs, and drug markets. Interactional theory predicts that "these youngsters are more likely to have serious and persistent careers" (p. 13).

Thornberry, Krohn, and colleagues (2003) tested their path model for males in the Rochester study. The overall results provided considerable empirical support. However, it needs to be tested in other sites and for applicability to girls, for there were too few female gang members in the sample to permit theory testing. In addition, younger samples should be used, because the youngest boys in the Rochester sample were about 13 years of age. Even stronger results might be seen.

Importance of Gang Membership in Developmental Trajectories

This section highlights research on the importance of the gang membership stage in the formation of youngsters' criminal trajectories. But this is never the first stage.

Longitudinal gang studies inevitably find that delinquency involvement typically precedes gang membership, with Loeber and fellow researchers' (2008) Pittsburgh Youth Study a primary example. In particular, alcohol and drug use, along with serious theft and violence precede gang joining. More serious drug involvement begins to occur in the same year as gang involvement, followed by drug dealing and gun carrying. Several studies show that gang members commit many more serious and violent acts while they are gang members than before joining and after leaving the gang. This is the most widely researched trajectory pattern, and Thornberry (1998) notes the finding that gang involvement increases youths' criminality as "one of the most robust and consistent observations in criminological research" (p. 147). Since Thornberry made this observation, this important finding has been further replicated in the United States (Gordon et al., 2004; Krohn & Thornberry, 2008) and also in Canada (Gatti, Tremblay, Vitaro, & McDuff, 2005; Haviland & Nagin, 2005; Haviland, Nagin, Rosenbaum, & Tremblay, 2008) and Norway (Bendixen, Endresen, & Olweus, 2006).

In Focus 5.1
Pittsburgh Youth Study: Age of Onset of Problem and Criminal Behaviors

Behavior (mean age in parentheses)

alcohol use (12.0)

serious theft (13.1)

tobacco use (13.2)

serious violence (14.6)

gang membership (15.4)

hard drug use (15.6)

marijuana use (15.7)

drug dealing (16.2)

gun carrying (17.3)

Source: Loeber, Farrington, Stouthamer-Loeber, White, & Wei, 2008, p. 150 (oldest cohort)

The following are noteworthy findings from studies that have analyzed the influence of gang membership on trajectory patterns.

The first studies to examine delinquency offending patterns among gang members along a developmental trajectory before, during, and after boys are active gang members were conducted in Rochester, New York (Thornberry et al., 1993), Denver, Colorado (Esbensen & Huizinga, 1993), Pittsburgh, Pennsylvania (Gordon et al., 2004), and Seattle, Washington (Hill, Chung, Guo, & Hawkins, 2002). These studies revealed the escalating crime effect of active gang involvement, which Thornberry (1998) called a "facilitation" effect.

Hill and colleagues' (1996) Seattle study showed that, even when youths engaged in violent behavior beforehand, involvement in violence escalated substantially during active gang membership. A second analysis of the Seattle study sample, led by Hill (2002), found different violence patterns that (irrespective of gang membership) characterize four trajectory groups: (1) nonoffenders, (2) desistors, (3) late escalators, and (4) chronics. Next, the researchers entered gang membership to see if violence changed in the four groups as a function of active gang involvement. Violence increased in each of the groups except for nonoffenders.

Using a similar approach in a Montreal study, Lacourse and colleagues (2003) identified three trajectory groups: (1) adolescents who never joined a gang, (2) a childhood onset group, and (3) an adolescent onset group. When gang membership was interjected, violence escalated most precisely during the period of gang membership in each trajectory group. The researchers also explored the effects on violence of movement in and out of a gang. At all ages and for all trajectory groups, transitions into a gang were associated with an increase in violent behaviors, and vice versa.

Next, Gordon's (2004) research team used more comprehensive controls for preexisting characteristics of boys as well as controls for age, season, and calendar year. This Pittsburgh study further substantiated earlier research, finding "a substantial elevation in delinquency when boys are active gang members" (pp. 80–81). In an added feature of this research, controlling for crime trends, the temporary increase fell to pregang levels once boys left gangs.

Two additional trajectory studies used the same Montreal data to refine trajectory modeling. Haviland & Nagin (2005) concluded that facilitation effects observed in the Rochester study are present in a very different ethnic and cultural setting in Canada, among white French youths. Haviland (2008) and fellow researchers next found that, in Montreal, gang membership was highly transient. "Boys who joined gangs at age 14 soon quit; others soon joined" (p. 432). As expected, this transitory process served to dissipate the effects of gang joining on violence.

The insights from these studies prompted Howell, Egley, et al. (2011) to explore whether cities' gang problem histories could be grouped in trajectories similar to individual offender careers. The effort was successful and the results are summarized in Chapters 7 and 8.

A Developmental Model of Gang Involvement

Readers are reminded that the developmental model of gang involvement is not a theory of gang formation. Those theories were reviewed in Chapter 4 (in addition, see the discussion of starter gangs in Chapter 9). Rather, this chapter explains gang joining as a developmental process. The main goal is to provide a framework to guide the development of successful communitywide prevention and intervention strategies. For this purpose, causal relationships are not made explicit and, no doubt, oversimplified.

The central theoretical question is this: Why are some youth at high risk of joining a gang? Howell & Egley (2005b) extended the age span of Thornberry and colleagues' (Thornberry, Krohn, et al., 2003) gang membership theory downward, so that the developmental model presented here (see Figure 5.3) encompasses antecedents of gang membership from birth through adolescence. This is important because, as Loeber and colleagues (2009) note, about equal numbers of risk factors associated with early disruptive and delinquent behaviors begin very early, at or close to birth, and in the elementary school years, with fewer first appearing during the middle and high school years. "Thus, the most salient risk window of children's exposure to risk factors is prior to adolescence" (p. 301).

Figure 5.3   Delinquency and Gang Theory

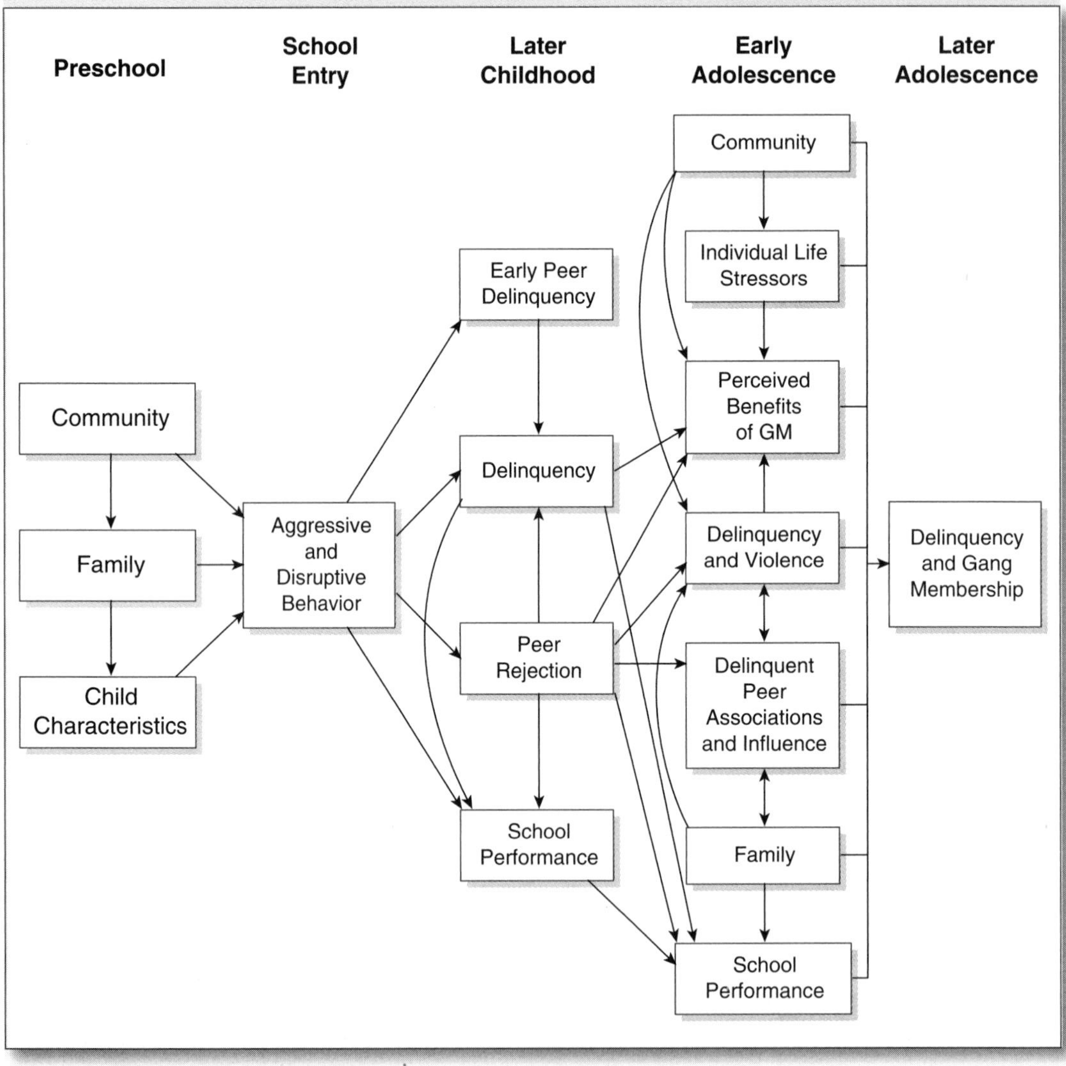

Source: Author

Perceived Benefits of Joining a Gang: Gang Attractions

Early gang research suggested that gang joining is a somewhat natural process. Thrasher (1927/2000) initially viewed the typical gang member as a rather healthy, well-adjusted, red-blooded American boy seeking an outlet for normal adolescent drives for adventure and expression. More recent gang studies have refuted this romanticized view of what gangs offer young candidates. As Reckless, Dinitz, and Murray (1956) noted in the juvenile delinquency field, two sets of forces are at work. The first set is called *pulls,* or gang attractions, and an opposite set of forces is called *pushes,* or risk factors that may propel youth toward gangs. The latter set is the main focus of this chapter.

A male gang leader expressed succinctly the main attractions of the gang in response to a question posed to him in the course of an interview. Hardman (1969) asked him what additional questions might be included in the interviews of other gang members to obtain a better understanding of the reasons why youths join gangs. He replied (quoted verbatim here):

> Ask them about feelin's of not bein' wanted at home; ask them about feelin' left out; ask them about the gang makin' up for some of these things they didn't get at home, y'know; ask them about the gang makin' them feel important, wanted, needed. Ask them about the feelin' of security that it gives you, always knowin' you got the guys backin' you up. Ask them about the feelin' of importance that they get from bein' in the gang. (p. 179)

Vigil (2002) helps further illustrate the attraction of gang membership as follows:

> What established gangs in the neighborhood have to offer is nurture, protection, friendship, emotional support, and other ministrations for unattended, un-chaperoned resident youths. In other words, street socialization fills the voids left by inadequate parenting and schooling, especially inadequate familial care and supervision. This street-based process molds the youth to conform to the ways of the street. (p. 10)

In an 11-city study, Esbensen, Deschenes, and Winfree (1999) compiled the following reasons youth reported for joining a gang, in the order of descending importance:

- For protection
- For fun
- For respect
- For money
- Because a friend was in the gang

Of these reasons, most studies relay that youth most commonly join gangs for the safety they believe the gang provides, particularly in urban areas. Gangs are often at the center of appealing social action, including hanging out, music, drugs, and opportunities to socialize with members of the opposite sex. In other words, the gang may be appealing because it meets a youth's social needs. Decker and Van Winkle (1996) found youth also reported, albeit far less frequently, more instrumental reasons for joining a gang such as drug selling or making money. J. W. Moore's (1978) and Vigil's (1988) early ethnographic studies noted the strong influence of friends and family members who already are part of the gang, especially for Mexican American youth. More recent student surveys also show this, and for all racial/ethnic groups (Peterson et al., 2004).

In Focus 5.2
San Diego Mexican American Gangs

The San Diego Mexican American gangs Sanders (1994) studied had a strong social orientation, focused on action events, status, and "kicking back." Action consisted of risk-taking activities. Status ensued from belonging to a gang that was feared or respected by other gangs in the community. Kicking back meant hanging out with other members, partying, drinking, and taking drugs. Although heavy drug users, they rarely were involved in trafficking (pp. 137–138).

These gangs were organized around cliques and barrios. A given barrio might have one or more gangs, although typically only one. Each gang consisted of age-grade cliques, each of which had one or more informal leaders. Sanders (1994) characterized these gangs as having more of the *influential* leader model of organization than the *horizontal* structure commonly found among Chicano gangs in Los Angeles. Strong solidarity grounded in the barrio identity held each gang and clique within it together. The continued existence of barrios gave stability to even inactive gangs.

Unlike the Black gangs described by Jankowski (1991) and Keiser (1969), the San Diego Black gangs (Crips and Bloods) that Sanders (1994) studied were not highly organized, but very large and divided into allied branches. Each of the branch territories was made up of sets. Their lack of organization was related in part to the fact that their territories were large, according to Sanders, because of their money-making enterprises. Each gang was made up of hard-core, affiliate, and fringe members. Because of the age cohorts, leadership structure, and limited number of hard-core members, these gangs were constantly in flux.

Sanders (1994) suggests that gangs ebb and flow from inactive neighborhood street groups to active fighting gangs with the incorporation of each new cohort of members. "As a new age cohort of boys comes into the gang, the level of activity rises. After a time, they either make their names, become more aware of the risks they are taking, or become cowed by the police or other gangs. When this occurs, the gang's activity subsides" (p. 61).

Source: Based on Sanders, 1994.

W. Miller suggests (2001) that, apart from personal reasons for joining a gang, media presentations make gangs seem very appealing. The "hip" lifestyle and sensational portrayals of gangs and their members have a significant influence, particularly on more susceptible youth. Previously, rap lyrics promoted prosocial values; that changed with gangsta rap, which promotes the gang culture popularized by movies such as *Colors, Boyz in the Hood,* and *Menace II Society* that transformed gangsters into folk heroes (Ro, 1996).

Romantic interests are another factor. Many female adolescents are attracted to gang life through their boyfriends. One study placed the research spotlight on San

Antonio, Texas, girls who never joined the gang (Petersen & Valdez, 2004, 2005; A. Valdez, 2007). They began hanging out with gang boys in childhood, just before age 12, and at the time of the study, 40% reported having a boyfriend in a gang and 80% said they had a good friend in a male gang. Gang associations led to the girls' involvement in delinquent and criminal activities, including holding drugs (55%), selling drugs (31%), and holding weapons (27%).

Relationship of Risk Factors to Gang Involvement

Loeber and colleagues (2009) found "data from prediction studies indicate that risk factors from each of the domains (individual, family, peers, schools, and neighborhoods) contribute to the explanation of why some individuals and not others progress from minor problem behavior, such as bullying, to physical fighting and then to violence" (p. 301). Moreover, a dose-response relationship is seen in available research "between an individual's exposure to an accumulation of risk factors across multiple domains and an increased probability of adverse outcomes" (p. 301). Importantly, "The dose-response relationship appears robust in that it applies to both genders, different ethnic or racial groups, households with different income levels, single- and two-parent families and different neighborhoods" (p. 301).

Recent youth gang research has produced four seminal findings with respect to the impact of risk factors on the likelihood of gang membership. First, Howell and Egley (2005b) note that risk factors for gang membership span all five of the risk factor domains (family, peer group, school, individual characteristics, and community conditions).

Second, risk factors have a cumulative impact; that is, the greater the numbers of risk factors the youth experience, the greater the likelihood of gang involvement. In a Seattle study, children under the age of 12 who evidenced as many as 7 of 19 measured risk factors were 13 times more likely to join a gang than children with none or only one risk factor (Hill et al., 1999). Esbensen and colleagues (2010) found a similar exponential relationship in a multiple-city survey beyond the accumulation of 6 of the particular risk factors measured in that study. As youth accumulated more of these risk factors, they were more likely to become involved with gangs as opposed to violence (52% of gang members experienced 11 or more risk factors, compared with 36% of violent offenders).

Third, the presence of risk factors in multiple developmental domains appears to further enhance the likelihood of gang membership. Rochester researchers, led by Thornberry, Krohn, and Lizotte (2003), examined seven risk factor domains, including (1) area characteristics, (2) family sociodemographic characteristics, (3) parent–child relations, (4) school, (5) peers, (6) individual characteristics, and (7) early delinquency. They found that 61% of the boys and 40% of the girls who had elevated scores in all seven risk factor domains were gang members. Thus, for optimal impact, gang prevention and intervention programs not only need to address multiple risk factors, they also need to address a number of risk factors in multiple developmental domains.

Fourth, general delinquency, violence, and gang involvement share a common set of risk factors (Esbensen et al., 2010).

A Review of Risk Factors for Gang Involvement

Researchers do not agree on the most important risk factors for gang membership because of different criteria employed in discovering these. Three credible lists of such risk factors have been generated. First, Howell and Egley's (2005b) review is detailed in the discussion in the next section. Second, Klein and Maxson's (2006) compilation was drawn predominantly from cross-sectional studies (14 of the 20 studies they reviewed are in this category). The cross-sectional studies measure both risk factors and outcomes at the same time, hence the causal ordering cannot be determined with certainty; what appears to be a predictor could well be an outcome of gang involvement. Third, Huizinga and Lovegrove (2009) compiled a short list of research–supported risk factors from an analysis of a number of longitudinal studies. This list was limited to factors that proved especially strong in at least two longitudinal study sites. This method is sound, but the drawback is that only 11 of more than 50 scientifically established risk factors in longitudinal studies met Huizinga and Lovegrove's stringent criteria. Consequently, Krohn and Thornberry (2008) find this listing problematic because research clearly shows that youth who have numerous risk factors in multiple domains are most likely to join gangs. Moreover, research has established that the prevalence of risk factors varies among study sites (which explain, in large part, the dissimilar lists from one longitudinal study to another). Reliance on a very narrow list such as that generated by Huizinga and Lovegrove's stringent criteria could have the unintended consequence of overlooking important factors that have not been well researched. For example, unsafe school environment and high rates of school suspensions and truancy and other punitive school policies and practices are potent risk factors in North Carolina cities (Graves, Ireland, Benson et al., 2010; Weisel & Howell, 2007) and in France (Debarbieux & Baya, 2008), but these factors are not yet on any lists of research-based gang risk factors because of their newness and the practical orientation of the school-related research.[1] Therefore, for strategic planning purposes, each community should examine a broad array of research–supported risk factors, and practical ones as well, to identify those that apply to a given community. Because the risk research has been conducted with a variety of racial and ethnic samples, there is every expectation that the research–supported risk factors apply across predominant racial/ethnic groups.

Table 5.2 contains risk factors for juvenile delinquency and gang involvement—almost all of which emanate from longitudinal quantitative studies. The risk factors for delinquency were identified in two national OJJDP study groups, one on child delinquents (Loeber & Farrington, 2001), another on serious and violent juvenile offenders (Loeber & Farrington, 1998), and those lists are updated here. The purpose for integrating them is to promote a holistic approach to delinquency and gang involvement. Note that the risk factors shown in Table 5.2 are organized by age level and that risk factors for gang involvement (almost all of which also predict general delinquency) are noted with an asterisk. Indeed, delinquency almost invariably precedes gang joining. Therefore, the theoretical model proposed here integrates developmental processes for delinquency involvement with gang membership. This developmental process is more likely to occur in disorganized communities where families, schools, and religious institutions are structurally weak or disorganized.

Table 5.2   Risk Factors for Delinquency, Violence, and Gang Involvement

Risk Factors Measured at Approximately Ages 0–2

Individual

- Conduct disorders (authority conflict/rebellious/stubborn/disruptive/antisocial)*
- Difficult temperament
- Hyperactive (impulsive, attention problems)*

Family

- Family poverty/low family socioeconomic status*
- Family violence (child maltreatment, partner violence, conflict)
- Having a teenage mother
- Maternal drug, alcohol, and tobacco use during pregnancy
- Parental criminality
- Parental psychiatric disorder
- Parental substance abuse
- Poor parent–child relations or communication
- Pregnancy and delivery complications

Risk Factors Measured at Approximately Ages 3–5

Individual

- Conduct disorders (authority conflict/rebellious/stubborn/disruptive/antisocial)*
- Lack of guilt and empathy*
- Low intelligence quotient
- Physical violence/aggression*

Family

- Family violence (child maltreatment, partner violence, conflict)
- Parental psychiatric disorder
- Parental use of physical punishment/harsh and/or erratic discipline practices

Risk Factors Measured at Approximately Ages 6–11

Individual

- Antisocial/delinquent beliefs*
- Lack of guilt and empathy*
- Early and persistent noncompliant behavior
- Early onset of aggression/violence*
- Few social ties (involved in social activities, popularity)
- General delinquency involvement*
- High alcohol/drug use*

(Continued)

(Continued)

- Hyperactive (impulsive, attention problems)*
- Low intelligence quotient
- Low perceived likelihood of being caught
- Medical/physical condition
- Mental health problems*
- Physical violence/aggression*
- Poor refusal skills*
- Victimization and exposure to violence

Family

- Abusive parents
- Antisocial parents
- Broken home/changes in caretaker*
- Family poverty/low family socioeconomic status*
- Family violence (child maltreatment, partner violence, conflict)
- High parental stress/maternal depression
- Living in a small house
- Parent proviolent attitudes*
- Parental use of physical punishment/harsh and/or erratic discipline practices
- Poor parental supervision (control, monitoring, and child management)*
- Poor parent-child relations or communication
- Sibling antisocial behavior*
- Unhappy parents

School

- Bullying
- Chronic absenteeism
- Frequent school transitions
- Frequent truancy/absences/suspensions, expelled from school, dropping out of school*
- Identified as learning disabled*
- Low academic aspirations*
- Low achievement in school*
- Low school attachment/bonding/motivation/commitment to school*
- Old for grade/repeated a grade
- Poor student-teacher relations
- Poorly defined rules and expectations for appropriate conduct
- Poorly organized and functioning schools/inadequate school climate/negative labeling by teachers*

Community

- Availability of firearms*
- Economic deprivation/poverty/residence in a disadvantaged neighborhood*
- Exposure to firearm violence
- Feeling unsafe in the neighborhood*

- Low neighborhood attachment*
- Neighborhood youth in trouble*

Peer

- Association with antisocial/aggressive/delinquent peers; high peer delinquency*
- Association with gang-involved peers/relatives*
- Peer alcohol/drug use
- Peer rejection*

Risk Factors Measured at Approximately Ages 12–17

Individual

- Antisocial/delinquent beliefs*
- Lack of guilt and empathy*
- Early dating/sexual activity/fatherhood*
- Few social ties (involved in social activities, popularity)
- General delinquency involvement*
- High alcohol/drug use*
- High drug dealing
- Illegal gun ownership/carrying
- Life stressors*
- Makes excuses for delinquent behavior (neutralization)*
- Mental health problems*
- Physical violence/aggression*
- Violent victimization*

Family

- Antisocial parents
- Broken home/changes in caretaker*
- Delinquent/gang-involved siblings*
- Family history of problem behavior/criminal involvement
- Family poverty/low family socioeconomic status*
- Family violence (child maltreatment, partner violence, conflict)
- Having a teenage mother
- High parental stress/maternal depression
- Lack of orderly and structured activities within the family
- Living in a small house
- Low attachment to child/adolescent*
- Low parent education*
- Parental use of physical punishment/harsh and/or erratic discipline practices
- Poor parental supervision (control, monitoring, and child management)*
- Poor parent-child relations or communication

(Continued)

(Continued)

School

- Bullying
- Frequent school transitions
- Low academic aspirations*
- Low math achievement test scores (males)*
- Low parent college expectations for child*
- Low school attachment/bonding/motivation/commitment to school*
- Poor school attitude/performance, academic failure*
- Poorly organized and functioning schools/inadequate school climate/negative labeling by teachers*

Community

- Availability and use of drugs in the neighborhood*
- Availability of firearms*
- Community disorganization*
- Economic deprivation/poverty/residence in a disadvantaged neighborhood*
- Exposure to violence and racial prejudice
- Feeling unsafe in the neighborhood*
- High-crime neighborhood*
- Low neighborhood attachment*
- Neighborhood physical disorder
- Neighborhood youth in trouble*

Peer

- Association with antisocial/aggressive/delinquent peers, high peer delinquency*
- Association with gang-involved peers/relatives*
- Gang membership

* Risk factors for gang membership

Sources: National Gang Center, OJJDP Strategic Planning Tool (2010 update); Howell & Egley, 2005b; Huizinga & Lovegrove, 2009

Table 5.2 also includes *hindering risk factors.* These are risk factors that hinder desistance from moderate and serious delinquency offending, and thus prolong delinquent careers (this pioneering research is reported in Stouthamer-Loeber et al., 2008). Key hindering risk factors for such offenses in the Pittsburgh Youth Study (measured in middle adolescence) are high alcohol use, high marijuana use, high drug dealing, gun carrying, and gang membership. White and fellow researchers (2008) report substance use, drug dealing, gun carrying, and gang membership as a particularly potent combination of hindering risk factors against desistance for both serious violence and serious theft.

The following discussion summarizes longitudinal research shown in Table 5.2. Thornberry's (2005) developmental model specifies four distinct developmental stages in the pathway to delinquency and gang involvement—preschool, school entry, childhood, and adolescence—and these are displayed in Figure 5.3. This discussion and Figure 5.3 also include attractions of the gang, or perceived benefits of joining, although this has not been conceptualized in research as a risk factor; it could be included, because gang attractions clearly elevate youths' risk for joining.

The Preschool Stage

In this first developmental stage, child characteristics and community and family deficits produce aggressive and disruptive behavior disorders by the time of school entry (Kalb & Loeber, 2003) and, in turn, according to Loeber and Farrington (2001), delinquency and school performance problems in later childhood. At birth—or beginning in the prenatal period for some infants—the family of procreation is the central influence on infants and children. During the preschool years, and especially in the elementary school period and onward, the array of risk factors expands, as some children are exposed to negative peer influences outside the home. Vaughn and colleagues (2009) describe a "severely impaired subgroup" (9.3%) of children identified in their analysis of data collected on a nationally represented cohort of more than 17,000 children drawn from approximately 1,000 kindergartner programs and collected by the National Center for Education Statistics.

Some children are also exposed to additional risk factors situated in schools or in the community at large during this period. When linked with certain family and child characteristics, Loeber, Farrington, and Petechuk (2003), as well as Tremblay (2003), find concentrated disadvantage impedes socialization of children. Loeber and Farrington (2001) identify important family variables in the preschool stage that include low parental education (social capital) and a host of family problems, including what Pogarsky, Lizotte, and Thornberry (2003) found as a broken home, parental criminality, poor family or child management, abuse and neglect, serious marital discord, and young motherhood. Keenan (2001) along with Loeber and Farrington (2001) contend that pivotal child characteristics during the preschool period include a difficult temperament and impulsivity, typically described as aggressive, inattentive, and sensation-seeking behaviors.

Kroneman, Loeber, and Hipwell (2004) strongly associate neighborhood poverty with problem behavior—specifically girls' disruptive behaviors—as young as age 5. Researchers led by both Hipwell (2002) and Wei (2005) conclude very young girls are seriously affected in the most disadvantaged and disordered neighborhoods. In these neighborhoods, girls—just like boys—"are exposed to a greater number of risk factors, including exposure to different forms of community and family violence and an increased likelihood of affiliation with deviant peers" (Kroneman et al., 2004, p. 117). Coleman (1988, 1990) finds a lack of "social capital" is a key by-product of what Sampson and Groves (1989) describe as "concentrated disadvantage" in impoverished, distressed, and crime-ridden communities. When combined with family and child deficits, concentrated disadvantage increases the odds of disruptive behavior disorders in children by the time of school entry and delinquency later during childhood, for both girls and boys (Hipwell et al., 2002; Keenan et al., 2010; Kroneman et al., 2004). Loeber, Farrington, and Stouthamer-Loeber's team (2003) indicated families with a harsh child punishment profile are overrepresented in such disadvantaged neighborhoods, and serious delinquency tends to occur more quickly in youngsters residing in these communities. Taken together, concentrated disadvantage at the community level, family problems, and certain child characteristics lead to early childhood problems (aggression and disruptive behavior), and each of these four variables, in turn, increases the likelihood of delinquency in childhood and gang membership in adolescence.

The School Entry Stage

Kalb and Loeber (2003) identify products of dysfunctional families as early childhood aggression and disruptive behaviors, including stubbornness, defiance and disobedience, and truancy after school entry. This is particularly true in disadvantaged communities. Harsh parenting, according to Hipwell and colleagues (2007, 2008), is a key factor at this stage for girls, leading to conduct problems. Aggressive and disruptive behaviors are likely to be followed by rejection by prosocial peers, thus opening the door to antisocial or deviant peer influences, which predict delinquent activity in later childhood and early adolescence. The link between physical aggression in childhood and violence in adolescence is particularly strong (Brame, Nagin, & Tremblay, 2001; Broidy et al., 2003).

Coie and Dodge (1998) propose a developmental model that emphasizes early peer rejection as an often-overlooked factor in the onset of child delinquency. Their model applies in particular to aggressive and disruptive children. Disruptive behavior disorders include conduct disorders, such as aggression and oppositional defiant disorder (Burke et al., 2002). Difficulties these children face in getting along with and being accepted by conventional peers may lead to increasing aggressiveness and more disruptive behavior. It appears that Coie and Dodge have identified a pivotal selection process that helps explain why aggressive and disruptive children are more likely to turn to delinquency and gang involvement: because their opportunities for prosocial peer relationships are limited.

It is important to note that most disruptive children do not become child delinquents, nor do most child delinquents engage in delinquency in adolescence. Loeber and Farrington (2001) report one-fourth to one-third of the disruptive children are at risk of becoming child delinquents, and about a third of all child delinquents later become serious, violent, and chronic offenders. However, Thornberry and Krohn (2001) explain, "the earlier the onset, the greater the continuity" (p. 297). Compared with late-onset delinquents, Wasserman and Seracini (2001) suggest child delinquents tend to come from dysfunctional families with one or more of the following characteristics: family disruption (especially a succession of different caregivers), parental antisocial behavior, parental substance abuse, mother's depression, and child abuse and neglect.

The Later Childhood Stage

In the third developmental stage, later childhood, other risk factors (explanatory variables) that prepare children for gang membership begin to come into play. Children who are involved in delinquency, violence, and drug use at an early age are at higher risk for gang membership than other youngsters (Craig et al., 2002). Hipwell and colleagues (2005) contend girls are more prone to earlier alcohol use than boys and greater physical aggression was associated with early alcohol use. More than one-third of the child delinquents in Krohn and researchers' (2001) Montreal and Rochester samples became involved in crimes of a more serious and violent nature during adolescence, *including* gang fights. Thornberry and Krohn (2001) conclude, "In brief, very early onset offending is brought about by the *combination and interaction* of structural, individual, and parental influences" (p. 295).

Coie and Dodge (1998) link peer rejection in the early school years to a tendency for greater susceptibility from the influence of deviant peers, including more aggressive youths. Aggressive and antisocial youths begin to affiliate with one another in childhood (Coie & Dodge, 1998), and Cairns and Cairns (1994) suggest this pattern of aggressive friendships may continue through adolescence. A Montreal study led by Craig (2002) suggests displays of aggression in delinquent acts at age 10 or perhaps younger may be a key factor leading to gang involvement. Peers rated gang members as significantly more aggressive than nongang members at ages 10 to 14. Warr (2002) identifies the negative consequence of delinquent peer associates as one of the most enduring findings in empirical delinquency studies. Lacourse and colleagues (2006) find associations with delinquent peers increase delinquency and the likelihood and frequency of physical aggression and violence, which in turn increases the likelihood of gang membership in early adolescence.

Thornberry and Krohn (2001) suggest weakened prosocial bonds as a result of delinquency may be an important interaction effect of this process. Poor school performance (poor grades and test scores) in later childhood is likely to result from prosocial peer rejection, child delinquency, and family problems. Other school-related variables that lead to gang involvement, according to studies led by Craig (2002) and Hill (1999), include low achievement in elementary school, low school attachment, and having been identified as learning disabled. Factors that weaken the student–school bond (commitment to school) in the later childhood stage contribute to delinquency and gang membership.

The poor school performance of children is one side of the coin; poor-quality schools (poorly organized and functioning) are the other side (Lyons & Drew, 2006). A contemporary indicator of poor-quality schools is zero tolerance policies that produce high suspension, expulsion, and dropout rates.

Longitudinal studies have not directly examined the effects of these specific student statuses as a risk factor for gang membership. However, several of them have examined school failure, periods not in school, teachers' negative labeling, and bad behavior and attitudes toward school, all of which predict gang involvement and generally long-term membership in gangs. In particular, Thornberry and Krohn's research team (2003) examined the effects of gang involvement on dropout rates, finding that gang membership (especially long term) greatly increased the odds of dropping out of school. In addition to alienating students from schools and teachers, thus weakening the student–school bond, zero tolerance policies release many youths from adult supervision during the day and after school, which Vigil (2002) suggests potentially exposes them to deviant influences on the streets, and a higher likelihood of delinquency involvement themselves.

The Early Adolescence Stage

The remainder of this developmental model incorporates only risk factors that predict gang involvement. Children who are on a trajectory of worsening antisocial behavior are more likely to join gangs during adolescence, and they tend to have more problems than nongang members (Esbensen et al., 2010; Howell & Egley, 2005b). Esbensen and Huizinga (1993) suggest gang entry might be thought of as the next developmental step in escalating delinquent behavior. Thornberry, Krohn, and colleagues (2003) also find future gang members are more likely to experience risk factors in multiple developmental domains, including community or neighborhood, family problems, school problems, delinquent peer influence, and individual characteristics. Each of these risk domains is considered next.

Community or Neighborhood Risk Factors

"Communities too have careers in delinquency" (Short, 1996, p. 224). As discussed in Chapter 1, the key conditions that gave rise to early gangs in U.S. history are the lack of a stable population and institutions. Most gang research indicates gangs tend to cluster in high-crime and economically disadvantaged neighborhoods. Subsequent research shows that this constellation of risk factors also predicts gang membership. As children grow older, and venture outward from their families, they are more and more influenced by community conditions. The key factors include residence in a disadvantaged neighborhood, a "culture of poverty" (Lewis, 1966), a high level of criminal activity, lots of neighborhood youth in trouble, and a ready availability and use of firearms and drugs. Bingenheimer, Brennan, and Earls (2005) claim exposure to firearm violence approximately doubles the probability that an exposed adolescent will perpetrate serious violence over the next two years. Other undesirable community conditions include feeling unsafe in the neighborhood, and low neighborhood attachment. All of these are conditions in which gangs and violence tend to thrive. In the worst conditions—where communities suffer from concentrated disadvantage—Morenoff, Sampson, and Raudenbush (2001) suggest the necessary *collective efficacy* (informal control and social cohesion) among residents to ameliorate its negative effects may be lacking.

Family Risk Factors

Studies show that both family structure (nonintact biological family) and poor quality of interactions and interrelationships among family members are important risk factors for gang membership. With respect to family structure, a broken home (absence of either or both biological parents) and multiple family transitions or caretaker changes are important predictors of gang membership. In Hill and fellow researchers' (1999) study, the most negative arrangement was one biological parent and other adults in the home.

Several family interactions and interrelationships predict gang membership. One of the most prominent risk factors is poor parental supervision (including control, monitoring, or management of family matters). These factors suggest the importance of highly structured family activities. Other family conditions compromise parental capacity to carry out their child development responsibilities, including low parent education, family poverty, or low family socioeconomic status, proviolent attitudes, and child maltreatment (abuse or neglect). A gang-involved family member has been well researched as a gang attraction, as noted previously, and also is seen as a risk condition in ethnographic studies (J. W. Moore, 1991) and student surveys (Decker & Curry, 2000).

School Risk Factors

Most studies of risk factors for juvenile delinquency and gang membership have examined only one segment of the school–student relationship: satisfactory academic performance. For example, Thornberry and Krohn's team (2003) concluded poor school performance on math tests predicts gang membership for males, as well as low attachment to teachers. Hill and colleagues (1999), and also Le Blanc and Lanctot (1998) found future gang members perform poorly in elementary school, and they have a low degree of commitment to and involvement in school. Modern-day studies of school experiences now include measures of "school climate" (Gottfredson, Gottfredson, Payne, & Gottfredson, 2005), "difficult" schools (typically characterized by higher levels of student victimization, sanctions, and poor student-teacher relations) (Debarbieux & Baya, 2008) and student "connectedness" to schools (Resnick, Ireland, & Borowsky, 2004). Although no studies of gang joining have employed these measures, the Denver Youth Survey examined negative labeling by teachers (as either bad or disturbed), which, as Esbensen and colleagues (1993) note, proved to be an important predictor of gang membership. Debarbieux and Baya's (2008) study of "difficult" schools found that the percentage of students who were gang members doubled in them. Gottfredson and Gottfredson (2001) also link feeling unsafe at school to gang involvement. In a survey of more than 10,000 junior high and high school students nationwide, Alvarez and Bachman (1997) found the key school-related situations that make students fearful of assault both at school and while going to and from school: recent victimization experiences, the presence of a violent subculture at the school (e.g., gang presence and attacks on teachers), and availability of drugs/alcohol were related to fear in both contexts. Vigil (1993) concurs that students who feel vulnerable at school may seek protection in the gang.

Peer Risk Factors

Based on their Montreal study, Lacourse and colleagues (2006) conclude, "Being part of a deviant peer group is associated with the onset, persistence, and aggravation of conduct problem symptoms during adolescence and substance use [and] also at increased risk of injury, incarceration, and even death" (p. 562). In addition, they found that "a behavior profile characterized by high hyperactivity, high fearlessness, and low prosociality is by far the best predictor of early affiliation with deviant peers" (p. 566). Furthermore, Thornberry and Krohn's team (2003) identified association with peers who engage in delinquency as one of the strongest risk factors for gang membership, particularly for boys. Both Craig's (2002) and Lahey's (1999) research teams found association with aggressive peers—whether or not they are involved in delinquency—during adolescence as a strong predictor of gang joining. "During early adolescence, however, friendships with aggressive peers may be a stepping stone to gang membership" (Lahey et al., 1999, p. 274). Lacourse and colleagues' (2003) Montreal study produced similar findings. Rejection by prosocial peers (unpopular) seems to be a key factor that pushes children into affiliations with delinquent groups and gangs (Haviland & Nagin, 2005; Thornberry, Krohn, et al., 2003).

Individual Risk Factors

Numerous studies show children on a trajectory of worsening antisocial behavior, including child delinquency, are more likely to join gangs during adolescence. Thornberry, Krohn, and colleagues (2003) include early involvement in delinquency, aggression or violence, alcohol or drug use, early dating, and precocious sexual activity as types of susceptible behavior. Additional antisocial behaviors among future gang members include aggression or fighting (Craig et al., 2002; Lahey et al., 1999).

Illegal gun ownership or carrying is another potentially important predictor of gang membership. Illegal gun ownership has not been analyzed as a risk factor, but analyses in the Rochester Youth Development Study show that gangs are more likely to recruit youths who own an illegal firearm and that gang involvement promotes the use of these weapons (Bjerregaard & Lizotte, 1995; Lizotte, Krohn, Howell, Tobin, & Howard, 2000).

Mental health problems increase risk for gang joining. Children who are child victims of abuse or neglect are more likely to join gangs (Fleisher, 1998; J. A. Miller, 2001; Thornberry, Krohn, et al., 2003). Other forms of violent victimization outside the home, such as assaults are also associated with gang joining (Taylor, 2008; Taylor, Freng, Esbensen, & Peterson, 2008). What the available research shows most clearly is that a constellation of risky behaviors, gang fights included, elevates one's risk of violent victimization,[2] and as Melde (2009) notes, for both genders. These risky behaviors include participating in gang or group fights, carrying a weapon, committing serious assault, selling drugs, and associating with delinquent peers. Youth characterized by any one of these risk factors in Pittsburgh and Denver were generally two to four times more likely to have been victims of violence than the group of youth who did not have the risk factor.

Given their typically disadvantaged social histories (Howell & Egley, 2005b), that of their parents (Gordon et al., 2004), and their early problem behaviors (Lahey et al.,

1999), it is not surprising that gang members often evidence more serious mental health problems than other youths. Thornberry and Krohn's (2003) research team found youngsters—particularly boys—who experience many negative or stressful life events also are more likely to join gangs. In this study, the measured events included failing a course at school, being suspended from school, breaking up with a boyfriend or girlfriend, having a big fight or problem with a friend, and the death of someone close.

Various clinical psychological conditions are prevalent among community adolescent samples, particularly adjudicated juvenile offenders (Schram & Gaines, 2005) and correctional populations (Di Placido, Simon, Witte, Gu, & Wong, 2006). Gang networks promote violence and support cognitive distortions (Goldstein & Glick, 1994; Goldstein, Glick, & Gibbs, 1998). In particular, Flannery and colleagues (2007) note exposure to violence at a very young age may cause or exacerbate mental problems. In addition, studies led by Esbensen (2010) and Peterson (2004) conclude violent victimization rates are higher for gang members than those for their nongang counterparts. In the most thorough psychological assessment of gang members reported to date, Davis and Flannery (2001) studied a Canadian sample of intractable prison inmates and found that nearly 9 out of 10 evidenced alcohol or other substance abuse and a variety of other mental problems including mood disorders, schizophrenia, anxiety disorders, adjustment disorders, mood disorders, and other personality disorders. Davis and Flannery noted that gang members in juvenile correctional facilities "often are admitted with histories of physical and sexual abuse, substance abuse, psychiatric disturbances, post-traumatic stress disorder, cognitive deficits, poor self-esteem, and other problems" (p. 37).

Considering the psychological findings, is it accurate to classify gang members as "sociopaths"? Yablonsky (1959, 1967) found gang members in general, and gang leaders in particular as "disturbed," "impulsive," and "sociopathic," lacking moral standards and empathy for others. Valdez, Kaplan, and Codina's (2000) study of San Antonio gangs and gang members contradicted Yablonsky's viewpoint. In a comparison of gang and nongang members, these researchers were able to classify only a very small proportion (4%) of the sample of exclusively Mexican American high-risk youth gang members as "sociopaths" as measured by the Hare Psychopathy Checklist (Hare, Hart, & Harpur, 1991). However, on the basis of this screening assessment, 4 out of 10 (44%) gang members were scored as possible psychopaths and in need of a clinical interview to determine the severity of their psychopathic problems.

Risk Factors for Girls

Briefly, research suggests that seven categories of risk factors are associated with gang joining among girls: (1) neighborhood conditions, (2) child physical and sexual abuse, (3) running away, (4) drug and alcohol abuse, (5) mental health problems, (6) violent victimization, and (7) juvenile justice system involvement. Each of these risk factors is discussed at length in Chapter 6.

Impact of Gangs on Participants

Most youths who join gangs have already been involved in delinquency and drug use. Once in the gang, they are quite likely to become more actively involved in delinquency, drug use, and violence, and they are more likely to be victimized themselves (Peterson et al., 2004; Taylor, 2008). Their problems do not end here. They are at greater risk of arrest, juvenile court referral, detention, confinement in a juvenile correctional facility, and, later, imprisonment. For the gang to have devastating consequences, it doesn't necessarily have to be a large formal gang. Loeber and Farrington (2011) find this trajectory often ends in murdering someone or being a homicide victim. Even low levels of gang organization have important consequences for involvement in crime and victimization (Decker et al., 2008; Esbensen, Winfree, et al., 2001; Taylor, 2008). Thornberry, Krohn, and colleagues (2003) assert that gang involvement has a way of limiting youngsters' life chances, particularly if they remain active in the gang for several years. Over and above embedding its members in criminal activity, the gang acts as a powerful social network in constraining the behavior of members, limiting access to prosocial networks, and cutting members off from conventional pursuits. These effects of the gang tend to produce precocious, off-time, and unsuccessful transitions that bring disorder to the life course in a cascading series of difficulties, including school dropout, early pregnancy or early impregnation, teen motherhood, and unstable employment. Long-term gang membership also leads to being arrested as a young adult for both males and females, and as Krohn and Thornberry (2008) note, other negative life-course transitions. Future studies are quite likely to show that the deeper the level of juvenile and criminal justice intervention, the greater the likelihood of gang involvement. For example, both secure juvenile correctional facilities and adult prisons surely have a higher proportion of gang members than found in probation caseloads (Olson & Dooley, 2006; Roush, Miesner, & Winslow, 2002; Weisel & Howell, 2007). Benda, Corwyn, and Toombs (2001) claim gang membership is a strong predictor of entry into the correctional system for adults. Several long-term follow-ups of gang members show anticipated continuation of lives in crime and other negative outcomes including high rates of homicide victimization (Loeber & Farrington, 2011; Thornberry, Krohn, et al., 2003).

Limitations of the Risk Factor Paradigm

A fundamental misunderstanding of the meaning of *risk factors* exists in criminological literature. Those who reject the study of risk factors—in particular Hughes (2006), Hughes and Short (2005), and Pitts (2008)—and sometimes degrade the use of them as what Hughes (2006) calls the "variables paradigm," assert that these factors ignore the "context," while contending that "all social facts are located in contexts" (Abbott, 1997, p. 1152). Hughes (2006) correctly points out that

> the formation and evolution of gangs in time and space need to be examined, with special consideration of the immediate and long-term effects of varying social forces and processes, including changes in relationships among members of individual gangs and

between gang members and others with whom they come into contact (e.g., police and members of both friendly and rival gangs). . . . Moreover, criminal and violent activities of gang members must be located within larger sets of behaviors and related to specific situational characteristics and processes of interaction. (p. 39)

Missing from Hughes's (also Short & Hughes, 2006a) otherwise excellent conceptualization are the developmental experiences and environmental conditions that brought gang members together in the incipient violence context. Cumulatively, risk factors mark the 10- to 20-year pathway (earlier contexts, that is) in which current gang participants first incubated—beginning initially in troubled families, then in unhealthy school climates, delinquent peer groups, and disadvantaged neighborhood settings—that brought them together in that particular neighborhood and at that point in time. Valid risk factors actually *describe* much of the context and the social facts in the context that preceded gang involvement. Indeed, several risk factor variables are sufficiently grounded in theories to provide empirically supported explanations for gang involvement (Thornberry, Krohn et al., 2003), although the testing work is yet in its infancy. But "the majority of theories about the causes of disruptive delinquent behavior are generally based on risk factors, although different terms for such factors have been used" (Loeber et al., 2009, p. 300). Moreover, some risk factors are associated with problem behaviors in specific contexts, including characteristics of schools attended and the neighborhood in which local children and adolescents reside or spend their time.

A free user-friendly system for gathering the necessary research-based risk factors for systematic assessment of neighborhood contexts of gang activity has been online for several years at the National Gang Center, the OJJDP Strategic Planning Tool. (www.nationalgangcenter.gov/About/Strategic-Planning-Tool). In addition, a gang problem assessment process guides communities in assessing key features of gang contexts. Ethnographic research can be combined with law enforcement survey data to generate city and neighborhood histories of gang activity (such as the Pittsburgh case study in Chapter 8). Other electronic tools are available at the National Gang Center, and detailed in Chapters 9 and 10, to facilitate rigorous neighborhood and community assessments of risk factors and other research tools assist them in assessing the existence of gangs and scope of their behaviors.

To be sure, there are important limitations of the risk factor paradigm. First, it cannot be used to explain the existence of gangs; rather it is individual focused, particularly with intervention in offender careers in mind. Second, the risk factor paradigm is not useful for examining juvenile and criminal justice system effects. Bernburg, Krohn, and Rivera (2006) state, "Teenagers who experience juvenile justice intervention are substantially more likely than their peers to become members of a gang in a successive period" (p. 81). Third, as Krohn and Thornberry (2008) note, measurement of risk factors in individuals cannot fully take neighborhood context into account. Fourth, the risk factor paradigm does not provide a grand "action theory" that integrates the components of the social sciences into a single theoretical framework. Talcott Parsons (1951) stands alone in having accomplished this feat. Theories, too, have their limitations.

Nevertheless, risk factor critics fail to grasp the vast practical utility of the public health model—the application of prevention science (Institute of Medicine, 2008)—in engaging communities in preventing delinquency, crime, and gang involvement. It is feasible for community stakeholders to manipulate successfully several risk and protective factors across a city or county and produce a material change in behavioral outcomes for youth, what amounts to an accomplishment of collective efficacy (Sampson et al., 1997). Some communities have demonstrated measurable success in such an initiative, for example, in addressing minor forms of problem behavior (delinquency and drug use) among low-risk youth (Hawkins, Oesterle, Brown et al., 2009), adolescent violence (Swenson, Henggeler, Taylor, & Addison, 2005), and gang homicide (Skogan, Hartnett, Bump, & Dubois, 2008).

Concluding Observations

The goal of this chapter was to show how youth become involved in gangs. Developmental models help to map the offense-based pathway to gang involvement. The developmental framework, in a simplified manner, depicts key stepping-stones in the gang member trajectory. Future theoretical work needs to incorporate these research–supported risk factors. Gang involvement is a key stage in the formation of youngsters' criminal trajectory.

> The gang life path has a set of unique roadways, turns, narrow choices, and seductive draws, all of which shape youths' personal and group identity. (Vigil, 2010, pp. 10–11)

But gang membership also has a way of limiting youngsters' life chances, that extends far beyond the period of active gang participation. Over and above embedding its members in criminal activity, in Rochester, New York, the gang acts as a powerful social network in constraining the behavior of members, limiting access to prosocial networks, and cutting members off from conventional pursuits. These effects of the gang tend to produce precocious, off-time, and unsuccessful transitions that bring disorder to the life course in a cascading series of difficulties, including school dropout, early pregnancy or early impregnation, teen motherhood, and unstable employment.

Several tentative conclusions can be drawn about studies to date on risk factors for gang membership. First, a caveat is in order. Only a relatively small number of longitudinal studies have investigated these, and as Krohn and Thornberry (2008) note, still fewer of them have used a common set of risk factors. Consequently, there are few replicated results. In addition, it must be emphasized that, as in the case of many other problem behaviors,

> gang membership does not seem to be a product of a few central risk factors; none exerts a massive impact on the likelihood of being a gang member. But, the accumulation of risk is strongly related to the chances of becoming a gang member. Gang members have multiple deficits in multiple developmental domains, each of which contributes in a small but statistically significant way to the chances of becoming a gang member. (p. 138)

Future gang members share several of the same risk factors seen in future serious and violent adolescent offenders, including association with delinquent peers, drug and alcohol use, school problems, and family problems. Thus, a continuum of responses consisting of age-appropriate services is needed to develop risk factors and service needs as they emerge with age. Gang involvement should routinely be assessed for gang involvement at each stage of juvenile justice and criminal justice system processing. Recommended procedures are detailed in Chapter 10.

DISCUSSION TOPICS

1. What are the key propositions of developmental or life-course theories?

2. How does Loeber's pathways model explain continued progression from less serious to more serious delinquent offenses?

3. Which developmental theory best explains gang membership?

4. Can any of the theories of gang existence in Chapter 4 be integrated with developmental theories to account for the existence of gangs and also explain why youth join them, in a single theory?

5. Which theories best account for youths who advance to the top of Loeber's pathways model, that is, become serious (property), violent, and chronic offenders?

RECOMMENDATIONS FOR FURTHER READING

Developmental and Life Course Theories

Caspi, A., Lahey, B. B., & Moffitt, T. E. (2003). *The causes of conduct disorder and serious juvenile delinquency.* London: Guilford Press.

Farrington, D. P. (2003). Developmental and life-course criminology: Key theoretical and empirical issues— The 2002 Sutherland Award Address. *Criminology, 41,* 221–255.

Farrington, D. P. (2006). Building developmental and life-course theories of offending. In F. T. Cullen, J. P. Wright, & K. R. Blevins (Eds.), *Taking stock: The status of criminological theory* (pp. 335–364). New Brunswick, NJ: Transaction Publishing.

Hawkins, J. D. (Ed.). (1996). *Delinquency and crime: Current theories.* New York: Cambridge University Press.

Le Blanc, M., & Loeber, R. (1998). Developmental criminology updated. In M. Tonry (Ed.), *Crime and justice: An annual review of research* (Vol. 23, pp. 115–198). Chicago: University of Chicago Press.

Melde, C., & Esbensen, F. (2011). Gang membership as a turning point in the life course. *Criminology, 49,* 513–552.

Patterson, G. R., Capaldi, D., & Bank, L. (1991). An early starter model for predicting delinquency. In D. J. Pepler & K. H. Rubin (Eds.), *The development and treatment of childhood aggression* (pp. 139–168). Hillsdale, NJ: Lawrence Erlbaum.

Thornberry, T. P. (Ed.). (1997). *Developmental theories of crime and delinquency.* New Brunswick, NJ: Transaction Publishing.

Protective Factors Against Gang Membership

Bjerregaard, B., & Smith, C. (1993). Gender differences in gang participation, delinquency, and substance use. *Journal of Quantitative Criminology, 9,* 329–355.

Esbensen, F., Huizinga, D., & Weiher, A. W. (1993). Gang and non-gang youth: Differences in explanatory variables. *Journal of Contemporary Criminal Justice, 9,* 94–116.

Hill, K. G., Howell, J. C., Hawkins, J. D., & Battin-Pearson, S. R. (1999). Childhood risk factors for adolescent gang membership: Results from the Seattle Social Development Project. *Journal of Research in Crime and Delinquency, 36,* 300–322.

Howell, J. C. (2004). Youth gangs: Prevention and intervention. In P. Allen-Meares & M. W. Fraser (Eds.), *Intervention with children and adolescents: An interdisciplinary perspective* (pp. 493–514). Boston: Allyn & Bacon.

Klein, M. W., Weerman, F. M., & Thornberry, T. P. (2006). Street gang violence in Europe. *European Journal of Criminology, 3,* 413–437.

Li, X., Stanton, B., Pack, R., Harris, C., Cottrell, L., & Burns, J. (2002). Risk and protective factors associated with gang involvement among urban African American adolescents. *Youth and Society, 34,* 172–194.

Maxson, C. L., Whitlock, M., & Klein, M. W. (1998). Vulnerability to street gang membership: Implications for prevention. *Social Service Review, 72*(1), 70–91.

Thornberry, T. P., Krohn, M. D., Lizotte, A. J., Smith, C. A., & Tobin, K. (2003). *Gangs and delinquency in developmental perspective.* New York: Cambridge University Press.

Wyrick, P. A. (2000). *Vietnamese youth gang involvement* (Fact Sheet No. 2000–01). Washington, DC: Office of Juvenile Justice and Delinquency Prevention.

Interactional Theory

Thornberry, T. P. (1987). Toward an interactional theory of delinquency. *Criminology, 25*(4), 863–891.

Thornberry, T. P., & Krohn, M. D. (2001). The development of delinquency: An interactional perspective. In S. O. White (Ed.), *Handbook of youth and justice* (pp. 289–305). New York: Plenum.

Thornberry, T. P., Krohn, M. D., Lizotte, A. J., Smith, C. A., & Tobin, K. (2003). *Gangs and delinquency in developmental perspective.* New York: Cambridge University Press.

Gang Involvement Increases and Delinquent and Violent Behavior

Bendixen, M., Endresen, I. M., & Olweus, D. (2006). Joining and leaving gangs: Selection and facilitation effects on self-reported antisocial behaviour in early adolescence. *European Journal of Criminology, 3,* 85–114.

Gatti, U., Tremblay, R. E., Vitaro, F., & McDuff, P. (2005). Youth gangs, delinquency and drug use: A test of selection, facilitation, and enhancement hypotheses. *Journal of Child Psychology and Psychiatry, 46,* 1178–1190.

Gordon, R. A., Lahey, B. B., Kawai, E., Loeber, R., Stouthamer-Loeber, M., & Farrington, D. P. (2004). Antisocial behavior and youth gang membership: Selection and socialization. *Criminology, 42,* 55–88.

Haviland, A. M., & Nagin, D. S. (2005). Causal inferences with group based trajectory models. *Psychometrika, 70,* 1–22.

Haviland, A. M., Nagin, D. S., Rosenbaum, P. R., & Tremblay, R. E. (2008). Combining group-based trajectory modeling and propensity score matching for causal inferences in nonexperimental longitudinal data. *Developmental Psychology, 44,* 422–436.

Hill, K. G., Chung, I. J., Guo, J., & Hawkins, J. D. (2002). *The impact of gang membership on adolescent violence trajectories.* Paper presented at the International Society for Research on Aggression, XV World Meeting, Montreal, Canada, July.

Krohn, M. D., & Thornberry, T. P. (2008). Longitudinal perspectives on adolescent street gangs. In A. Liberman (Ed.), *The long view of crime: A synthesis of longitudinal research* (pp. 128–160). New York: Springer.

Melde, C., & Rennison, C. M. (2008). The effect of gang perpetrated crime on the likelihood of victim injury. *American Journal of Criminal Justice, 33*, 234–251.

Thornberry, T. P., Lizotte, A. J., Krohn, M. D., Smith, C. A., & Porter, P. K. (2003). Causes and consequences of delinquency: Findings from the Rochester Youth Development Study. In T. P. Thornberry & M. D. Krohn (Eds.), *Taking stock of delinquency: An overview of findings from contemporary longitudinal studies* (pp. 11–46). New York: Kluwer Academic/Plenum Publishers.

Gang Attractions

Decker, S. H., & Curry, G. D. (2000). Addressing key features of gang membership: Measuring the involvement of young members. *Journal of Criminal Justice, 28*, 473–482.

Decker, S. H., & Van Winkle, B. (1996). *Life in the gang: Family, friends, and violence.* New York: Cambridge University Press.

Melde, C., Taylor, T. J., & Esbensen, F. (2009). "I got your back": An examination of the protective function of gang membership in adolescence. *Criminology, 47*, 565–594.

Thornberry, T. P., Krohn, M. D., Lizotte, A. J., Smith, C. A., & Tobin, K. (2003). *Gangs and delinquency in developmental perspective.* New York: Cambridge University Press.

Vigil, J. D. (1993). The established gang. In S. Cummings & D. J. Monti (Eds.), *Gangs: The origins and impact of contemporary youth gangs in the United States* (pp. 95–112). Albany: State University of New York Press.

Longitudinal Studies of Gang Members

Currently, four longitudinal adolescent delinquency studies are examining subsamples of gang members, (1) the Rochester Youth Development Study, (2) the Pittsburgh Youth Study, (3) the Denver Youth Survey, and (4) the Seattle Social Development Project. The following are the gang research reports to date on the four studies (see the References at the back of the book for full entries):

Rochester Youth Development Study: Thornberry, Krohn, Lizotte, & Chard-Wierschem (1993); Bjerregaard & Smith (1993); Lizotte, Tesoriero, Thornberry, & Krohn (1994); Bjerregaard & Lizotte (1995); Lizotte, Howard, Krohn, & Thornberry (1997); Thornberry & Burch (1997); Thornberry (1998); Lizotte, Krohn, Howell, Tobin, & Howard (2000); Thornberry & Porter (2001); Thornberry, Krohn, Lizotte, Smith, & Tobin (2003); Thornberry, Lizotte, Krohn, Smith, & Porter (2003); Krohn & Thornberry (2008); Tobin (2008).

Pittsburgh Youth Study: Stouthamer-Loeber & Wei (1998); Lahey, Gordon, Loeber, Stouthamer-Loeber & Farrington (1999); Gordon, Lahey, Kawai, Loeber, Stouthamer-Loeber, & Farrington (2004); Stouthamer-Loeber, Wei, Loeber, & Masten (2004); Loeber, Farrington, Stouthamer-Loeber, White, & Wei (2008); Loeber & Farrington (2011).

Denver Youth Survey: Esbensen & Huizinga (1993); Esbensen, Huizinga, & Weiher (1993); Huizinga (1997); Huizinga & Schumann (2001); Huizinga, Weiher, Espiritu, & Esbensen (2003); Huizinga & Lovegrove (2009).

Seattle Social Development Project: Battin, Hill, Abbott, Catalano, & Hawkins (1998); Battin-Pearson, Thornberry, Hawkins, & Krohn (1998); Hill, Chung, Guo et al. (2002); Hill, Hawkins, Catalano, Kosterman, Abbott, & Edwards (1996); Hill, Howell, Hawkins, & Battin-Pearson (1999); Hill, Lui, & Hawkins (2001); Kosterman, Hawkins, Hill, Abbott, Catalano, & Guo (1996).

Notes

1. Reports of research of this sort are classified as grey literature because they were not published in traditional professional journals or books.

2. Participants were asked if they had received injuries from an assault or robbery (e.g., been attacked by someone with a weapon or by someone trying to seriously hurt or kill them). Injuries were considered serious if they involved a cut or bleeding, being knocked unconscious, or hospitalization.

Girls and Gangs ❖

Introduction

Girls have always been active participants in U.S. gangs (Asbury, 1927). The early street gangs in New York City typically had female auxiliaries (many members of which were highly skilled thieves and fighters) comparable to what Sante (1991) describes as the "farm leagues for boys" (p. 309).[1] Female gangsters who had ties with the older Five Points gangs helped by serving as lookouts and decoys. Younger members of some of the female gangs acted independently at times, hiring themselves out as errand runners, lookouts, or spies in brothels. Otherwise, the younger gangs spent a great deal of their free time emulating the largely adult gangs, practicing the roles to which they aspired in the mean streets.

Alongside the typical roles as lookouts and decoys, a couple of women are renowned in Asbury's (1927) historical accounts of early gangs for their fighting ability, Hell-Cat Maggie and Battle Annie, both of whom were active in male gangs in the 1850s. Hell-Cat Maggie fought along the Bowery gang's front line in many of the great battles. Distinguished by her filed teeth and artificial nails constructed of brass, "When Hell-Cat Maggie screeched her battle cry and rushed biting and clawing into the midst of a mass of opposing gangsters, even the most stout-hearted blanched and fled" (p. 30). Battle Annie and other females often positioned themselves on the outskirts of the gang battles, "their arms filled with reserve ammunition, their keen eyes watching for a break in the enemy's defense and always ready to lend a hand or tooth in the fray" (p. 29). Thrasher (1927/2000) reports that females were excluded from male gangs, but they were sometimes active in mixed-gender ("immoral") gangs that pursued sexual pleasures.

The preceding introduction encapsulates typical early characterizations of female gangs as auxiliaries to male gangs, as tomboys, or as sex objects, and these images, until recently, dominated the literature on female gang members. Moore and Hagedorn (2001) explain, "Even when describing female gang members as tomboys, researchers emphasized that the females' motivations were focused on males" (p. 1). Bowker (1978b) adds that as auxiliaries to male gangs, it is assumed that females were much like a "little sisters" group to them. In fact, Moore and Hagedorn (2001) point out that "most early reports focused on whether female gangs were 'real' gangs or merely satel-lites of male groups" (p. 1). These authors suggest, "In retrospect, the early skepticism about whether female gangs were 'real gangs' seems odd. It seems to have been based on a very narrow view of what a gang really is. Gangs—male and female alike—differ greatly from one another" (p. 2).

Females in Early Gang Studies

Historically, as Moore and Hagedorn (2001) note, "Much of the research on gangs has ignored females or trivialized female gangs" (p. 1). Thus, Chesney-Lind and Hagedorn (1999) pose the important question, "Were there really so few female gangs, or did sexist researchers just not see them? We don't know for sure" (p. 7). It would be under-standable if few females in New York City found the early street gangs to be particularly enticing, given their penchant for the nasty street fights that often originated as brawls in seedy barrooms, although a few female auxiliaries clearly were visible. The possibil-ity remains that these very public events masked more covert gang activity among girls.

In the first gang study, of more than one thousand Chicago gangs, Thrasher (1927/2000) documented only five or six gangs composed entirely of girls and even fewer mixed ones (p. 81). Curry (1999) asserts that it was Thrasher who "laid the foun-dation for the dismissal or minimization of 'woman's place' in the world of gang activ-ity" (p. 134). Thrasher (1927/2000) opined: "They lack the gang instinct, while boys have it" (p. 80). His more reasoned two-part explanation for limited female involve-ment is as follows:

> First, the social patterns for the behavior of girls, powerfully backed by the great weight of tradition and custom, are contrary to the gang and its activities; and secondly, girls, even in urban disorganized areas, are much more closely supervised and guarded than boys and are usually well incorporated into the family group or some other social structure. (p. 80)

Curry (1999) asserts that Thrasher was not the only gang researcher to overlook the active involvement of girls. Spergel's (1964) reliance on male sources in his New York City study, according to Curry, left him supporting Thrasher's minimalist perspective of female gangs. In the tradition of Riis's (1902/1969) early colorful descriptions of gangs in the city, other widely cited journalists described female gangs similarly (Bernard, 1949; Rice, 1963; Hanson, 1964). However, Chesney-Lind and Hagedorn (1999) note that these accounts have been questioned for accuracy. Popular accounts of female gangs as well as males are often tainted with sensationalism (e.g., Sikes, 1996).

Breakthrough Female Gang Studies

Pioneering research on female crime involvement set off a firestorm in criminology and specifically in the gang arena. Adler (1975a, 1975b) contended that some important differences in the organization of female gangs and in members' behavior occurred in the period from 1945 to 1975. The key change, she argued, was the evolution from predominantly auxiliary status—as weapon and drug carriers, lookouts, and girlfriends—to greater autonomy and independent activity. Adler contended that the women's "liberation hypothesis" created a female crime wave in the 1970s. This claim has been soundly discredited. Rather, Chesney-Lind (1993, 1999) insists that what occurred was a "surge of media interest" in crimes committed by females in that period, particularly slanted stories that stereotyped females as "liberated crooks."

Chesney-Lind (1999) relays a second "media crime wave" of the same sort noted in newspaper and magazine articles published in the 1990s. In "the first crime wave, the liberated 'female crook' was a white political activist, a 'terrorist,' a drug using hippie" (p. 309). In the second crime wave, the "demonized woman is African American or Hispanic, and she is a violent teenager" (p. 309) and, according to Esbensen and Tusinski (2007), often a gang member. Based on her meticulous review of the evidence, Chesney-Lind's (1999) assessment is that "in both instances, there was some, small amount of truth in the image found in the articles. [But] girls and women have always engaged in more violent behavior than the stereotype of women supports; girls have also been in gangs for decades" (p. 309).

Several seminal female gang studies undertaken in the late 1970s and 1980s moved research on female gangsters beyond the auxiliary role, including research on Chicano, Black, Puerto Rican, and members of other race/ethnicities. Studies of Mexican American gangs or gang members were carried out in Los Angeles (Harris, 1988; Moore & Long, 1987; Moore, 1991; Quicker, 1983), and in Chicago (Horowitz, 1983), and on Puerto Ricans in New York City (Campbell, 1984/1991, 1987). Studies of Black female gangs or gang members were made in Philadelphia (Brown, 1977, 1999), Los Angeles (Bowker, 1978a; Bowker & Klein, 1983), and in Boston (W. B. Miller, 1973, 1980). In addition, Shacklady-Smith (1978) interviewed girls in three British gangs in Bristol, UK. These diverse studies supported two general themes: first, that girls exhibited some independence from boys and, second, a tendency to engage in fighting and more serious criminal activity was observed.

By and large,

> by the mid-seventies descriptions of girl gang roles and activities were less likely to be restricted to the traditionally female (subordinate) role. Gang girls were more often depicted as being actively involved in conflict situations, which, in the past, were believed to be male-dominated; e.g., gang feuds, individual and gang fights. (Fishman, 1999, p. 67)

Interestingly, Campbell's participant observation research (1984/1991, 1987, 1999) drew the most attention with respect to these two themes. Campbell's 1984 book, *The Girls in the Gang: A Report From New York City* was pivotal. Curry (1999) explains, "Without a doubt, an essential contribution to contemporary thinking on female involvement in gangs was provided by Anne Campbell" (p. 137). What is particularly remarkable is that her subjects consisted of only three female gang members (two of whom were Puerto Rican) from four different gangs (one woman was a member of two gangs). Yet Campbell's presentation of their compelling life histories ranked on a par with the early Chicago school studies. Make no mistake—despite the fact that the girls' gangs were auxiliaries to male gangs—these were criminally active women, involved in selling drugs, intergang warfare, organized crime, prostitution, domestic violence, and a variety of property crimes. More important, Campbell insisted that gang girls simply had not been studied as assiduously as boys.

> Instead they had been stereotyped as promiscuous sex objects—segregated in "ladies auxiliary" gangs—or as socially maladjusted tomboys, vainly trying to be "one of the boys." The stereotypes appeared in the social work literature and were strongly imbedded in much of the [gang] research literature. (Moore & Hagedorn, 1996, p. 205)

Modern-Day Studies of Female Gang Members

Joan Moore (1991) returned to the barrios of East Los Angeles to see what had changed, if anything, in the Mexican American gangs that she described in her earlier (1978) widely acclaimed study. For her follow-up study, Moore chose two gangs, White Fence and Hoyo Maravilla, that were representative of those she found in the 1970s. These gangs operated within two neighborhoods, Boyle Heights and Maravilla, and both of these large gangs had been in existence for 45 years when Moore returned. Because she was interested in gang behaviors of both females and males, Moore oversampled cliques within the two gangs that had allied female cliques. She then randomly selected samples of 51 female and 106 male members of the two gangs. They were interviewed in 1986–1987. Moore's study is arguably the most important female gang study of the 20th century, for the rigor of her research design, the methodology she used (blending ethnographic and quantitative methods), and her discoveries on female gang involvement. Her research reveals the striking evolution of female gangs to a greater independent status and also the more active criminal involvement of female gang members.

A brief description of the setting within which the White Fence and Hoyo Maravilla gangs thrived, and their structures, is instructive. During the period of Moore's two studies, Los Angeles had become a "Chicano capital" (1991, p. 11). Boyle Heights and Maravilla competed with one another as an area of first settlement, and because the larger communities were poor and marginalized, the gangs that began to emerge there in the 1930s and 1940s were very attractive to both boys and girls. These gangs started as friendship groups of adolescents, bound by a norm of loyalty, who shared common interests, with a more or less clearly defined territory in which most of the members lived. "They were committed to defending one another, the barrio, the families, and the gang name in the status-setting fights that occurred in school and on the streets" (p. 31).

White Fence soon became quite violent, constantly fighting older boys from other gangs. El Hoyo Maravilla was a barrio invaded by the marauding servicemen in the zoot suit riots of 1941, and the boys who lived there fought back, soon developing a tradition of fighting one-on-one with rival gang members. Both gangs crystallized during the late 1930s, consisting of rather highly structured age-graded cliques with separately named cliques that are part of the larger gang (pp. 26–31, 139). Moore

(1991) describes this as "the beginning of institutionalization" (p. 31), ensuring orderly recruitment of replacement members through the formation of younger cliques every three to five years, most notably in the male Hoyo Maravilla gang (p. 29). In the two gangs, about half of the male cliques had associated female cliques. Between the mid-1970s and the late 1980s, the Hoyo Maravilla gang had 18 male and 8 female cliques, and the White Fence gang had 14 male and 7 female cliques. At the beginning of Moore's follow-up study, Hoyo Maravilla had 241 members and White Fence had 238 just in the cliques that she sampled (p. 141). Although most of the girls' cliques drew their names from the boys' gangs with which they were affiliated, others had social histories (Table 6.1).

Moore made several important discoveries in comparing female gang members of the 1970s with those of the 1950s through intensive interviews. First, more of the gang members were female than expected. Although gang cliques with female participants were oversampled, it was surprising that about one-third of the total membership of both very large gangs was female (p. 136).

Second, there were more independent female cliques in the 1970s than in the 1950s—particularly in the Hoyo Maravilla gang. "Most [Hoyo Maravilla female cliques] claimed no linkage to any specific gang . . . in the neighborhoods in Maravilla" (p. 27). In addition to their weak ties to the boy's cliques, the girls' gangs were also much less bounded with particular barrios than the boys' gangs. In contrast, White Fence girls "were more like the stereotyped notion of girl gangs" in functioning as auxiliaries to male cliques (p. 29). Moore also documented very young starter girl gangs, but they were not organized at all, only small cliques of friends and relatives.

Table 6.1 Names and Beginning and Ending Dates for Gang Cliques*

Hoyo Maravilla Gang	*Dates*	*White Fence Gang*	*Dates*
Originals	1935–1945	Originals	1944–1952
Cherries	1939–1950	Monsters Lil White Fence (girls)	1946–1954
Vamps (girls)	Unknown	Cherries WF Cherries (girls)	1947–1960
Jive Hounds	1943–1953	Tinies	1949–1961
Cutdowns Las Cutdowns (girls)	1946–1956	Spiders Chonas (girls)	1953–1960
[Big] Midgets	1950–1955	Midgets	1957–1966

*Partial listing, only earliest cliques

Source: Moore, 1991, p. 28

Third, the newer female cliques had more rowdy girls who fought, drank, or used drugs heavily. A surprisingly large proportion (80%) of the women defined themselves as *loco*—compared with 65% of men—although few women admitted to *muy loco* behaviors (p. 63). As a White Fence woman recalled: "I mean things didn't really matter. I mean nobody could explain anything. First we would just go and do it. We wouldn't even think, you know. . . . And then we would go and get in trouble and we'd think about it later on" (p. 63).

Moore found a gaping discrepancy between male and female gang members' attitudes regarding how women were treated. When she asked both male and female gang members about the role of women in the gang, half of the male members claimed that female members were "possessions." This response not only referred to the sexual exploitation of females but also reflected the males' general demand to be in charge. The other half of the male members felt that female members were respected and treated like family. However, about two-thirds of the female members vehemently denied that they were treated like possessions. To be sure, Moore found a continuing strong tendency among males to view female gang members as sex objects, contradicting the more independent posture that Campbell (1984/1991) had suggested.

Moore's findings soon were substantiated in other studies of females in a variety of locations across the United States. First, a series of longitudinal studies of male and female gang members among large random samples revealed far greater female participation in gangs, as Moore's research had suggested. The first such premier studies were the Rochester Youth Development Study and the Denver Youth Survey. These were the first of three major longitudinal studies of random samples of high-risk youth in which substudies of gang members were embedded, and the first large-scale U.S. studies to include representative samples of girls.[2] The self-report measure of gang involvement used in these two studies revealed that gang members in large urban samples of adolescents were about as likely to be females as males. The Denver Youth Survey found that females constituted between 20% and 46% of the gang members during a 4-year study period, and 18% of the boys and 9% of the girls self-identified themselves as gang members (Esbensen & Huizinga, 1993; Esbensen et al., 1993). More startling, in the Rochester Youth Development Study, a *larger* proportion of female (22%) than male (18%) adolescents self-reported gang membership up to age 15 (Bjerregaard & Smith, 1993). In both study sites, female gang members evidenced a higher prevalence rate for delinquency involvement than both nongang girls and nongang delinquent boys, and a higher incidence rate for all types of offenses than for nongang boys in Denver and Rochester.

Put simply, girls in these cities had positioned themselves between gang boys and nongang delinquents with respect to offender careers measured over the early adolescent years. These findings immediately drew widespread attention in the gang research community, and they prompted a number of studies that dissected female versus male involvement. It remained to be seen whether or not these findings on high-risk inner-city samples would be replicated elsewhere. The burning issues that these findings stimulated were twofold: First, could female gang member crime involvement also be as serious as for gang-involved males in other less high-crime places? Second, what was the composition of the gangs to which females belonged, and did this matter?

Level of Female Gang Involvement and Seriousness of Crimes

Several important discoveries have been made over the past decade or so regarding the level of female gang involvement and seriousness of crimes (Bjerregaard, 2002a). First, female involvement in gangs promotes delinquency involvement at a higher level than if females associate with highly delinquent peers not involving gangs. Gang membership facilitates delinquency over and above the effect of delinquent peers for females as well as males (Battin-Pearson et al., 1998). Comparing the 15-year-old girls who are gang members with nonmembers who are in the highest quartile of delinquent peers shows that the female gang members in Rochester still self-report significantly more involvement in general delinquency, violent delinquency, drug selling, and drug use. Gang involvement has the same facilitating effect for girls as for boys, and in Rochester, Thornberry, Krohn, and colleagues (2003) suggest this effect is particularly strong with respect to violent delinquency and drug selling.

Second, gang members of both sexes are significantly more likely to have participated in delinquency, including serious delinquency and substance abuse, and to have committed these acts at much higher frequencies than nonmembers. Thus, gang membership promotes delinquency and substance use across both sexes (Krohn & Thornberry, 2008; Thornberry, Krohn et al., 2003).

Third, delinquency among girl gang members, just as among boys, is higher than among nongang members (Deschenes & Esbensen, 1999; J. Miller & Decker, 2001). For example, an 11-city survey of eighth graders undertaken in the mid-1990s found that delinquency among girl gang members is up to five times higher than among boys who are not members of gangs (Esbensen & Winfree, 1998).

Fourth, female gang members commit similar crimes to those male gang members commit, but both Bjerregaard (2002a) and Haymoz and Gatti (2010) conclude that a smaller proportion of girls participate in serious, violent offenses. However, studies vary somewhat with regard to the types of crimes committed; that is, some indicate that girls commit less violent crimes than boys, while others show no difference. But the most consistent finding is that, among early adolescents, the crimes girl gang members committed were similar to those by boys, including assault, robbery and gang fights (Esbensen et al., 2010), and that a smaller proportion of girls are involved (Bjerregaard, 2002b). Interestingly, Bjerregaard found that within organized gangs, larger proportions of girls than boys were involved in some offenses.

Fifth, Bjerregaard (2002a) suggests that female gang members are becoming more extensively involved in the more serious and violent offenses. In the Rochester study, 29% of females in the sample were gang members at some point during the middle and high school period, and they accounted for virtually all of the entire sample of females' serious delinquencies (88%), for nearly two-thirds (64%) of all female violent offenses, and for almost 8 out of 10 female drug sales (Thornberry, Krohn et al., 2003). The

11-city survey found that more than 90% of both male and female gang members reported having engaged in one or more violent acts in the previous 12 months (Table 6.2). The researchers found that 75% of female gang members reported being involved in gang fights, 37% had attacked someone with a weapon (Esbensen et al., 2010). Ness (2010) recently observed another common situation in some cities when a gang sister or group of them is victimized—or rolled on—by others.

Table 6.2 Gang Member's Annual Prevalence and Individual Offending Rates (IORs) by Sex

Violent Act	Male		Female	
	Prevalence (%)	IOR (mean)	Prevalence (%)	IOR (mean)
Hit someone	79	7.5 [a]	78	6.6
Attacked someone with a weapon	53 [a]	5.2	37	4.3
Robbed someone	33 [a]	6.0	15	5.2
Participated in a gang fight	81	6.6	75	5.9
Shot at someone	35 [a]	4.6	21	4.1
General violence	94	19.6 [a]	94	14.2
Serious violence	85	14.0 [a]	81	9.8

[a] $p < .05$, gang boys versus gang girls; chi-square test for prevalence; t-test for IOR

Source: Esbensen, Peterson, Taylor, & Freng, 2010, p. 84

In Focus 6.1
Does the Code of the Street Apply to Girls and Young Women?

Several researchers suggest that young women also adopt the street code that Anderson (1998, 1999) first described for men on inner-city Philadelphia streets, the main principle of which is that one must display a mean persona to ward off potential victimizers. What keeps one poignantly on guard is the assumption that trouble can materialize at any moment. Among the largely Black and Mexican American communities in west and north Philadelphia, respectively, where Ness (2010) interviewed and observed girls' behavior, the main "capital" a girl has is her measure of control, her reputation for handling herself. Girls who did not measure up were considered "punks" and being so labeled "was the equivalent of being labeled a nobody," the most severe put-down (p. 49). Another scholar, Adamshick (2010), observed a similar set of behavioral expectations among girls in an alternative school in the Midwest. The three top themes that emerged in her interpretation of instances of girl-to-girl aggression in this study are (1) to protect oneself, (2) a part of the search for self, (3) as a means to enhancing attachment and friendship. In particular, "fighting served the purpose of marking one's territory or space, and attempting to overcome powerlessness" (p. 546). In contrast, in an analysis of National Longitudinal Survey of Youth data, Park, Morash, and Stevens (2010) found gang presence (exposure) and hopelessness to be strong predictors of assault for boys and girls alike. As Garot (2010) notes, Anderson's code is very much limited to "dynamics of face and identity that arise in fights among young people" (p. 122). There are many other contributing factors.

Main sources: Batchelor, 2009; Jones, 2004, 2008; Morash & Chesney-Lind, 2009; Ness, 2004; Nurge, 2003

Last, Moore (2007a) suggests that female gangs began to proliferate along with male gangs in the mid-1980s and that that the older female gangs—which were in existence long before the mid-1980s growth period—are more likely now than in the past to be involved in drug dealing, and also somewhat more likely to project an assertive stance toward the male gangs and the members with which they are associated. Moore is also convinced that these changes in the essential features of female gangs represent the sort of shift that is typically associated with "economic restructuring," that is, further marginalization of the inner city since the 1980s, which created more criminal as well as gang opportunities for women.

The Question of Increasing Female Gang Involvement

Although the question of increasing female gang involvement cannot be answered with certainty, the limited available evidence suggests that this is the case (Peterson, forthcoming). Only one nationally representative annual survey gauges female gang membership—the National Youth Gang Survey; however, the respondents in this survey are law enforcement agencies, which typically estimate that only about 10% of the gang members they observe are female.[3] This relatively low estimate should come as no surprise because girls who join gangs tend to leave them at an earlier age than boys and hence have less street presence than boys. In a national school survey, Gottfredson and Gottfredson (2001) found gang joining among boys peaked at the 10th grade, but 2 years earlier for girls, at the 8th grade. Related to this point, law enforcement seems to pay less attention to very young gang members altogether because their attention is properly focused on older more serious and violent gang members.

Nevertheless, the proportion of females among active gang members is undoubtedly larger than in the past century. In the 1997 National Longitudinal Survey of Youth, male versus female differences in the proportion who joined gangs was not as large as previous research had suggested. Snyder and Sickmund (2006) report the male-to-female ratio in this national sample was approximately 2:1 (11% of males versus 6% of females). In a nine-city purposive sample, Esbensen and colleagues (2008) found almost equal proportions of boys (9%) and girls (8%) self-reported gang membership.

Also, Black and Hispanic boys have only slightly higher gang membership rates (10.2% each) than girls. By comparison, in known gang problem areas, surveys of students and other adolescent samples show that approximately one-third of all gang members are female (Bjerregaard, 2002a). For example, in Gottfredson and Gottfredson's (2001) nationwide student survey, 35% of the self-identified gang members were girls.

The Importance of Gang Gender Composition

Only in the past decade or so has widespread attention been given to the gender composition of the gangs in which females participate (Peterson, Miller, & Esbensen, 2001). Several studies indicate that mixed gender gangs are quite common. Curry's (1998) study in three cities found that more than half (57%) of the girls described their gang as mixed sex, about a third said they belonged to gangs that were affiliated with male gangs, and only 6% said their gang was autonomous. Student respondents in Esbensen and fellow researchers' (2008) nine-city student sample of male and female gang youths, mostly ages 12 to 15, classified the members of their gang as predominantly (54%) half male and half female, regardless of racial/ethnic composition.[4] Only 10% of the boys and 4% of the girls said their gang was same gender. These findings are consistent with several other studies that show considerable gender-mixed gang activity.

Interestingly, in the first definitive study of gender mixing, Peterson and colleagues (2001) found that boys and girls in the aforementioned 11-city student survey—from a sample of male and female gang youths attending public schools in the mid-1990s—provided differing accounts. "Approximately 45% of male gang members described their gangs as having a majority of male members; 38% said their gangs were sex-balanced; 16% were in all-male gangs; and just under 1% (two cases) reported being in gangs that were majority-female" (p. 423). In contrast, more than half (54%) of the girls described their gangs as sex-balanced, followed by 30% in majority-male gangs. The greater mix in female gangs would prove to have noteworthy implications for involvement in criminal activity. Peterson and colleagues (2001), as well as Miller and Esbensen (2001), found the females in all- or majority-female gangs exhibited the lowest delinquency rates, and females in majority-male gangs exhibited the highest delinquency rates (including higher rates than males in all-male gangs).

In general, these studies suggest that criminal activity and violence tend to increase as the proportion of males in the gang increases. Why this is so is not well understood. It is worth noting that in Giordano's (1978) and Warr's (1996, 2002) studies of peer influence on delinquency, the cross-sex groups of delinquents have higher rates. In particular, Warr (1996) found in an analysis of national data that, unlike males, delinquent offenses reported by females are significantly more likely to occur in mixed-sex groups. In 2007, Haynie, Steffensmeier, and Bell's national study confirmed this, finding that "among females, the odds of engaging in violence are greatest when adolescent girls are enmeshed in a highly violent friendship network comprised of a greater proportion of male friends."

Studies to date in several sites strongly suggest that girls are more likely to be involved in serious offenses when they are members of more organized gangs

(Bjerregaard, 2002a). In reference to gangs in general, research on information provided by Esbensen and fellow researchers' (2001) 11-city sample of students found that members of gangs that are somewhat "organized" (i.e., have initiation rites, symbols or colors, established leaders, specific rules, and engage in illegal activities) self-report higher rates of delinquency and involvement in more serious delinquent acts than other youths. There may well be few differences between male and female gangs in this regard. In fact, Esbensen, Deschenes, and Winfree (1999) note that this study found, if anything, the gangs in which girls were involved were slightly more organized than those to which boys belonged. Equally important, Decker, Katz, and Webb's (2008) Arizona study found that the more organized the gang (having initiation rites, established leaders, and symbols or colors), the more likely members were to be involved in violent offenses, drug sales, and violent victimizations—even at low levels of organization.

According to reports of self-nominated gang members in a study conducted in California, Illinois, Louisiana, and New Jersey, both male and female members of young gangs considered their gangs to be relatively well organized; and in one report on this study, Bjerregaard (2002a) found more females reported "having larger gangs, a gang name, a leader, regular meetings, a turf, and special clothing" (p. 91). Interestingly, males were significantly more likely to report being involved in gangs that engaged in criminal activity, but "when one examines the gender patterns among those reporting membership in organized gangs, a different pattern emerges. The gender differences disappear" (p. 91). Once she applied the "organized" requirement to self-identified gang involvement, Bjerregaard (2002a) found that "female prevalence rates of gang delinquency are similar to or higher than those reported by the males. . . . This finding is contrary to much of the prior research which reported that females were less likely to be involved in both criminal offenses and personal offenses and that females sometimes acted as inhibitors to such offenses occurring (Campbell, 1990, Fishman, 1995; Klein, 1995)" (p. 93).

Girl Gangs Outside of the United States

Much of the research on girls' experiences with gang involvement elsewhere in the world is relatively recent (Batchelor, Joe-Laidler & Myrtinen, 2010; J. W. Moore, 2007a). In her international literature review of girl gang studies, Moore (2007a) found virtually no literature on female gangs outside of the United States. She contends that this is partly because gangs of either gender appear to be recent phenomena in Europe. In England, for example, Moore suggests that studies completed there since the 1970s strongly indicate that that the reported gangs are not really organized in American-style gangs—with few exceptions (such as the Manchester gang described in Klein, Kerner, Maxson, et al., 2001). Rather, they appear to be bonded with each other in loose and independent friendship cliques. These may well be the groups that surveyed students referenced in two reports that suggested female gang involvement, first in Smith and Bradshaw's (2005) Scotland study and second, in Sharp, Aldridge, and Medina's (2006) national survey in England and Wales. Both surveys suggested substantial female gang involvement. Not all observers are in agreement, however, with respect to the seriousness of gang problems on the European gang scene. Hallsworth and Young (2008) exhort officials and practitioners "to be more skeptical about gang talk and gang talkers" (p. 191). Specifically, the authors suggest, "unless you have good reason:

- Refrain from doing gang talk to your friends.
- Refrain from doing gang talk to your enemies.
- Refrain from doing gang talk to yourself. (p. 192)

It may well be that gang culture is spreading across Europe in conjunction with family migration; not, as Peterson, Lien, and Van Gemert (2008) assert, gangs themselves. Girls' involvement has been reported in various types of gangs in Europe, in the UK (Batchelor, 2009), Germany (Bruhns & Wittman, 2002), Norway (Natland, 2006), Scotland (Brawshaw, 2005), Russia (Salagaev, Shaskin et al., 2005), Italy and Switzerland (Haymoz & Gatti, 2010), and the Netherlands (Erickson, Butters, Cousinea, Harrison, & Korf, 2006); Central America, including Nicaragua (Rodgers, 2006) and Guatemala (Winton, 2007); as well as Hong Kong (Li & Joe-Laidler, 2009) and New Zealand (Dennehy & Newbold, 2001). In the UK studies, female membership appears comparable with male participation, particularly in the younger age categories, but according to Batchelor (2009), girls report involvement delinquent offenses at a much lower rate than their male peers.

Landmark Studies of Females in Gang Contexts: Risks and Revictimization

J. W. Moore's (1991) follow-up interviews with former and still-active gang members documented the influence of family factors on gang involvement among females. Despite the fact that traditional Mexican families are intensely opposed to any kind of street involvement for girls, many Mexican girls did join gangs, usually the sisters of the boys who were centrally involved. Those girls who participated in gangs tended to come from families characterized by extensive poverty, unhappiness, drug and alcohol abuse, incest, and other family problems. Three-quarters of the gang girls ran away from home at least once, and nearly half of those who ran did so repeatedly. They mainly were escaping severely troubled families. Almost one-third of the female gang members said they had been molested by a family member, most commonly the father, but also uncles, brothers, and grandfathers. A near majority of the molested girls had been victims of repeated sexual encounters. One of Moore's particularly important findings in this context is that more women than men who joined gangs came from troubled families.

Moore (1991) displayed a level of sensitivity to gendered victimization of girls and women in her Los Angeles gang study that remains unparalleled. Her main discovery is that even though family values prohibited the girls' involvement, it appeared that many of them had been propelled by family, gang, and community dynamics to join a gang (Moore & Hagedorn, 1996). Those who joined tended to view the gang as "a group that could sympathize with them, welcome them, and in some cases shelter them" from abusive family members and otherwise troubled families (p. 208). The gangs themselves had different standards for girls and boys. Whereas boys who joined were not necessarily viewed as deviants, girls inevitably were, as they were labeled as "bad" girls. In accordance with community standards, parents in conventional families would not permit their children to play with others from "underclass" families. Thus, girls from "bad" families were stigmatized. In these ways, "to put it succinctly, there was a self-selection process in gang recruitment that revolved around gender" (p. 207). Moore and Hagedorn's (1996) comparison of females' status in Milwaukee during the early 1990s, versus Moore's subjects in the late 1980s, led to this main conclusion: "Times may have changed but gendered exploitation persists" (p. 184). This prophetic statement continues to be salient.

In an intriguing gang study conducted approximately two decades later, A. Valdez (2007) found girls in San Antonio in the same plight as Moore's young women. In this study, Valdez achieves a level of sensitivity that reaches the very high bar that Moore set in her Los Angeles study. His research setting is the Mexican barrios on the West Side of San Antonio, where there are 2,945 housing units in eight public housing projects, where some of the city's poorest families live. Valdez chose to move his research spotlight off the gangs themselves and onto 150 randomly selected girls who associated with male gang members but never joined their gangs.[5] Because gangs were ever-present in the West Side barrios, hanging out with gang members became "a daily routine in the lives of many of these adolescent females" (p. 3).

Valdez discovered that girls in these neighborhoods who were continuously exposed to high-risk situations and dangers unwittingly place themselves at an even higher risk for violent victimization, having the effect of further elevating them "beyond risk" (or at extreme risk) and practically assuring extremely multiple bad outcomes. "Their involvement in risky behavior—such as the perpetration of physical fights, early onset of sexual behavior, substance use, and/or other delinquent behavior—is characteristic of growing up in these environments" (p. 3). Valdez isolated the major risk factors that elevated risk for the girls he studied.

First, the family context was extremely detrimental for the females who would later be gang associates. The use of drugs and alcohol, violent acts, and criminal behavior were "normalized" within the family context of the girls in his study, and 8 out of 10 of the girls had a family member who used drugs. More than 6 out of 10 had someone in the house with a drinking problem, typically the father. Many of the young women felt the impact of family violence in their everyday lives. More than 7 out of 10 of them had family members involved in criminal activity. More than half of them had witnessed their parents physically fighting in the home.

Second, peer relationships increased the girls' risk for violent victimization, particularly their gangs' associations. Each of the female gang associates was "distinctly integrated into the male gangs through their relationships with the male gang members" (p. 87). The girls typically first associated with the gang members before age 12. When interviewed by Valdez, 43% said they currently had a boyfriend in a gang, and 81% said they had a good friend in a gang. The longer the duration of gang affiliation, the greater the girls' participation in male-gang delinquent activities. Not surprisingly, "The data indicate that many of the girls had a history of incarceration that occurred after they became affiliated with the gang" (p. 86).

Third, individual exposure to risky behaviors including delinquency, violence (particularly physical fights), substance use, and sexual relations was extensive among the female gang associates. Based on his lengthy study of girls who associated with male gangs in San Antonio, A. Valdez (2007) developed the following typology of girl types among the associates to categorize nongang females:

- *Girlfriends* are defined as a type of female gang associate who is a current steady partner of a male gang member. Relationships range from being a male gang member's "main chick" or *santita* (saint) to the teenage mother of his child. This type of female is least involved in everyday gang activities, and to some extent these girls were shielded from male gang members (e.g., not sexually harassed), out of respect.
- *Hoodrats* are a more complex type of female gang associate. Although often (but not necessarily) she is sexually promiscuous, she is often seen hanging out and partying with the guys and generally is a heavy polydrug and alcohol user. Male gang members often refer to this type of female as *bitch, shank, player,* and *whore.* She normally does not develop an emotional relationship with any of the boys, yet among the four female types, she is most actively involved in everyday gang activities.
- *Good girls* include childhood friends of many of the male gang members, often having attended the same schools, and having parents who interacted with each other. In time, these relationships came to be based on mutual respect. Males characterize these females as "nice girls." Compared with the other types of female gang associates, this type tends to have conventional lifestyles, very infrequent involvement in criminal activities, and also limited involvement in everyday gang activities.
- *Relatives* refer to girls who are close relatives to gang members, typically sisters and cousins. These kinship ties accorded this female type special status within the social network. For example, if one of these girls were dating a gang member, she would be given special status as his main chick and also as a homeboy's sister or cousin. This type of female gang associate also has limited involvement in everyday gang activities.

Valdez's (2007) classification scheme is very useful, for no other gang research has identified distinguishable subgroups of girls who hang around boys actively involved in gangs. However, some researchers have used other terms to characterize them, including *associates* (Curry et al., 2002), *wannabes* and youth who *kick it* (participate in gang social activities) (Garot, 2010). But Valdez's classification scheme is far more explicit than others and should prove very useful in distinguishing girls' levels or degrees of gang involvement for research, prevention, and intervention purposes.

All types of girls in Valdez's typology were involved in physical street fights, ranging from pushing and shoving to more violent attacks that usually resulted in injuries. Hoodrats were most actively involved in everyday gang activities and were "more likely to participate in illegal activities in association with male gang members" (p. 98). Some of the Hoodrats were involved in more serious activities such as drug dealing and weapon sales, but often independently of the male gang. In addition, by virtue of their association with a particular male gang, tensions were created with other girls associated with rival gangs. Valdez observed that the female gang associates in the barrios on the West Side of San Antonio "are continuously exposed to high-risk situations and dangers, which are exacerbated when associating with gangs" (p. 180).

A. Valdez (2007) emphasizes violent victimization as a predominant result of high-risk girls' exposure to gangs. Victimization at the hand of boyfriends typically begins with a verbal instigation. Importantly, adolescent girls "are prey for male adolescents and older adults in a street culture that promotes hyper masculinity, sexual conquest, sexual aggression, and sexual objectification of women" (p. 111). Unfortunately, these girls view the gang life as an opportunity for them "to gain autonomy and independence, not only from family oppression, but also from cultural and class constraints" (p. 113). For girls in these circumstances, the gang is very likely to be viewed as a refuge for girls who have been victimized at home. Paradoxically, these vulnerable girls are placing themselves at greater risk of violent victimization as a direct result of their association with gang members, including sexual and physical victimization by boyfriends, and also by other male gang members in the context of the gang such as at parties (pp. 114–128).

In Focus 6.2
Jealousy (Precursor) and Acts of Instigation

1st Act of Verbal Instigation: He says, "My friends told me they saw you talking to some guy at school today."

2nd Act of Verbal Instigation: She says, "I wasn't talking to anybody, your friends are lying."

3rd Act of Verbal Instigation: He screams, "Don't lie to me. You're a ho!"

1st Act of Physical Confrontation: He slaps her across the face.

2nd Act of Physical Confrontation: She pushes him away from her.

3rd Act of Physical Confrontation: He grabs her violently by the arm.

4th and Final Act of Physical Confrontation: She pushes him away from her and storms off.

Source: A. Valdez, 2007, p. 141

A Pathway to Serious Offender Careers for Girls

This section proposes nine possible stepping-stones in girls' pathways to serious, violent, and chronic juvenile offender careers, including gang involvement. These stones are traversed in a developmental process. Researchers sometimes fail to recognize the cumulative requirement of the social development process.[6] Empirical evidence of this process is abundant (detailed in Chapter 5). The model proposed here applies to girls who grow up in disadvantaged neighborhoods and communities, where Hipwell and colleagues (2002) report girls disproportionately experience multiple problems. In the representative Pittsburgh community sample of 5- to 8-year-old girls, nearly 16% of the girls living in disadvantaged neighborhoods[7] scored high on a Disruptive Behavior Index, compared with 9% of girls in the moderately disadvantaged neighborhoods, versus only 5% of those in the advantaged neighborhoods. Hence, girls who scored deviant on several domains relative to their peers were overrepresented threefold in disadvantaged neighborhoods. The stepping-stones proposed here apply mainly to girls similarly situated.

> A small group of girls, just like boys, experience an early onset of disruptive behavior and their behavior shows a relatively stable pattern. These girls may be particularly at risk for negative long-term outcomes, including teenage pregnancy, dropping out of school, involvement in delinquency [and gang participation]. (Kroneman et al., 2004, p. 119)

Of course, girls cannot join gangs if these groups are not already established in neighborhoods. This is not a theory of gang formation; rather a suggested series of stepping-stones that might apply to young girls at high risk of becoming persistent serious and violent offenders. Gang participation can become an interlocking step in this pathway. The potentially cumulative stepping-stones are (1) neighborhood conditions, (2) child physical and sexual abuse, (3) school problems, (4) running away, (5) drug and alcohol abuse, (6) mental health problems, (7) gang membership, (8) violent victimization, and (9) juvenile justice system involvement. Not all girls at high risk can be expected to traverse each of these stepping-stones. Supporting research indicates that risk factors and protective factors vary from one community or neighborhood to the next (Sampson & Graif, 2009; Wikstrom & Loeber, 2000)—and, according to Thornberry (2005), they vary in influence from one point in time to another from birth to adolescence—and so do individual girls' responses and adaptations to them. Unfortunately, little research on girls' risk factors for gang joining exists and, as Peterson and Morgan (2010) note, a more standardized approach in the examination of sex differences is needed.

Hipwell and colleagues (2002) conducted the first major longitudinal experimental study of a large representative sample of young girls. Their findings suggest that

> owing to sex-stereotyping socialization processes, it is likely that the cumulative effects of risk factors may be more extensive for girls than for boys [and] this notion is, in fact, supported by the higher rates of co-occurring behavioral, emotional, and social problems among disruptive female adolescents compared with their male counterparts. (p. 115)

This analysis suggests that risk factors for girls' deviance may begin to show negative effects as young as ages 5 to 8. The central proposition here is that the gendered nature of these risk factors and experiences may well have a greater impact on girls than on boys for the most severe delinquency outcomes—that is, serious, violent, and chronic juvenile offender careers. For gang girls, involvement in multiple problem behaviors is commonplace, particularly sexual activity, substance use, and violence.

Community conditions are the first stepping-stone. Children who grow up in neighborhoods characterized by concentrated poverty are at elevated risk for multiple negative outcomes, including poor physical and mental health, risky sexual behavior, delinquency, and gang involvement (Sampson, Morenoff, and Gannon-Rowley 2002). Neighborhood poverty is strongly associated with problem behavior—specifically girls' disruptive behaviors—as young as age 5 (Hipwell et al., 2002). Hipwell and colleagues (2002) explain that a subgroup of females at an early age display a range of disruptive disorders, particularly among girls in the most disadvantaged neighborhoods" (p. 99). In these neighborhoods, girls—just like boys—"are exposed to a greater number of risk factors, including exposure to different forms of community and family violence and an increased likelihood of affiliation with deviant peers" (Kroneman et al., 2004, p. 117). Park, Morash, and Stevens (2010) contend that gang presence in neighborhoods is a key contextual factor, particularly where gangs have been active in neighborhoods for several years and friends or siblings have been in gangs for multiple years. However, for recent migrants, Vigil (2008) claims some effects of disadvantaged neighborhoods may not manifest themselves for a year or more as families and the children themselves become more and more marginalized.

Child abuse and sexual exploitation is the second stepping-stone. Girls are more likely than boys to suffer physical abuse up to about age 4, and they are about twice as likely as boys to suffer sexual abuse throughout childhood and adolescence (Finkelhor, Turner, Ormrod, & Hamby, 2009). There are similar differences when abuse and neglect are considered together. For example, Ireland, Smith, and Thornberry (2002) found in the Rochester study that by age 18, 16% of the males and 27% of the females had substantiated cases of maltreatment. Chesney-Lind (1989, 1997) concludes sexual abuse victimization leads girls to run away from home. Finkelhor and colleagues (2009) examined national survey data to show that girls are more likely than boys to be sexually victimized: 7% of girls reported a sexual victimization within the past year, and nearly one in eight (12%) reported being sexually victimized during their lifetimes. Sexual victimizations are more common for girls and also strongly concentrated nationwide in the 14- to 17-year-olds. These girls had sexual assault rates of 8% (per year) and 19% (lifetime), completed and attempted rape rates of 6% (per year) and 14% (lifetime), and rates of sex assault by a known adult of 2% (per year) and 8% (lifetime). Some research suggests that the sexual victimization of girls who affiliate with gangs may begin much earlier and that the sexual abuse experiences of girls are also closely linked to a history of running away, which may expose them to greater risk for multiple victimizations (Moore, 1991; Moore & Hagedorn, 2001).

Experiencing school problems is the third stepping-stone. But research on school-related problems as a precursor for girls' gang involvement is a glaring gap. Most research on this topic addresses males' school problems and exclusion of racial/ethnic minorities, particularly how they are propelled along the "school-to-prison pipeline" by zero tolerance policies, extremely high dropout, suspension, and expulsion rates (Howell, 2009). Make no mistake, these factors apply to girls as well. In addition, Peterson (forthcoming) tags low commitment to school as a distinguishing risk factor for girls' gang joining.

The fourth stepping-stone for girls on the pathway to delinquency is running away. The proportion of girls (11%) who run away from home is about the same as that for boys (10%) (Snyder & Sickmund, 1999, p. 58). The relationship between running away, or being thrown away, and subsequent delinquency is well established: Kaufman and Widom (1999) found that childhood victimization (sexual or physical abuse and neglect) increases the runaway risk and that both childhood victimization and running away increase the likelihood of juvenile justice system involvement. Theirs and other studies, particularly Tyler and Bersani (2008), show that running away is correlated with subsequent high-risk outcomes. Life on the streets is risky for homeless children, leading to substance abuse, and associations with deviants, risky sexual behaviors, and violent victimization (Fleisher, 1995, 1998). In a study of homeless and runaway adolescents in four midwestern states, Whitbeck, Hoyt, and Yoder (1999) found that such street experiences also amplified the effects of early family abuse on victimization and depressive symptoms for young women. Moreover, in a subsequent analysis of only the girls in the initial study sample, Tyler, Hoyt, and Whitbeck (2000) found that girls who leave home to escape sexual abuse are often sexually victimized on the streets. Tyler and colleagues explain the pathway as follows: Exposure to dysfunctional and disorganized families early in life places youths on a trajectory of early independence; this early independence on the streets, along with the homeless or runaway lifestyle, exposes them to dangerous people and places; and this environment and the absence of conventional ties put them at increased risk of sexual victimization (p. 238).

Drug abuse is the fifth stepping-stone. There is evidence that the trend for earlier initiation of alcohol use is more pronounced among girls than among boys, and Hipwell and colleagues (2005) note that girls may be more vulnerable to persistent use, alcohol abuse, and dependence at an early age. Keenan's (2010) research team detail this likelihood:

> PGS data from the two oldest cohorts suggest that African American girls may be more likely, compared to European American girls, to experiment with alcohol during childhood, but that European American girls appear to be at greater risk, relative to African American girls, for initiation of alcohol use during adolescence, and possibly for progression to more regular alcohol use following initiation of alcohol use during adolescence. (p. 9)

Another study led by Huizinga (2003) suggests that, in adolescence, prior delinquency involvement is a stronger indicator of drug use among girls than vice versa, but nearly half of the persistent serious female delinquents are drug users.

Mental health problems are the sixth stepping-stone. Shufelt and Cocozza (2006) report girls at significantly higher risk (80%) than boys (67%) for a mental health disorder, with girls demonstrating higher rates of internalizing disorders than boys. In another study, Teplin and colleagues (2006) collected a random sample of a large population of juvenile detainees, and found that three-quarters of females had a psychiatric diagnosis compared with only two-thirds of males. Boys and girls may also differ with respect to the impact of depression on violent behavior. In Chicago, Obeidallah and Earls (1999) found that, compared with nondepressed girls, mildly to moderately depressed girls had higher rates of aggressive behavior and crimes against people. Thus, depression may be a central pathway through which girls' serious anti-social behavior develops, so it may be a precursor to delinquency and violence. Early pubertal maturation might be a precursor for the onset of both depression and anti-social behavior among girls (Keenan, Hipwell, Chung et al., 2010), so this may be an important screening point.

Youth gang involvement is the seventh stepping-stone. Thornberry, Krohn, and colleagues (2003) assert that child abuse, early dating, and precocious sexual activity increase the risk of gang involvement for both girls and boys. Several studies suggest that family conflict and child victimization in the home may have greater importance as risk factors for gang membership for girls than for boys (Moore & Hagedorn, 2001). In a study of gang girls, J. A. Miller (2001) found that girls with serious family problems often choose gangs as a means of avoiding chaotic family life and meeting their own social and developmental needs, and unsafe conditions at home helped drive them into the streets and into gangs. Many studies conclude that some gangs provide refuge and support for girls. Indeed, as Joe and Chesney-Lind (1995) suggest, the gang may serve as an alternative family.

Violent victimization is the eighth stepping-stone. The contributing factors to girls' violent victimization are well documented in the preceding discussion, including family, peer, and neighborhood contexts. It must be said, and as Morash and Chesney-Lind (2009) concur, that much of girls' violence is in response to neighborhood conditions prompting them to assault others to ward off victimization. Many of the fearful girls join gangs, according to Melde, Taylor, & Esbensen (2009), for protection and to address their fear of victimization, and as Carbone-Lopez, Esbensen, and Brick (2010) assert, in hopes of preventing some of the repeated harassment they experience in social settings. In doing so, they increase their risk of revictimization.

Juvenile justice system involvement is the ninth stepping-stone on girls' pathway to delinquency. Thornberry, Krohn and colleagues (2003) report girls who are actively involved in gangs become the most serious, violent, and chronic juvenile offenders of all girls. Gang membership increases the odds of continued involvement in criminal activity throughout adolescence and into adulthood for both young men and young women. However, girls are far more likely than boys to enter the juvenile justice system via arrests for running away from home, and Valdez (2007) concludes this group of runaway girls also has high rates of substance abuse problems and gang involvement. Snyder and Sickmund (2006) found that although only a little more than one-fourth (29%) of juveniles arrested in 2003 were girls, they represented more than half (59%) of the juvenile arrests for running away. Once girls are placed on probation—typically for status offenses, such as running away or liquor law violations—Acoca (1999) says, "Any subsequent offense [becomes] a vector for their greater involvement in the juvenile justice system" through probation violations and new offenses (p. 7). Girls are far more likely than boys to be held in detention centers for minor offenses, particularly for probation and parole violations (American Bar Association & National Bar Association, 2001). Girls are also more likely than boys to be detained for contempt of court, increasing the likelihood of their return to detention centers, providing more opportunities for further victimization. In short, gender disparities exist in all stages of juvenile and criminal justice systems. Incarceration, ironically, often leads to further victimization. Although female sexual abuses at the hands of staffs within juvenile correctional facilities have been exaggerated in some reports (Beck, Harrison, & Guerino, 2010), Acoca (1998) believes that when girls are placed in these facilities, they are often subjected to further victimization in the forms of emotional abuse; emotional distress from isolation; physical abuse, threats, and intimidation from staff; and unhealthy conditions of confinement. In addition, Roush, Miesner, and Winslow assert gang-involved girls receive few specialized services in juvenile correctional facilities, and Petersen (2004) links this to forming friendships with other gang-involved girls in correctional gang subcultures.

Research is needed on the sequencing of the developmental pathway steps suggested here (and others that are research supported) to determine the most common combinations that girls follow. Welcomed illumination of girls' pathways to delinquency and gang involvement is forthcoming from the first major U.S. longitudinal study focused exclusively on girls (the Pittsburgh Girls Study, $N = 2,451$). It is underway in the Pittsburgh Life History Studies (directed by Rolf and Magda Loeber). Headed by Alison Hipwell, it was launched in 2000.

The Moving to Opportunity Experiment

The preceding discussion of girls' pathway to gang involvement and violence provides a propitious opportunity to consider ways of deterring girls from gang involvement and violent victimization. Although specific programs for girls are presented in Chapters 9 and 10, a large-scale program is reviewed here to take advantage of readers' focus on the previous research. Chapter 4 described the negative effects of community disorganization that allow gangs to flourish. Following the failure of many federal programs to overcome the effects of poverty, the U.S. federal government undertook a major experiment to change poverty-stricken families' plight by moving them out of public housing settings altogether! Outcomes of this dramatic undertaking are intriguing and thought provoking.

The Moving to Opportunity for Fair Housing Demonstration (MTO) was a unique effort to try to improve the life chances of very poor families with children by helping them leave the disadvantaged environments that contribute to undesirable outcomes (Popkin, Leventhal & Weismann, 2008). The U.S. Department of Housing and Urban Development (HUD) launched MTO in 1994 in five cities: Baltimore, Boston, Chicago, Los Angeles, and New York—some of the nation's poorest, highest-crime communities with "distressed public housing" (p. 1). In this $70 million HUD program, housing subsidies were offered to families in high-poverty areas, giving them a chance to move outward to lower-poverty neighborhoods and better opportunities. Participation in MTO was voluntary. Those families that volunteered were randomly assigned to one of three treatment groups (Orr et al., 2003):

- The *experimental group* was offered housing vouchers that could only be used in low-poverty neighborhoods (where less than 10% of the population was poor). Local counseling agencies helped the experimental group members to find and lease units in qualifying neighborhoods.

- The *Section 8 group* was offered vouchers with no geographical restriction and no special assistance. (Section 8 is a federal tenant-based housing voucher program for public housing tenants who are forced to move because of rehabilitation or demolition of their public housing unit.) To use the voucher, experimental group families were required to move to census tracts with poverty rates below 10% in the 1990 Census.
- Finally, *control group* members were not offered vouchers but continued to live in public housing or receive other project-based housing assistance.

Almost half of the families assigned to the experimental group "leased up" to low-poverty areas with program vouchers, as did more than three-fifths of the families in the Section 8 group. This family relocation program was based on the belief that "moving to a better neighborhood might benefit adolescents in several ways" (Popkin et al., 2008, p. 1):

- Providing better monitoring of behavior to reduce the threat of violent crime and disorder
- Offering stronger institutional resources for youth, notably high-quality schools, youth programs, and health services
- Providing access to more positive peer groups
- Promoting changes in parents' well-being and behavior because of increased opportunities and social pressures

Orr and colleagues' (2003) follow-up research on the MTO families about 5 years after they moved revealed that those who relocated were very satisfied with their situation. From the families' perspectives, the principal benefit of the move was a substantial improvement in housing and neighborhood conditions. Families who moved with program vouchers largely achieved the single objective that loomed largest for them at baseline: living in a home and neighborhood where they and their children could feel and be safe from crime, violence, gangs, and drugs.

As expected, Popkin, Leventhal, and Weismann (2010) reveal that women and girls in the experimental families cited a general sense of safety as their biggest gain. "An issue of particular concern for many mothers and girls was the pressure for early sexual initiation—especially what they viewed as older men and boys preying on very young girls" (Popkin et al., 2008, p. 3). And the majority of those in the experimental group believed they had attained their goal. "These girls—and their mothers—often spoke about what had happened to their friends who still lived in public housing and how they felt they had avoided that fate" (p. 2). Popkin and fellow researchers (2010) found the diminished pressure to succumb to sexual advances as especially significant for very young girls who, in high-poverty neighborhoods, began experiencing sexual harassment and pressure from older boys and men during early adolescence, typically by ages 12 or 13.

The experimental family girls reported significantly less psychological distress and anxiety than did girls in the control group. Moreover, they were less likely to report marijuana use or smoking and were less likely to be arrested (for both violent and property crimes) than control group girls. Perhaps more important, the girls who moved experienced lower exposure to gangs and drug trafficking (Popkin et al., 2008), and both mothers and girls achieved a "dramatic reduction" in "female fear" (Popkin et al., 2010, p. 735). "Compared with their counterparts still living in high-poverty neighborhoods, female experimental-group participants reported less harassment from men and boys, less pressure to engage in sexual behavior, and, as a result, said they were less fearful" (Popkin et al., 2008, p. 2). Indeed, the change was dramatic. "Nearly all the girls in the experimental group and their mothers described feeling confident that they were safe from those types of risks in their new neighborhoods, while nearly the same proportion in high poverty described living with harassment" (p. 2).

Both adolescent girls and their mothers who had moved in lower-poverty neighborhoods were keenly aware of the dangers they had left behind, and expressed enormous relief that they felt less stressed and scared. The difference was significant:

> In contrast, those who were still living in—or who had moved back to—high-poverty communities spoke of their fears, the often extreme strategies they used to protect themselves (or their daughters), and the consequences the girls had faced—pregnancy, sexually transmitted diseases, intimate partner violence, and sexual assault. (Popkin et al., 2010, p. 735)

Interestingly, although adolescent girls benefited in important ways from moving to better neighborhoods, Popkin and colleagues (2008) found the boys who relocated seemed to have not benefited at all, and their behavior worsened in some respects. In contrast to the favorable outcomes for girls,

> adolescent boys in the experimental group reported *more* behavior problems, were *more* likely to smoke, were *more* likely to be arrested for property crimes, and—perhaps most surprising—were *no less* likely to be arrested for violent crimes than their counterparts in the control group. (Popkin et al., 2010, p. 718)

Needless to say, "these findings have been very controversial," particularly the absence of an explanation for why boys seem to have fared so poorly (p. 1). Explanations for the disparate outcomes for boys and girls come from in-depth interviews with a random sample of teenagers in the Baltimore and Chicago sites (Clampet-Lundquist, Edin, Kling, & Duncan, 2006). Many of the boys in experimental families "seemed to lack the set of neighborhood navigation skills that would allow them to avoid neighborhood locations where trouble was likely to occur and to distinguish between good and bad influences when forming friendships" (p. 37). Of particular importance, many of the experimental boys maintained "risky ties" to friends and relatives in the public housing out of which their families had moved. Consequently, Popkin's research team (2010) found they "must go to great extremes to avoid becoming involved in gang activity" (p. 736). In contrast, control group boys who remained behind seemed more adept at using strategies that enabled them to avoid neighborhood trouble in their more familiar community surroundings.

Experimental group girls were able to benefit more from low-poverty neighborhoods than boys, in part, because they engaged in activities including peer interactions in prosocial school networks that facilitated acclimation into low-poverty neighborhoods (Clampet-Lundquist et al., 2006). The protective role of mothers also seemed important. They found it easier "to monitor and protect their daughters when they live in lower-poverty communities where harassment and sexual pressure are not as pervasive" (Popkin et al., 2010, p. 736). For families and girls who cannot move from harmful community, family, and school conditions, programs are discussed in Chapters 9 and 10.

Park and colleagues (2010) conducted a national study of gender-sensitive predictors of assaultive behavior in the late adolescence period that suggests several important targets for intervention (risk factors) and protective measures that should be taken for "beyond risk" girls. This study also supports the research on San Antonio girls in addressing several of the stepping-stones to gang involvement and violent behavior. Early-age runaway girls (before age 12 or 13) reported significantly more assaults than other girls in the national study (Park et al., 2010).[8] Additionally, for each year of school attended, girls' reports of assault were much lower, on average. With greater gang presence (exposure), the more likely the girls assaulted others. Low hope for the future (expectations of arrest and victimization) also significantly predicted more assaults. Parental monitoring significantly predicted fewer assaults for girls. Finally, Park and colleagues (2010) found girls with low violence levels had more parental monitoring and stronger ties to school and religious institutions. For the entire sample of girls and boys, the hopelessness indicator was positively correlated with self-reported assaults, gang presence, and years of gang membership.

Concluding Observations

Girls have been participants in U.S. gangs from the earliest gang records. Many years passed, however, before gang researchers took note of their active involvement. Still, the early research minimized their involvement to auxiliary roles with strong emphasis on sexual gratification. But female gang studies undertaken in the late 1970s and 1980s moved research on female gangsters beyond these subservient and auxiliary roles.

Modern-day studies of female gang members reveal the more active involvement of girls in gangs and several important discoveries regarding the level of gang involvement and seriousness of crimes including the following:

- Female gangs began to proliferate along with male gangs in the mid-1980s.
- More of the gang members are female than expected.
- Gang membership promotes delinquency and substance use across both sexes.
- Delinquency among girl gang members, just as among boys, is higher than among non-gang members.
- Female gang members commit similar crimes to those committed by male gang members, but a smaller proportion of girls participate in serious, violent offenses.
- Some evidence suggests that female gang members are becoming more actively involved in more serious offenses.
- Little is known about girls' experiences with gang involvement elsewhere in the world, and the few studies are relatively recent, though more are underway.

Several possible stepping-stones occur in girls' pathways to serious, violent, and chronic juvenile offender careers, including (1) neighborhood conditions, (2) child physical and sexual abuse, (3) school problems, (4) running away, (5) drug and alcohol abuse, (6) mental health problems, (7) gang membership, (8) violent victimization, and (9) juvenile justice system involvement. These risk factors provide an excellent backdrop for considering how to change conditions to maximize successes in the life-course of girls who are not privileged to move away from them. A continuum of prevention programs and strategies is recommended in Chapter 9.

DISCUSSION TOPICS

1. Why was gang research so slow to notice the active involvement of girls in gangs?

2. Explain the more serious criminal activity of modern-day female gangs. Do these factors also apply to male gangs?

3. Why does the presence of males in female gangs elevate the criminal involvement of the group as a whole?

4. What are the main attractions of gangs in inner-city areas?

5. Class exercise: Develop a strategy for insulating girls from community conditions that lead to gang involvement and violent victimization when their family does not move out.

RECOMMENDATIONS FOR FURTHER READING

Proliferation of Female Gangs in the 1980s

Brotherton, D. C. (1996). The contradictions of suppression: Notes from a study of approaches to gangs in three public high schools. *Urban Review, 28,* 95–117.

Fleisher, M. S. (1998). *Dead end kids: Gang girls and the boys they know.* Madison: University of Wisconsin.

Moore, J. W. (1991). *Going down to the barrio: Homeboys and homegirls in change.* Philadelphia: Temple University Press.

Moore, J. W., & Hagedorn, J. M. (1999). What happens to girls in the gang? In M. Chesney-Lind & J. Hagedorn (Eds.), *Female gangs in America: Essays on girls, gangs, and gender* (pp. 176–186). Chicago: Lake View Press.

Moore, J. W., & Hagedorn, J. M. (2001). Female gangs. *Juvenile Justice Bulletin. Youth Gang Series.* Washington, DC: U.S. Department of Justice, Office of Juvenile Justice and Delinquency Prevention.

Taylor, C. S. (1993). *Girls, gangs, women, drugs.* East Lansing: Michigan State University Press.

High Proportions of Female Gang Members

Bjerregaard, B. (2002a). Operationalizing gang membership: The impact measurement on gender differences in gang self-identification and delinquent involvement. *Women and Criminal Justice, 13,* 79–100.

Bjerregaard, B., & Smith, C. (1993). Gender differences in gang participation, delinquency, and substance use. *Journal of Quantitative Criminology, 9,* 329–355.

Esbensen, F., Brick, B. T., Melde, C., Tusinski, K., & Taylor, T. J. (2008). The role of race and ethnicity in gang membership. In F. V. Genert, D. Peterson, & I. Lien, *Street gangs, migration and ethnicity* (pp. 117–139). Portland, OR: Willan.

Esbensen, F., & Deschenes, E. P. (1998). A multi-site examination of gang membership, Does gender matter? *Criminology, 36,* 799–828.

Esbensen, F., & Huizinga, D. (1993). Gangs, drugs, and delinquency in a survey of urban youth. *Criminology, 31,* 565–589.

Esbensen, F., Winfree, L. T., He, N., & Taylor, T. J. (2001). Youth gangs and definitional issues: When is a gang a gang, and why does it matter? *Crime and Delinquency, 47,* 105–130.

Moore, J. W. (1991). *Going down to the barrio: Homeboys and homegirls in change.* Philadelphia: Temple University Press.

Thornberry, T. P., Krohn, M. D., Lizotte, A. J., Smith, C. A., & Tobin, K. (2003). *Gangs and delinquency in developmental perspective.* New York: Cambridge University Press.

Early Sexual Victimization of Girls Who Affiliate With Gangs

Cepeda, A., & Valdez, A. (2003). Risk behaviors among young Mexican American gang associated females: Sexual relations, partying, substance use, and crime. *Journal of Adolescent Research, 18,* 90–106.

Joe, K. A., & Chesney-Lind, M. (1995). Just every mother's angel: An analysis of gender and ethnic variations in youth gang membership. *Gender & Society, 9,* 408–430.

Miller, J. A. (2001). *One of the guys: Girls, gangs and gender.* New York: Oxford University Press.

Moore, J. W. (1991). *Going down to the barrio: Homeboys and homegirls in change.* Philadelphia: Temple University Press.

Moore, J. W., & Hagedorn, J. M. (1996). What happens to girls in the gang? In C. R. Huff (Ed.), *Gangs in America,* (2nd ed., pp. 205–218). Thousand Oaks, CA: Sage.

Schalet, A., Hunt, J., & Joe-Laidler, K. (2003). Respectability and autonomy: The articulation and meaning of sexuality among the girls in the gang. *Journal of Contemporary Ethnography, 32,* 108–143.

Valdez, A. (2007). *Mexican American girls and gang violence: Beyond risk.* New York: Palgrave Macmillan.

Yoder, K. A., Whitbeck, L. B., & Hoyt, D. R. (2003). Gang involvement and membership among homeless and runaway youth. *Youth and Society, 34,* 441–467.

Level of Female Involvement and Seriousness of Crime

Bjerregaard, B. (2002a). Operationalizing gang membership: The impact measurement on gender differences in gang self-identification and delinquent involvement. *Women and Criminal Justice, 13,* 79–100.

Esbensen, F., Winfree, L. T., He, N., & Taylor, T. J. (2001). Youth gangs and definitional issues: When is a gang a gang, and why does it matter? *Crime and Delinquency, 47,* 105–130.

Haymoz, S., & Gatti, U. (2010). Girl members of deviant youth groups, offending behavior and victimisation: Results from the ISRD2 in Italy and Switzerland. *European Journal on Criminal Policy and Research, 16,* 167–182.

Peterson, D. (Forthcoming). Girlfriends, gun-holders, and ghetto-rats? Moving beyond narrow views of girls in gangs. In S. Miller, L. D. Leve, & P. K. Kerig (Eds.), *Delinquent girls: Contexts, relationships, and adaptation.* New York: Springer.

Gender-Mixed Gang Activity

Fleisher, M. S. (1998). *Dead end kids: Gang girls and the boys they know.* Madison: University of Wisconsin.

Miller, J. A. (2001). *One of the guys: Girls, gangs and gender.* New York: Oxford University Press.

Miller, J. A., & Brunson, R. (2000). Gender dynamics in youth gangs: A comparison of males' and females' accounts. *Justice Quarterly, 17,* 419–448.

Valdez, A. (2007). *Mexican American girls and gang violence: Beyond risk.* New York: Palgrave Macmillan.

Pittsburgh Girls Study (*N* = 2,451)—Pittsburgh Life History Studies

Hipwell, A. E., Keenan, K., Bean, T., Loeber, R., & Stouthamer-Loeber, M. (2008). Reciprocal influences between girls' behavioral and emotional problems and caregiver mood and parenting style: A six year prospective analysis. *Journal of Abnormal Child Psychology, 36,* 663–677.

Hipwell, A. E., Loeber, R., Stouthamer-Loeber, M., Keenan, K., White, H. R., & Kroneman, L. (2002). Characteristics of girls with early onset of disruptive and antisocial behavior. *Criminal Behaviour and Mental Health, 12,* 99–118.

Hipwell, A. E., Pardini, D. A., Loeber, R., Sembower, M. A., Keenan, K., & Stouthamer-Loeber, M. (2007). Callous-unemotional behaviors in young girls: Shared and unique effects relative to conduct problems. *Journal of Clinical Child and Adolescent Psychology, 36,* 293–304.

Hipwell, A. E., White, H. R., Loeber, R., Stouthamer-Loeber, M., Chung, T., & Sembower, M. (2005). Young girls' expectancies about the effects of alcohol, future intentions and patterns of use. *Journal of Studies on Alcohol, 66,* 630–639.

Keenan, K., Hipwell, A. E., Chung, T., Stepp, S., Stouthamer-Loeber, M., McTigue, K., & Loeber, R. (2010). The Pittsburgh girls studies: Overview and initial Findings. *Journal of Clinical Child and Adolescent Psychology, 39,* 506–521.

Revictimization

Fleisher, M. S. (1998). *Dead end kids: Gang girls and the boys they know.* Madison: University of Wisconsin.

Hunt, G., & Joe-Laidler, K. (2001). Situations of violence in the lives of girl gang members. *Health Care for Women International, 22,* 363–384.

Miller, J. A. (2001). *One of the guys: Girls, gangs and gender.* New York: Oxford University Press.

Miller, J. A. (2008). *Getting played: African American girls, urban inequality, and gendered violence.* New York: New York University Press.

Moore, J. W. (1991). *Going down to the barrio: Homeboys and homegirls in change.* Philadelphia: Temple University Press.

Moore, J. W. (1994). The *chola* life course: Chicana heroin users and the barrio gang. *International Journal of Addictions, 29,* 1115–1126.

Moore, J. W., & Hagedorn, J. M. (1996). What happens to girls in the gang? In C. R. Huff (Ed.), *Gangs in America,* (2nd ed., pp. 205–218). Thousand Oaks, CA: Sage.

Moore, J. W., & Hagedorn, J. M. (1999). What happens to girls in the gang? In M. Chesney-Lind & J. Hagedorn (Eds.), *Female gangs in America: Essays on girls, gangs, and gender* (pp. 176–186). Chicago: Lake View Press.

Moore, J. W., & Hagedorn, J. M. (2001). Female gangs. *Juvenile Justice Bulletin. Youth Gang Series.* Washington, DC: U.S. Department of Justice, Office of Juvenile Justice and Delinquency Prevention.

Valdez, A. (2007). *Mexican American girls and gang violence: Beyond risk.* New York: Palgrave Macmillan.

Moving to Opportunity Experiment

Aldridge, J., Shute, J., Ralphs, R., & Medina, J. (2009). Blame the parents? Challenges for parent-focused programmes for families of gang-involved young people. *Children & Society,* 1–11.

Briggs, X., Popkin, S. J., & Goerin, J. (2010). *Moving to opportunity: The story of an American experiment to fight ghetto poverty.* Oxford, UK: Oxford University Press.

Notes

1. Key fighting female auxiliary gangs of the late 1800s were the Lady Locusts, the Lady Barkers, the Lady Flashers, the Lady Truck Drivers, and the Lady Liberties of the Fourth Ward (Sante, 1991, p. 220).

2. The third causes and correlates study, the Pittsburgh Youth Survey (directed by Rolf and Magda Loeber), also investigates gang members (see Further Reading), but the study of girls was undertaken later. In this study, subjects are 3,241 girls between age 5 and 8. Currently, the girls, their parents, and teachers are participating in the ninth assessment, when the girls are ages 14–17. Research on gang-involved females is forthcoming.

3. See the National Youth Gang Survey online analysis: http://www.nationalgangcenter.gov/Survey-Analysis/Demographics#anchorgender

4. The students were attending 15 public schools in a cross section of nine survey cities in 2002– 2004. The numbers of Black and multiracial groups were too small to be valid.

5. To be eligible, the young women needed to be Mexican American, between 14 and 18 years of age, and know of a male in one of the 27 male gangs in Valdez's previous study on gang violence in San Antonio (see Petersen & Valdez, 2005).

6. See Johansson & Kempf-Leonard, 2009, as an example.

7. Based on multiple indicators including poverty from the 1990 U.S. Census, and parental perceptions of racial strife, prostitution, assaults, burglary, and drug dealing.

8. Each year from 1997 to 2001, respondents were asked (yes or no) about gang membership in the last year. Gang presence or exposure was measured by affirmative responses to questions about (a) how many years youth had lived in a neighborhood with gangs and (b) how many years they had friends or siblings who had been in a gang between 1997 and 2001. These two items were combined to form an index of gang presence in the youth's environment (Park et al., 2010, p. 319).

7

✦

National Gang Problem Trends

1996 to 2009

Introduction

There is no uniform nationwide timeline of street gang emergence in the United States. As discussed in Chapter 1, each of the four major geographic regions saw uneven development of gangs in the history of this country. Serious gangs first emerged on the East Coast in the 1820s, followed by the Midwest (Chicago) and West (Los Angeles) regions a century later, and in the South after another half century. In the more recent era, gang activity in America appears to exhibit a more uniform pattern. The National Gang Center (NGC) has tracked U.S. gang activity since its first systematic National Youth Gang Survey (NYGS) in 1996. This is the first study in any country that surveys a nationally representative sample of authoritative respondents annually regarding the prevalence of gang activity.

This chapter first reviews the history of gang growth in the United States over the past half century. Next, trends in U.S. cities' histories of gang activity are examined using a new analytical tool that groups cities with similar gang problems. Next, factors that contribute to widespread gang activity are considered. Last, an overview of gang activity in other countries is provided.

An Overview of Nationwide Gang Activity in the Modern Era

Gang Growth From the 1960s to the 1990s

W. Miller (2001) combined data from several sources to reveal sharp increases among counties and cities that reported gang problems at any time between the 1970s, 1980s, and the early 1990s (Figure 7.1). Despite the limitations of these data, a large proportion of jurisdictions that experienced gang activity in the mid-1990s said that their gang problem first emerged in the late 1980s and early 1990s, suggesting a sharp increase in the emergence of gangs in that period with a peak around the mid-1990s (Figure 7.1).

Figure 7.1 Cumulative Numbers of Gang Cities in the United States, 1970–1995, by Decade

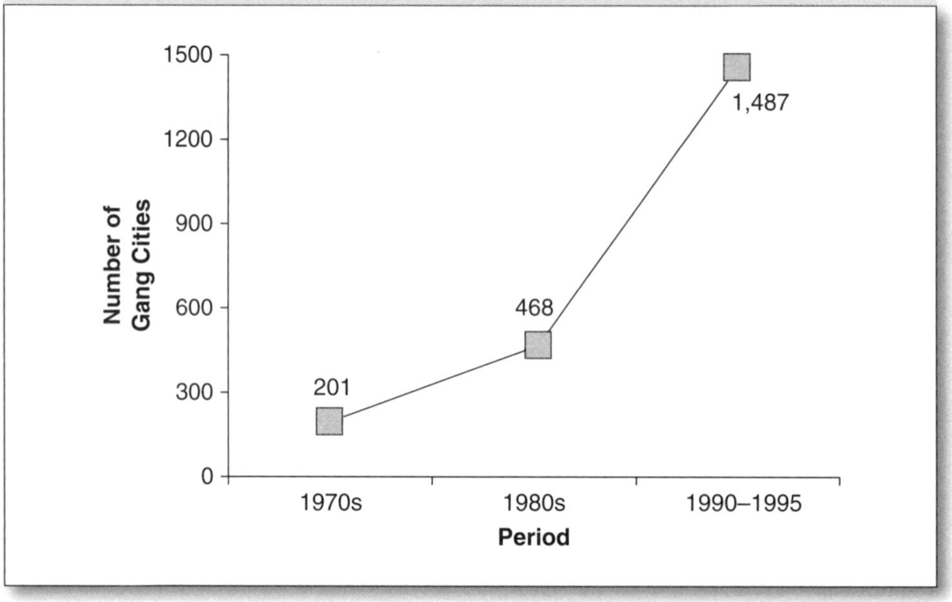

Source: W. Miller, 2001, p. 15

Given consistent growth of gang activity early in U.S. history, and evidence that gangs were present in almost 200 cities by the end of the 1970s, it is surprising that between 1967 and 1973, three major federal commissions,[1] each presenting comprehensive reviews of a wide range of major crime problems in the country, reached the following misleading conclusions (W. Miller, 1975, p. 75):

- Youth gangs are not now or should not become a major object of concern in their own right.
- Youth gang violence is not a major crime problem in the United States.
- What gang violence does exist can fairly readily be diverted into "constructive" channels, primarily through provision of services by community-based agencies.

In a sharp contradiction of these assumptions, W. Miller (2001) summarizes his findings in the following manner:

Youth gang problems in the U.S. grew dramatically between the 1970s and the 1990s, with the prevalence of gangs reaching unprecedented levels. This growth was manifested by steep increases in the number of cities, counties, and states reporting gang problems. Increases in the number of gang localities were paralleled by increases in the proportions and populations of localities reporting gang problems. There was a shift in the location of regions containing larger numbers of gang cities, with the Old South showing the most dramatic increases. The size of gang problem localities also changed, with gang problems spreading to cities, towns, villages and counties smaller in size than at any time in the past. (p. 42)

Miller's (1982/1992) study also produced another very important conclusion regarding the locations of gangs. "More seriously criminal or violent gangs continued to be concentrated in slum or ghetto areas, but in many instances the actual locations of these districts shifted away from central or inner-city areas to outer-city or suburban communities outside city limits. [Yet] there was little evidence of any substantial increase in the proportions of middle-class youth involved in seriously criminal or violent gangs" (p. 78). The assumption was that gangs were migrating outward from "gangland," in downtown areas of New York City, in the transition zones where Thrasher found them in Chicago. Rather, "what has happened . . . is that slums or ghettos have shifted away from the inner city-to suburban outer-city, ring-city, or suburban areas—often to formerly middle- or working-class neighborhoods. Special concentration occurs in housing project areas. The gangs are still in the ghettos, but these are often at some remove from their traditional inner-city locations" (p. 76). This important discovery is also reflected in national gang surveys begun in the mid-1990s, which are discussed in a later section of this chapter.

Explanations of Gang Growth in the 1980s and 1990s

Factors that account for the growth in serious gang activity in large U.S. cities during this period are discussed in some detail in Chapter 8. Here, possible contributing factors to more general emergence of gang activity across the United States in the 1980s and 1990s are discussed briefly.

A large influx of immigrant groups arrived in U.S. communities after the mid-1960s. The Immigration and Nationality Act of 1965 ended the national quotas on foreigners in the United States. Bankston (1998) notes this led to a shift in immigration to the states, from European origins to Central and South America and Asia. The new groups, according to W. Miller (2001), consisted largely of Asians (Cambodians, Filipinos, Koreans, Samoans, Thais, Vietnamese, and others) and Latin Americans (Colombians, Cubans, Dominicans, Ecuadorians, Mexicans, Panamanians, Puerto Ricans, and others). But not all of these groups have experienced similar problems in assimilation in America's communities. Bucerius (2010) contends those affected most "disproportionately originate from Mexico, Latin America and the Caribbean countries, including El Salvador, Guatemala, Haiti, the Dominican Republic, and the West Indies" (p. 235). By the late 1980s, according to Portes and Zhou (1993), the children of many American-born or Americanized parents among the new immigrants, dubbed "the new second generation," had reached adolescence. According to Rumbaut's Children of Immigrants Longitudinal Study, in 2005, children of post-1965 immigrants had grown to more than 30 million with a median age of eighteen (as cited in Zhou, Lee, Vallejo et al., 2008, p. 37). Gang joining tends to commence with second-generation youth (Telles & Ortiz, 2008).

However, as W. Miller notes, "There is little consensus as to what has caused the striking growth in reported youth gang problems during the past 25 years. It is unlikely that a single cause played a dominant role; it is more likely that the growth was a product of a number of interacting influences" (p. 42). These include:

- Drugs—This popular explanation centers on the growth of the drug trade, and there is little doubt that this was one important factor in the gang growth, but most gang involvement is at the street level, in individual sales, and not drug trafficking, according to law enforcement survey responses.
- Gang names and alliances—In the 1980s, the pattern of adopting a common name and claiming a federated relationship with other gangs expanded enormously. The most prominent of these were the Los Angeles Crips and Bloods that originated in the 1960s. In succeeding years, hundreds of gangs adopted their names. Similar loose alliances of People and Folk "nations" formed in Chicago.
- Migration—Those who also support the drug-trade explanation also particularly favor attributing the spread of gangs to migration of gangs themselves.
- Government policies—Permissive policies that followed the civil rights movement and urban riots of the 1960s allowed more ganging. In addition, the exodus of better educated and more prosperous city residents led to a reduction in the anti-gang influences the previous residents had provided in inner-city areas.
- Gang subculture and the media—Contributions of the media to increased popularity of the gang culture is of central importance to recent youth gang growth in the United States from the mid-1970s onward. (pp. 42–46)

This latter factor—the diffusion of gang culture—may well be the most important accounting for growth in reported youth gang problems in the 1980s and early 1990s, for reasons that W. Miller (2001) succinctly articulates:

The lifestyle and subculture of gangs are sufficiently colorful and dramatic to provide a basis for well-developed media images. For example, the Bloods/Crips feud, noted earlier, caught the attention of media reporters in the early 1990's and was widely publicized. Gang images have served for many decades as a marketable media product—in movies, novels, news features, and television drama—but the 1980s saw a significant change in how they were presented.

In the 1950s, the musical drama *West Side Story* portrayed gang life as seen through the eyes of adult middle-class writers and presented themes of honor, romantic love, and mild rebellion consistent with the values and perspectives of these writers. In the 1990s, the substance of gang life was communicated to national audiences through a new medium known as "gangsta rap." For the first time, this lifestyle was portrayed by youthful insiders, not adult outsiders. The character and values of gang life described by the rappers differed radically from the images of *West Side Story*. Language was rough and insistently obscene; women were prostitutes ("bitches," "ho's," and "sluts") to be used, beaten, and thrown away; and extreme violence and cruelty, the gang lifestyle, and craziness or insanity were glorified. Among the rappers' targets of hatred, scorn, and murder threats were police, especially black police (referred to as "house slaves" and "field hands"); other races and ethnic groups; society as a whole; and members of rival gangs.

The target audience for gangsta rap was adolescents at all social levels, with middle-class suburban youth constituting a substantial proportion of the market for rap recordings. The medium had its most direct appeal, however, for children and youth in ghetto and barrio communities, for whom it identified and clarified a set of values, sentiments, and attitudes about life conditions that were familiar to them. The obscene and bitterly iconoclastic gangsta rappers assumed heroic stature for thousands of potential gang members, replacing the drug dealer as a role model for many. Gangsta rap strengthened the desire of these youth to become part of a gang subculture that was portrayed by the rappers as a glamorous and rewarding lifestyle. (pp. 45–46)

Klein (2002) describes the result as "general diffusion of street gang culture—the dress and ornamentation styles, the postures, and the argot of gang members—to the general youth population of the country" (p. 246). Now, "most young people in America recognize the look, the walk, and the talk of gang members. Many mimic it in part or in whole. Many try it out as a personal style. Play groups, break-dancing groups, taggers, and school peer groups experiment with gang life" (p. 246). Gang culture is intertwined with the general youth subculture.

Nationally Reported Youth Gang Activity From the Mid-1990s

The Office of Juvenile Justice and Delinquency Prevention established the National Gang Center (NGC), formerly called the National Youth Gang Center, in 1994. Since 1996, the NGC has conducted an annual gang survey of a nationally representative sample of law enforcement agencies. The agencies included in the two nationally representative NYGS samples are detailed in Table 7.1.

Table 7.1 National Youth Gang Survey Samples

1996–2001 NYGS Sample (Former Sample)	2002–Present NYGS Sample (Current Sample)
All police departments serving cities with populations of 25,000 or more ($N = 1,216$)	All police departments serving cities with populations of 50,000 or more ($N = 624$)
All suburban county police and sheriffs' departments ($N = 661$)	All suburban county police and sheriffs' departments ($N = 739$)
A randomly selected sample of police departments serving cities with populations between 2,500 and 24,999 ($N = 398$)	A randomly selected sample of police departments serving cities with populations between 2,500 and 49,999 ($N = 543$)
A randomly selected sample of rural county police and sheriffs' departments ($N = 743$)	A randomly selected sample of rural county police and sheriffs' departments ($N = 492$)

Hereafter, *larger cities* refers to cities with populations of 50,000 or more, and *smaller cities* refers to cities, towns, and villages with populations between 2,500 and 49,999. Finally, *study population* refers to the entire group of jurisdictions that the current sample represents, that is, all jurisdictions served by county law enforcement agencies and all jurisdictions with populations of 2,500 or more served by city (e.g., municipal) police departments.

In the NYGS sample for 2002 to present, 63% of the agencies were also surveyed from 1996 to 2001, permitting an ongoing longitudinal assessment of gang problems in a large number of jurisdictions. The average annual survey response rate is approximately 85% for the entire sample, as well as within each area type. Of the respondents in the current sample, 99% have provided gang-related information in at least one survey year.

Survey recipients are asked to report information solely for *youth gangs,* defined as "a group of youths or young adults in your jurisdiction that you or other responsible persons in your agency or community are willing to identify as a 'gang.'" Motorcycle gangs, hate or ideology groups, prison gangs, and exclusively adult gangs are excluded from the survey. Using this definition, the NYGS measures youth gang activity as an identified problem by interested community agents, specifically law enforcement officials.

Across survey years, questionnaire items have been designed to evaluate the soundness of this definitional approach. Respondents in the 1998 NYGS overwhelmingly defined gangs in terms of their involvement in criminal activity (National Youth Gang Center, 2000). Although some U.S. gang researchers have questioned the reliability and validity of information that law enforcement officers provide on gangs (Moore & Hagedorn, 2001; Sullivan, 2006), Decker and Pyrooz (2010b) and Katz and Fox (2011) proved the NYGS survey to have good reliability and validity in independent tests. *Reliability* refers to the extent to which results are consistent over time and an accurate representation of the total population under study. *Validity* refers to whether the research truly measures the phenomenon that it was intended to measure. Katz, Webb, and Schaefer (2000) suggest that law enforcement gang intelligence databases can provide a useful estimate of city gang activity. Pyrooz, Fox, and Decker (2010) note the police frequently interact with gangs and gang members, with ample opportunities to observe, document, and address emerging gang trends on the street.

Based on nationwide law enforcement reports in 2009, Egley and Howell (2011) estimate there were 28,100 gangs and 731,000 gang members throughout 3,500 jurisdictions in the United States. Table 7.2 provides percentage changes in these key indicators of gang activity from 2002 to 2009. The number of jurisdictions with gang problems increased 21% from 2002 to 2009, and the estimated number of gangs increased 29% in the same period. The estimated number of gang members, which has averaged over 750,000 across survey years, decreased slightly from 2008 to 2009, but remains unchanged from the 2002 total. The prevalence rate of gang activity increased slightly from 2008 to 2009, to 34.5%, meaning that just over one in three localities reported gang activity in 2009 (Figure 7.2). Over the entire survey period, three trends are apparent in the level of gang activity: (1) a sharp decline throughout the late 1990s, (2) a sudden upturn beginning in 2001 and continuing until 2005, and (3) a relative leveling off thereafter.

Table 7.2   Percentage Changes in Nationwide Gang Estimates from 2002 to 2009

	2002–2009	*2005–2009*	*2008–2009*
Gang-problem jurisdictions	20.7	<1.0	5.1
Gangs	28.9	5.3	<1.0
Gang members	<1.0	−7.4	−5.6
Gang homicides (cities over 100,000 population only)	2.1	7.2	10.7

Source: Egley & Howell, 2011

Figure 7.2   Prevalence of Gang Problems in Study Population, 1996 to 2009

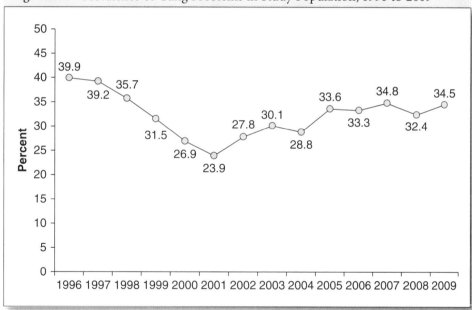

Source: National Gang Center

Figure 7.3 shows the prevalence of gang activity within each the four NYGS subsamples, (1) police departments serving cities with populations of 50,000 or more, (2) suburban county police and sheriffs' departments, (3) police departments serving cities with populations between 2,500 and 49,999, and (4) rural county police and sheriffs' departments. Gang activity in all subsamples increased somewhat from 2002 to 2009. Each subsample follows a similar trend over time but at distinctly different levels.

Larger cities consistently exhibit the highest prevalence rates of gang activity among the four groups, followed, in order, by suburban counties, smaller cities, and rural counties. The rates of reported gang activity in suburban counties are closest to rates for larger cities because of the relatively large populations in suburban counties (i.e., a higher capacity to sustain gang activity, Egley et al., 2006), the shifting of previous inner-city slums and ghettos to ring-city or suburban areas (Miller, 1982/1992), and the growing popularity of gang culture in these areas (W. Miller, 2001). Mirroring the overall prevalence trend, each of the subsamples shows uniform declines in the late 1990s, reaching a low point in 2001 and then steadily increasing before leveling off somewhat in recent years. Again, larger cities' gang problems are considered in Chapter 8.

Figure 7.3 Reports of Gang Problems: 1996 to 2009, by Area Type

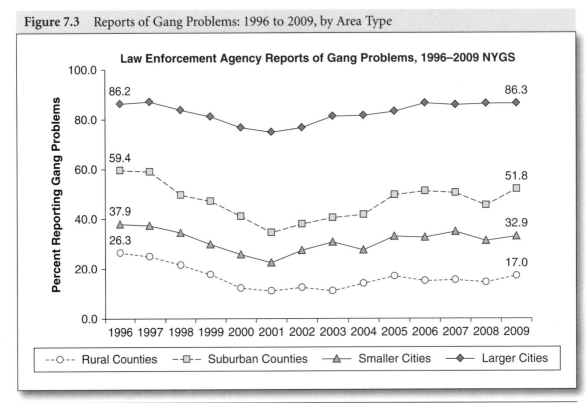

Source: National Gang Center

Egley and Howell (2011) report the 2009 survey respondents also estimated year-to-year changes regarding other gang-related crimes and violence in their jurisdictions. For the gang-related offenses of robbery, aggravated assault, drug sales, and firearms use, most frequently reported by respondents was "no substantial change" (i.e., neither significant increase nor decrease) from 2008 to 2009 in the number of offenses

committed. Among agencies reporting gang problems in 2009, half (49.8%) characterized their gang problems as "staying about the same," the highest percentage ever
recorded in the NYGS. Agencies reporting a fluctuating pattern of gang activity over
the past 5 years were more likely to characterize their gang problems as "getting worse"
than were agencies consistently reporting gang activity.

The next section examines localities' histories of reported gang activity. For the
first time, analysts are able to display groups of cities in terms of the consistency with
which they report gang activity.

Patterns in U.S. Localities' Histories of Gang Activity

Trajectory models are used here to group jurisdictions that share similar trends in the
outcome of interest (gang activity in this instance) and to graphically illustrate those
patterns over the 14-year period. For example, some jurisdictions may report a consistent presence of gangs while others could experience no gang activity over time, rapid
increases over time, rapid decreases, fluctuating presence of gang activity, or other
more complex trends between 1996 and 2009.

Figure 7.4   Trajectory Model: Presence of Gang Activity, 1996 Through 2009

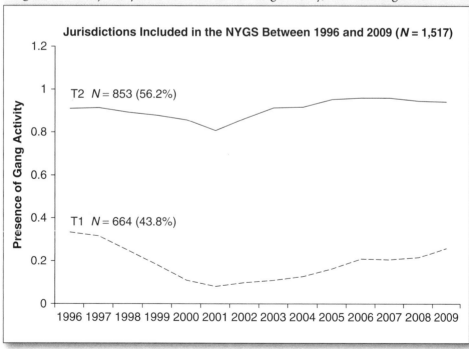

Source: Howell et al., 2011

The trajectory model in Figure 7.4 displays *trends in the presence of gang activity* across 1,517 jurisdictions—the number of U.S. localities surveyed every year in the NYGS between 1996 and 2009 (for samples, see Table 7.1). Of the total, 664 (43.8%) of the jurisdictions fall in the first trajectory (T1). This group exhibited a relatively lower prevalence of gang activity in 1996, which declined precipitously until 2001 before experiencing some growth that continued through 2009. By contrast, more than half of the jurisdictions (T2, $N = 853$; 56.2%) reported a near-chronic presence of gang activity across the time period. Thus, this trajectory model reveals that a small majority (56%) of all respondents reporting gang activity have a persistent gang problem that, apart from the minor deviation in 2001, has remained virtually constant over time. In other words, just over half of all localities that reported gang activity across the 14 survey years experienced near-chronic gang activity.

In summary, the 14-year gang problem prevalence trend reported in the NYGS (Figure 7.2) shows two peaks in reported gang activity, in 1996, and in 2007–2009. This view of gang activity also suggests a sharp drop in the late 1990s and early 2000s. However, the trajectory analysis shows that more than half of the localities represented in Figure 7.2 rather consistently experienced gang problems across the 14 years. In Chapter 8, trajectory models are used to gain insights into patterns with respect to serious gang problems in large cities.

Perpetuation of Gang Presence

In addition to social, economic, and cultural conditions that give rise to gangs, two important factors perpetuate their presence: prison gangs and gangs' histories. Each of these factors is discussed next. Attention is also called to an often overlooked area of gang problems in Indian Country.

Prison Gangs

The growth of prison gangs and the concomitant effects of returning inmates can affect local gang activity. This development is examined in more detail in the next chapter, but it is important to introduce the topic here because the potential that returning gang inmates have for sustaining local gang activity could be substantial. Owing to the dearth of prison gang studies, the relationship between prison gangs and gangs on the street is not well understood. Jacobs (2001) contends that "the worst-case scenario is that prison gangs serve to strengthen gangs on the street. There is fluid communication between the gang members outside and inside prison. Criminal schemes are hatched in prison and carried out on the streets and vice versa" (p. vi). In one study, Curtis (2003) reported middle-age returnees who attached themselves to street-corner groups found themselves challenged by youngsters eager to make a reputation on the street. These youngsters soon had stitches and broken bones to prove their lack of respect for returning *original gangsters*. In another study that researched the dynamic relationship between prison gangs and youth gangs, Valdez and colleagues (2009) found Mexican American prison gangs gradually came to dominate and dissolve the local gangs in a large southwestern city.

According to Fleisher (2006a), the Gypsy Jokers were the first known American prison gang formed in the 1950s in Washington state prisons. Lyman (1989) defines a prison gang as "an organization which operates within the prison system as a self-perpetuating criminally oriented entity, consisting of a select group of inmates who have established an organized chain of command and are governed by an established code of conduct" (p. 48). Lyman also noted an important distinction between prison gangs and street gangs. "The prison gang will usually operate in secrecy and has as its goal to conduct gang activities by controlling their prison environment through intimidation and violence directed toward non-members" (p. 48). Toch (2007) asserts that validating gang membership is very difficult, even within prisons. Al Valdez (2007) adds that most core members of prison gangs have committed violent crimes and the street gangs are a central cadre from which the prison gangs recruit. Interestingly, gang members who once were bitter rivals on the street become ethnic allies within the prison. Members of racial and ethnic groups stick together because "once inside, protection comes from members of your own race" (p. 239).

The term *prison gang* traditionally meant both prison gangs, such as Mexican Mafia, and prison counterparts of street gangs (hereafter, prison street gangs), such as Gangster Disciples (Fleisher, 2006b). Now, Decker (2007) contends prison gangs have increasingly stronger ties to street gangs. The other major prison gangs are *Nuestra Familia* (our family), Aryan Brotherhood (a White supremacist group), the Texas Syndicate, and Mexikanemi (known also as the Texas Mexican Mafia) (Fleisher & Decker, 2001). Yet another one, the Netas, recently moved into the U.S. prison system from Puerto Rico prisons. Other well-represented street gang members inside prisons are Crips, Bloods, Vice Lord Nation, 18th Street, and Latin Kings (National Alliance of Gang Investigators, 2005).

The Mexican Mafia and Nuestra Familia supergangs emerged in California prisons. Al Valdez (2007) tells the intriguing story of their origins. A power struggle among inmates within the prison system in the 1950s caused prison officials to separate the violent and defiant factions. The 13 "worst of the worst" wards of the California Youth Authority were sent to the Duel Vocational Institute located in Tracy, California. Because these inmates were juveniles, they were housed in a special section of the adult correctional facility. Between 1956 and 1957, these young offenders formed the "gang of gangs," commonly known as the Mexican Mafia or *La Eme*.

Texas is also known for the Texas Syndicate and Mexican Mafia, which have been the two best-known Mexican prison gangs in Texas for decades (Vogel, 2007). A third major Texas gang, Tango Blast, formed inside Texas's state prisons during the early 1990s, established to shield inmates from other prison gangs. Many older members claim the word *Tango* is an acronym standing for "Together Against Negative Gang Organizations." However, the most common translation of Tango is "hometown clique," in specific reference to the Houston-based parent gang, Houstone Tango Blast. Vogel (2007) notes one unique feature of Tango Blast as its organization, divided up by cities or hometowns for expansive protection; another is that no lifelong commitment is required.

It is surprising that there are no reliable national data on the prevalence and membership of prison gangs. Based on self-reported information from surveyed inmates, the Bureau of Justice Statistics (Beck et al., 1993) estimated that only about 6% of inmates in adult state prisons were gang members; however, other more realistic estimates range upward from 25% (Olson & Dooley, 2006). Indeed, the first collection of research articles was published on them just a decade ago in a special issue of *Corrections Management Quarterly*. The few established experts agree that prison gangs got bigger and more entrenched in the 1980s and 1990s. "Prison gangs are more structured than street gangs and have much stronger leadership. The rank-and-file membership often has several gradations, making prison gangs look much like organized crime groups" (Decker, 2007, p. 395).

Among prison inmates, Griffin and Hepburn (2006) find gang members more than twice as likely to commit an assault within the first 3 years of their imprisonment than nongang members, and Olson and Dooley (2006) claim they also have higher recidivism rates following release. Transferred juvenile gang members are also significantly more likely to assault fellow inmates in adult prison compared to nongang members (Tasca, Griffin, & Rodriguez, 2010). As Tasca and colleagues suggest, it is likely that youth who are gang members import norms from the streets into institutions that promote violence to gain power and respect, as well as use violence to deal with conflict.

Gangs' Histories

Street gang histories and gang lore serve to perpetuate gang activity. Prominent gang names are often copied in distant cities, and this, along with glorification of gang life, past exploits, and acclaimed victorious battles, serves to bolster recruitment and member retention.

In Focus 7.1
Regional Trends in Gang Activity as Viewed by the FBI and Police Agencies

The following documentation lists gang activity trends in large cities and states during the past few years compiled by the U.S. Federal Bureau of Investigation and state and local police agencies. This information suggests that gangs are often very adept in perpetuating gang culture.

Northeast region. New York City is no longer the epicenter of serious street gang activity in the Northeast as was the case in the early 1900s. Gradually, gang activity in this region expanded to include other East region and New England states, particularly Pennsylvania, New Jersey, and Connecticut (FBI, 2009). According to the FBI's intelligence reports, "The most significant gangs operating in the East Region are Crips, Latin Kings, MS-13, Ñeta, and United Blood Nation" (p. 16). "The most significant gangs operating in the New England Region are Hells Angels, Latin Kings, Outlaws, Tiny Rascal Gangster Crips, and UBN" (p. 17).

Central region. In the Midwest region, traditional Chicago gangs still have the strongest presence. According to the Chicago Crime Commission (2006), in 2008, the largest street gangs in Chicago include the Gangster Disciple Nation (GDN), Black Gangsters/New Breeds (BG), Latin Kings (LKs), Black P. Stone Nation, Vice Lords (VLs), the Four Corner Hustlers and the Maniac Latin Disciples

(MLDs) (p. 11). The most recent chapter in Chicago gang history is the proliferation of gangs outside the city. By 2006, 19 gang turfs were scattered around Chicago, throughout Cook County (p. 119). Next, gangs began emerging in the larger region surrounding Chicago on the north, west, and south sides. The FBI (2009) includes other cities in this region that have extensive gang activity as Cleveland, Detroit, Joliet, Kansas City, Minneapolis, Omaha, and St. Louis (p. 18).

Pacific region. Street gangs in Los Angeles remain legendary. Los Angeles is now said to be "the gang capital of the world" (The Advancement Project, 2007, p. 1). The Los Angeles Police Department (2007) recently designated the 11 most notorious gangs in the city: 18th Street Westside (Southwest area), 204th Street (Harbor area), Avenues (Northeast area), Black P-Stones (Southwest, Wilshire areas), Canoga Park Alabama (West Valley area), Grape Street Crips (Southeast area), La Mirada Locos (Rampart, Northeast areas), Mara Salvatrucha (Rampart, Hollywood, and Wilshire areas), Rollin 40s (Southwest area), Rollin 30s Harlem Crips (Southwest area), and Rolling 60s (77th St. area).

Southern region. The most significant gangs operating in the Southeast region (Deep South states) are said to be Crips, Gangster Disciples, Latin Kings, Sureños 13, and United Blood Nation (FBI, 2009). According to the FBI, the increased migration of Hispanic gangs into the region has contributed significantly to gang growth (p. 20). In the Southwest region (Texas, Oklahoma, New Mexico, Colorado, Utah, and Arizona), the most significant gangs are Barrio Azteca, Latin Kings, Mexikanemi, Tango Blast, and Texas Syndicate. Among 25 major Houston gangs, the Tango Blast, Houstone Tango Blast, and Latin Disciples are said to be the main regional gangs that are Houston-based (Houston Multi-Agency Gang Task Force, n.d.).

Sources: Chicago Crime Commission (2006); FBI (2009); Los Angeles Police Department (2007)

Gang Activity in Indian Country and Among Native American Youth

Given the similar geographical segregation, and economic, social, and cultural marginalization to which Native American Indians have been subjected in comparison with Blacks, Mexican Americans, and other minority groups, it is reasonable to expect that gang activity would flourish in Indian Country. Unfortunately, only limited information is available on the seriousness of gang activity therein. *Indian Country* is defined in 18 U.S.C.§1151 (1948) as including (1) land within Indian reservations, (2) dependent Indian communities, and (3) Indian allotments. Reservation trust lands refer to areas that have been set aside and the federal government recognize as being held in trust for a particular federally recognized tribe. A variety of federal treaties, regulations, and acts over the years have established these trust areas and have established laws governing sovereign Indian nations.

Gangs made up of Native American youth are believed to have originated in 1990–1991, according to Al Valdez (2007). He bases this period specification on oral reports made to him by educators, tribal police officers, council members, and Native Americans themselves. A national survey of gang problems in Indian Country conducted in 2000 revealed that 23% of Indian Country respondents (tribal leaders and law enforcement agencies) reported having active youth gangs in their communities during 2000. A majority of 70% responded that there was no gang activity in their communities, and

7% could not make a determination (Major, Egley, Howell, Mendenhall, & Armstrong, 2004). This report refers to each respondent tribe, reservation, and Alaska Native village as a *community,* which includes a wide range of settings—pueblos, rancherias, villages, towns, and rural settlements. Communities also include people who have been recognized by the U.S. government as a tribe or tribal community, but who do not occupy tribal trust, tribally owned, or Indian allotment lands. Communities are the people and land together or tribal community viewed as a group.

An analysis of gang membership in the National Longitudinal Study of Adolescent Health (Glesmann, Krisberg, & Marchionna, 2009) relays that Native American youth (15%) are about twice as likely to be gang members as any other race/ethnicity. Al Valdez (2007) reports that "recent interviews with educators who work at Native American schools or at schools with Native American populations have confirmed an increase in gang dress, gang graffiti, gang crime and the gang mentality since 1997" (p. 423).

Gang Activity Patterns in Other Countries

The purpose of this overview is to give readers a snapshot of gang activity in countries other than the United States. First, information is provided on two unique and highly publicized U.S. gangs that also have a presence in Mexico and Central America. Next, gang trends in Canada are examined. Last, European gang activity is reviewed.

Transnational Gangs and Gangs in Mexico and Central America

If one is using the term *transnational* broadly to mean multinationality of membership, then transnational gangs are not a new phenomenon in the United States. Franco (2008b) explains that so-labeled U.S. gangs

> have included [those] composed of members of Asian, Russian, African, Serbian, Bosnian, Jamaican, and other races, ethnicities, and nationalities. Some of these transnational gangs have evolved into highly organized and sophisticated criminal enterprises known for influencing government officials and the judiciary in the countries in which they operate. In the U.S., the most well-known example of this type of crime syndicate is the Mafia, or La Cosa Nostra. (p. 2)

For a gang to be considered transnational, Franco suggests that it should have more than one of the following characteristics: criminally active as well as operational in multiple countries, activities of a gang's members must be at the direction of gang leaders in another country, highly mobile, and involvement in sophisticated crimes that transcend national borders. However, "much of the literature characterizes . . . gangs as transnational merely because they are present in more than one country" (p. 2).

The notorious 18th Street gang (M-18) and Mara Salvatrucha (MS-13) Los Angeles gangs have been called transnational gangs.

18th Street gang. According to Valdez (2006b), the 18th Street gang (also called *Calle* 18, or *Barrio* 18) emerged in the 1960s. This gang was created because a local Mexican gang, Clanton Street, rejected all youths who could not prove 100% Mexican ancestry. "As a result, the kids from the Clanton Street neighborhood who were denied membership because of their tainted ancestry formed their own gang. They became the original 18th Street gang" (p. 146). To date, Al Valdez (2007) claims it "is probably one of the largest Hispanic street gangs in the country and it has become established nationally and internationally" (p. 145).[1]

Because it broke the racial barrier in accepting largely immigrant youths and those of mixed racial backgrounds, the 18th Street gang quickly grew enormously. "Predominately composed of Hispanics, some cliques of the 18th Street gang have even recruited African Americans, Asians, Whites, and Native Americans for membership. . . . Another unique aspect about the 18th Street gang is that, although it was primarily turf-oriented, some gang members traveled to other areas and states to establish cliques and start illegal activities" (Al Valdez, 2007, p. 146).

This gang epitomizes the tenacity of the Mexican American gang phenomenon in Los Angeles. "Like most gangs," claims Al Valdez (2007), "18th Street is involved in many types of criminal activities including auto theft, car-jacking, drive-by shootings, drug sales, arms trafficking, extortion, rape, murder, and murder for hire" (p. 147). The 18th Street gang is also said to have connections with the Mexican Mafia prison gang, according to Al Valdez.

Mara Salvatrucha. As Central American gang expert Cruz (2010) explains, the word *mara* in the Salvadoran vernacular commonly refers to any group of people, and is widely synonymous with *folks* and also is slang for *gang*. According to Manwaring (2005), *trucha* is also a slang term for *shrewd*. Thus, Mara Salvatrucha specifically refers to a gang of shrewd Salvadorans. Generally speaking, the term *maras* (and also *pandillas*) can be used to denote Central American gangs. McGuire (2007) clarifies that a purely Salvadoran gang term, *pandilla*, is distinct from *mara*. "In English the *maras* are frequently referred to as Central American gangs. This is somewhat of a misnomer, since the commonly understood origin of these gangs is not in Central America, but within the political boundaries of the United States" (p. 4).

Mara Salvatrucha (also called MS-13; the numeric *13* refers to *M* as the 13th letter of the alphabet) originated among Salvadoran refugees, who, according to McGuire (2007) were "fleeing civil wars [1979–1992] in which the United States–backed forces were known for committing human rights abuses." (p. 4). McGuire details this movement as follows:

> As the now standard account of the origin story goes, Central American gangs began to form in Los Angeles in the 1980s as the large numbers of war refugees and their children made their way into previously Mexican and Mexican-American dominated neighborhoods. Upon arrival in Los Angeles, these Central Americans encountered Mexican and Mexican-American gangs, which, as documented by Vigil and others, had long existed in the L.A. area. In part to defend themselves, they formed their own gangs based on national identities, like *Mara Salvatrucha*, which was originally linked to specifically Salvadoran immigrants. (p. 4)

Many Salvadoran youths faced a choice: either join the large Mexican American 18th Street gang or, for protection, form a gang of their own. Those who chose the latter option formed what became known as Mara Salvatrucha, and the rivalry between the two groups began (Al Valdez, 2007).

Within 2 years after the civil war ended in the early 1990s, Cruz (2010) recounts 375,000 Salvadorans voluntarily returned home, and another 150,000 Central Americans were deported to their home countries in a 3-year period during the mid-1990s. This "reverse migration" of youth to Central America facilitated contact between the new gang culture transported from Los Angeles gangs and older local maras in Central America. Within 5 years, gangs in Central America had adopted the culture of gangs in Los Angeles, particularly imitating the MS-13 and 18th Street gangs. According to the Washington Office on Latin America (2010), "What began in the 1980s as a series of small differentiated local gangs became by 1993 two larger and generally loosely associated trans-national groups of gangs extending between the U.S. and Central America" (p. 4).

Cruz (2010) relays that by 1996, 85% of youth gang members in El Salvador said they belonged to either the MS-13 or 18th Street gangs. Other gang cultures were also diffused from Los Angeles to the region but Cruz (2010) contends most local youth mimicked these two gangs, particularly in El Salvador, Guatemala, and Honduras. Although "press reports and some current and former Central American officials have blamed MS-13 and other gangs for a large percentage of violent crimes committed in those countries . . . some analysts assert that those claims may be exaggerated" (Seelke, 2008, p. 4). Local gangs (maras) and a variety of criminal groups already were active in the region. Thus "other gang experts have argued that, although gangs may be more visible than other criminal groups, gang violence is only one part of a broad spectrum of violence in Central America" (Thale, 2007, as cited in Seelke, 2008, p. 4).

U.S. gang involvement in drug trafficking along the U.S.–Mexico border. Even less information is available on specific crimes that can be attributed to the 18th Street and MS-13 gangs. A 2006 assessment of gang activity in Mexico and Central American countries conducted by the U.S. Agency for International Development (USAID) supported the following general conclusion: "Gangs such as MS-13 and 18th Street conduct business internationally, engaging in kidnapping, robbery, extortion, assassinations, and the trafficking of people and contraband across borders" (p. 6). Ribando (2005) relays MS-13 members are reportedly being contracted on an ad hoc basis by Mexico's warring cartels to carry out revenge killings, and regional and U.S. authorities have confirmed gang involvement in regional drug trafficking. But the question remains as to the extent to which the U.S.-based gangs are involved with drug cartels in the region. The United Nations Office on Drugs and Crime (UNODC) in 2007 concluded that "while some drug trafficking may involve gang members, the backbone of the flow seems to be in the hands of more sophisticated organized crime operations" (p. 17). Other observers agree that "organized crime is now the main cause of the bloodshed. Central America forms a bridge between Columbia, the world's biggest cocaine producer, and Mexico, which is the staging post for the world's biggest market for the drug—the United States" (*The Economist*, 2011, p. 45).

It is difficult to determine at this time the extent of U.S. gang or Central America maras' involvement in drug trafficking in the Mexico–Central America region. Figure 7.5 shows the five major drug-trafficking cartels, their sphere of operations, most commonly trafficked drugs, and affected counties. Various Mexican drug cartels also operate in the region, including *La Familia Michoacana* (LFM), Los Zetas, Tijuana, and the Beltran-Leyva Organization, and they have been battling among themselves, and with other drug-trafficking organizations (DTOs) and Colombian cartels over the past several years for lucrative smuggling routes (Manwaring, 2009a, 2009b). Among these drug cartels, Burton and West (2009) assert the LFM has drawn the most attention in the U.S. government because of its vigilante origins and acclaimed extensions into the states, as far north as Chicago. In addition, homicides related to drug trafficking in Mexico more than doubled recently, from 2,275 in 2007 to 5,207 in 2008. In particular, Grayson (2009) asserts the Michoacana family, or *La Familia,* is an increasingly important contributor to the extremely bloody drug wars among Mexican cartels. Evidence is lacking with respect to U.S. and Central America gang involvement in these wars. It also is difficult to draw a bright line between traditional street gangs, paramilitary organizations, and other criminal groups that operate in Mexico, Central America, and South America (Manwaring, 2009b).

Figure 7.5   Major Drug Cartels

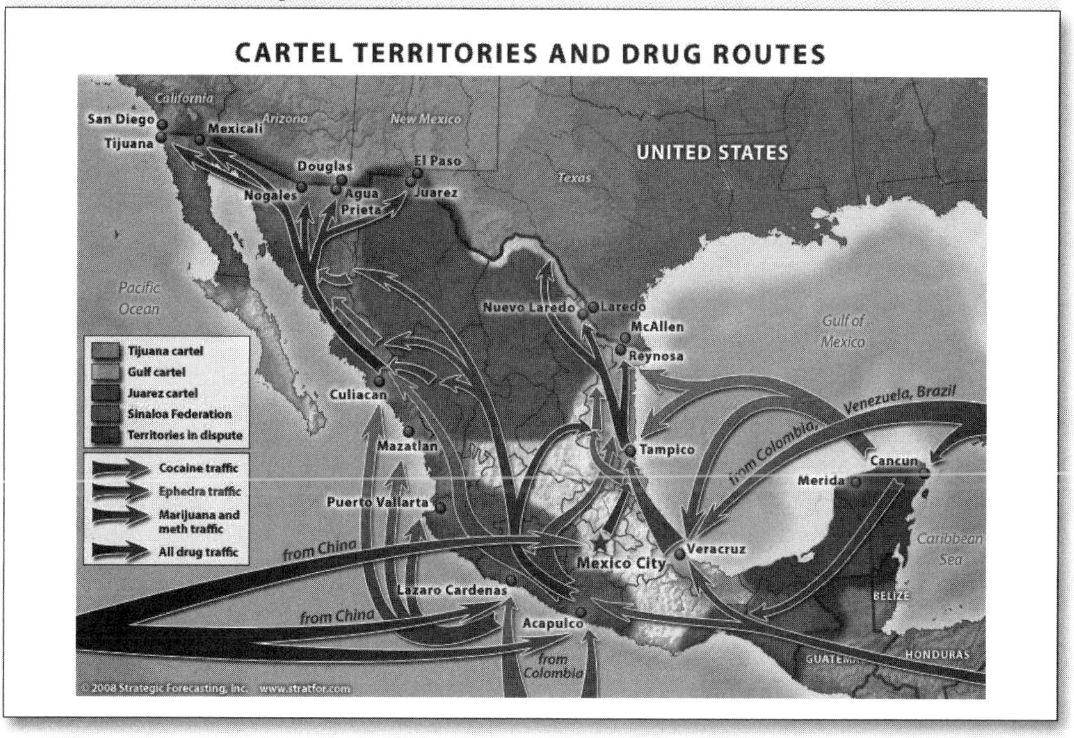

Source: Burton & West, 2009. ©Stratfor Global Intelligence. Reprinted with permission.

Still, there are gaps in knowledge of U.S. gang involvement in drug trafficking with Mexican DTOs. Burton and West (2009) explain, "The exact nature of the relationship between Mexican cartels and U.S. gangs is very murky, and it appears to be handled on such an individual basis that making generalizations is difficult. Another intelligence gap is how deeply involved the cartels are in the U.S. distribution network" (p. 7). No doubt, some U.S. gangs play an active role in street-level drug marketing north of the border. "However, the U.S. gangs do not constitute formal extensions of the Mexican DTOs. Border gangs developed on their own, have their own histories, traditions, structures and turf, and they remain independent" (p. 4). Their involvement in narcotics is similar to that of a contractor who provides certain services, such as labor and protection, while drugs move across gang territory, but drug money is not usually their sole source of income.[2] "Linkages to gangs located in Central America are not needed to source these drugs, and it appears both are mainly sourced from the Mexican drug trafficking organisations that control U.S. drug markets" (UNODC, 2007, p. 64). The UNODC report goes further in questioning this linkage:

> The maras are often referred to as "transnational" in their character, as groups exist with the same name in different countries. Since some mareros are former deportees, it would [be] odd if there were not some communication between these groups. But the specter of "mega gangs," responding to a single command structure and involved in sophisticated trafficking operations, does not, at present, seem to have been realised, at least insofar as drug trafficking is concerned. It is likely that the gang members are preoccupied with more local, neighbourhood issues. (p. 64)

Reports to Congress have suggested an MS-13 presence in a relatively large number of states (Franco, 2008a). However, the size and strength of MS-13 gangs outside Los Angeles have been questioned. The Washington Office on Latin America (WOLA) conducted a brief evaluation and analysis of the characteristics, both local and transnational, of Central American gangs in the Washington, DC, area (McGuire, 2007). This study revealed that 18th Street, at that time, did not "have a strong presence in the DC area, though small cliques may form and disband occasionally. Mara Salvatrucha, perhaps the most well known of the gangs that exist in Central America and the U.S., does have a presence in the Washington area" (pp. 1–2). McGuire drew four main conclusions. First, despite sensationalist media coverage of gang violence, "Central American youth gangs are a relatively minor security problem in the DC area" (p. 2). Although the gang problem is significant for particular communities, "relative to other public security threats in the area, Central American gangs are not a high-priority concern for area law enforcement" (p. 2). Second, based on her interviews of former gang members, "widely held stereotypes about gangs and gang members are oversimplified and often inaccurate" (p. 2). Last, McGuire's study concluded that "while there are cases of Central American gangs attacking random citizens, the overwhelming majority of their crimes are perpetrated against rival gangs, or against other Latinos in their communities" (p. 29).

MS-13 and 18th Street may well be the strongest, most influential, and most dangerous street gangs in Central American countries, but not in the United States (Howell & Moore, 2010). In Central America, these gangs threaten to destabilize neighborhoods, and in Mexico, some gangs have links to narcotics-trafficking cartels that go head-to-head with the military. At the present time, Howell and Moore (2010) believe political and governmental conditions in these countries are more conducive to gang development and expansion than in the United States.

Howell and Moore (2010) also note that the extent of collusion among U.S. gangs, DTOs, and other criminal organizations along the U.S.–Mexico border is not clear. Nevertheless, this intermingling is not a welcomed development for MS-13 and 18th Street gangs that already are considered to be among the most dangerous in this country. The involvement of the Mexican Mafia and other prison gangs in drug trafficking and other criminal activity in the West–Southwest region is also an unwelcomed development of great concern, along with the peripheral involvement of local U.S. gangs along the U.S.–Mexico border. These situations represent formidable challenges to U.S. public safety in the West and Southwest regions, gang policies, and programs.

Gangs in Canada

The scope and seriousness of Canada's gang problem is a matter that is widely debated in Canada. Three gang assessment exercises have produced differing impressions of the seriousness of gang problems over the past decade. In 2002, the core U.S. National Youth Gang Survey questions were replicated in Canada, and these were supplemented with a few other survey items (Chettleburgh, 2003). An estimated 434 gangs and 7,071 gang members were reported.

Chettleburgh (2003) points out the Canadian youth gangs depicted in the 2002 survey were broadly distributed throughout the country in police service jurisdictions that served 65% of the population. Similar to American gangs, almost half of Canadian youth gang members are young people under the age of 18. However, the racial/ethnic mix of gangs in the two countries widely differed. The largest proportions in Canada are African Canadian/Black (25%), followed by First Nations (indigenous peoples) (22%) and Caucasian/White (18%). In contrast, the most broadly represented racial/ethnic groups in American gangs in 2002 were Hispanic/Latino (47%), African American/Black (36%) and White (10%) in the 2002 National Youth Gang Survey. Another distinction is the greater prevalence Chettleburgh reports of hybrid gangs in Canada. Most important, however, is the far lower homicide rate among Canadian gangs versus American gangs, as reported in Chettleburgh (2003). However, almost half of the Canadian gangs were reportedly active in other violent crimes including assault and drug trafficking, and often in collaboration with organized crime groups in the latter area. In addition, "The movement of gang members from one jurisdiction to another, in addition to the return of gang-involved youth or adult inmates from Canadian correctional facilities, appears to be having an impact in a large number of Canadian jurisdictions" (p. 2).

The second Canadian gang assessment, made by the Criminal Intelligence Service Canada (2006) estimated more than 300 gangs in Canada with approximately 11,000 gang members, and overwhelmingly comprised of adult members, 20 to 30 years of age. This federal law enforcement agency also reported more extensive involvement of Canadian gangs in drug trafficking and with organized crime groups (particularly outlaw motorcycle gangs and Italian or Asian criminal groups). It also reported an increase in gang-related violence, "often related to street gang expansion, recruitment, and encroachment on other criminal groups' territory" (p. 23). In addition, this gang assessment notes extensive gang activity within low-income housing areas and inside Canadian federal and provincial correctional institutions, while "occasionally influencing gang activity outside institutions" (p. 25).

The older gangs described in this Canadian report resemble in some respects the largely adult criminal gangs that first emerged on the streets of New York City and Chicago. Aside from the ages of their members, the main parallels are their connections with organized crime groups, some money laundering, and involvement in the illicit sex trade. The most active adult criminal gang areas in Canada are in the Ontario (particularly Toronto) and Quebec provinces (especially Montreal). Considerable gang activity is also reported in the Alberta, Sask, and Manitoba provinces, along with southern British Columbia (the Vancouver area). In the Quebec province, gangs sometimes form alliances and rivals occasionally fight one other in the U.S. fashion over territorial encroachment and interpersonal disputes, sometimes with firearms, as Chettleburgh (2007) describes in the preceding discussion.

In a third Canadian gang assessment, Wortley and Tanner (2008) set out to gain a deeper understanding of the experiences and social circumstances that push and pull youth toward gangs. The key reasons a representative sample of gang youth in Toronto gave correspond closely to the factors U.S. gang members most frequently mention, detailed in Chapter 5. Wortley and Tanner (2008) pinpoint the experiences and social circumstances that promote gang involvement:

- Neighborhood, peer, and family influences
- Protection
- Support and companionship
- Status and respect
- Money
- Racial injustice (p. 198)

Across Canada, neighborhood residential instability—measured by family mobility, home renting, and single parenthood—predicts youth gang membership (Dupere, Lacourse, Wilms, Vitaro, & Tremblay, 2007). However, as in the American communities, some minority groups are more disadvantaged than others, particularly the Aboriginal youngsters in Canada (Deane, Bracken, & Morrissette, 2007). It is often argued that Vigil's (2002, 2006) multiple marginality theory applies to this and other minority groups in Canada. It is not surprising, then, that a similar proportion of Canadian and U.S. youngsters join gangs. In the Canadian National Longitudinal Survey of Children and Youth, gang involvement was measured at age 14. Dupere and colleagues (2007) found 6% of the nationally representative sample self-identified as gang members, a slightly lower proportion than in the United States (8%).

In sum, essentially the same conditions lead to gang joining in Canada as in the United States. Each country's gangs emerge from social, cultural, political, and economic conditions. In this respect, Canada's gang problem appears similar to the U.S. gang situation except on an infinitely smaller scale. Chettleburgh (2003) notes that the

Canadian population is roughly 11% of the U.S. population, and Canada's estimated youth gang membership is less than 1% of the U.S. figure. There are, however, some important differences. The street gangs of Canada seem more intertwined with organized crime groups. More racial/ethnic mixing of gang members occurs in Canada. Another key difference is the extremely low level of gang-related homicide in Canada. Largely on this basis, some well-informed observers protest that some Canadian media and government officials overstate gang problem seriousness in the country as a whole (Wortley & Tanner, 2006, 2008).

Gangs in Europe

Although the first active gangs in Western civilization—which Pike described in 1873 as gangs of highway robbers—may well have existed in England during the 17th century, it does not appear that these gangs had the features of modern-day, serious street gangs. Murray (2000) asserts more violent gangs formed in the early 1900s as the historic Troubles in Northern Ireland gave rise to two violent gangs in the Bridgetown area, known as the Billy Boys (a Protestant gang) and the Norman Conks (from Norman Street, a Roman Catholic residential area), in concert with the sectarian conflicts. Fifty years later, according to Boyle (1977), Glasgow, Scotland, had more stereotypical gangs—much like American forms, but Sanders (1994) notes these had dissipated by the early 1980s.

The presence of modern-day gangs in Europe prompted establishment of the Eurogang Programme in 1997, in which European and American researchers have undertaken a collaborative effort to research street gangs in Europe. Presently, Klein (2008) tabulates more than 60 gang-involved cities recognized in Europe, and the number continues to grow. Three comprehensive books have been published on European gangs in the past decade. These three volumes are impressive and quite accessible: The first one, written by Klein, Kerner, Maxson, and Weitekampf (2001), explores known gang problems in Europe. In the second volume, Decker and Weerman (2005) introduce the Eurogang Programme participants' agreed-upon gang definition and feature the near-decade of research on the status and nature of gangs in Europe. The third volume—on street gangs, migration, and ethnicity—features Van Gemert, Peterson, and Lien's (2008) compiled papers from the Eurogang Programme's first thematic focus on multiethnic gang issues. In a fourth summary report, Klein, Weerman, and Thornberry (2006) synthesize the results of Eurogang studies to date in 12 countries and compare the seriousness of European gangs with U.S. counterparts. This review suggests that level of violence in European gangs is less serious than within U.S. gangs and that "differences may be attributable to the recentness of the European gang development, the lower levels of firearms availability, and lower levels of gang territoriality in Europe" (p. 413). There are notable exceptions, of course. In South Manchester, Bullock and Tilley (2002) claim gang members carrying firearms is common, about 60% of shootings are thought to be gang related, and these are perpetrated by loosely turf-based gangs.

The following are more detailed research observations from Klein and colleagues' (2006) review, concerning the relationship between European street gangs and violent behavior. First, the frequency, severity, and lethality of youth violence are, generally speaking, lower in European countries than in the United States. Second, despite this differential, gang membership appears to have the same worsening effect on the behavior for gang-involved European youth as it does for U.S. youth. Third, as in the states, the researchers' qualitative reports of European gangs suggest that the nature of violence is varied—though typically limited to physical fighting—and similarly motivated. Fourth, again, as in American communities, gang membership in Europe facilitates violence over and above the baseline impact of association with delinquent peers. "Overall, perhaps what is most impressive about these results is their near universality . . . across a diverse set of European countries that probably have more differences than similarities [and these findings are] observed in both quantitative and qualitative studies and in both comparative and single-country studies" (p. 434).

Klein and colleagues (2006) caution that some important limitations apply to this summary: "These conclusions derive largely from unplanned comparisons and from single studies brought together after-the-fact so that patterns have been inferred rather than directly tested" (p. 434). The next phase of Eurogang Programme research is multisite comparative research on street gangs.

In another excellent summary of the Eurogang Programme research, Decker and Weerman (2005) report on studies in six European countries (the Netherlands, Germany, Russia, Scotland, Italy, Norway) and Scandinavia. The following are important findings from these studies. In general, Decker and Weerman (2005) caution that "it is not correct to speak of 'European gangs' as if all gangs in Europe were the same or shared a number of common features that distinguish them from American counterparts" (p. 306). Just as is the case in the United States, street gangs and troublesome youth groups in Europe vary considerably with respect to key features—including demographic characteristics, background, social, and economic elements—as well as structural components, particularly the degree of organization. Although some of the European gangs "have existed for quite a while and have developed structures and leaders, many others (probably most in Europe) are shorter lived and more loosely organized" (p. 305). Again, as in the states, territorial claiming tendencies in European gangs vary. There also is less preoccupation with gang identity among European gangs. "There are many groups that have no names for themselves and do not see themselves as gangs; others do self-identify as gangs and have dangerous names based on locality or inspired by American gang culture" (p. 306).

Future reports from the Eurogang Programme bear close watch. This is an important collective effort involving a network of researchers guided by a common definition and consistent research methods.

Concluding Observations

Overall, two predominant gang trends are evident across the 14 years of the National Youth Gang Survey. First, there is remarkable stability in reported gang activity in the very large cities. Few significant changes are seen in cities with populations greater than 50,000 persons (and adjacent suburban areas), where nearly two-thirds of all gangs and 8 out of 10 gang members are found. In addition, 8 out of 10 of these cities reported gang activity each year from 2005 to 2009. Of utmost concern, however, is the consistent increase from 2002 in reported homicides across 167 very large cities, with populations greater than 100,000. This trend is examined in more detail in Chapter 8. Second, gang activity is highly unstable in smaller cities, towns, villages, and rural counties. These localities have very small gangs, largely transitory gangs, few gang members, and little violence. To be sure, few gangs survive in these areas.

Recent developments have extended and expanded the scope and dangerousness of three U.S. street gangs—MS-13, 18th Street, and the Mexican Mafia—in particular. First, the funneling of major drug-trafficking routes from air transport and sea-crossing to the overland route via Central America and Mexico has opened more lucrative drug-trafficking opportunities to U.S. gangs along the border and within the Southwest and West regions. Second, expanded and intensified interactions with Mexico and Central American countries over the past 20 years or so have contributed to the growth of the MS-13 and 18th Street gangs. Third, the prison-based Mexican Mafia is said to control many of the Hispanic gangs in Southern California.

The European gang situation remains clouded, both in terms of prevalence and seriousness. Similarly, the scope and seriousness of Canada's gang problem is a matter that is widely debated. The United States remains unique in the scope and seriousness of gang activity, although the situation is complicated by gangs in Mexico and Central America in their violent capacities. Some Central American gangs threaten to destabilize neighborhoods, and in Mexico, some gangs have links to narcotics-trafficking cartels that go head-to-head with the military. At the present time, political and governmental conditions in these countries are more conducive to gang development and expansion than in the United States.

DISCUSSION TOPICS

1. Why do you suppose the presence of gang activity has an ebb and flow pattern?

2. What does this ebb and flow pattern have to do with the seriousness of gang crime?

3. How can prison gangs become involved in drug trafficking on the streets?

4. What are the prospects for American street gang involvement in drug trafficking in Central America?

5. How could street gangs in any country develop the capacity to operate transnationally?

RECOMMENDATIONS FOR FURTHER READING

Explanations of Gang Growth in the 1980s and 1990s

Alonso, A. A. (2004). Racialized identities and the formation of black gangs in Los Angeles. *Urban Geography, 25,* 658–674.

Chesney-Lind, M., & Hagedorn, J. (Eds.). (1999). *Female gangs in America.* Chicago: Lake View Press.

Cummings, S., & Monti, D. J. (Eds.). (1993). *Gangs: The origins and impact of contemporary youth gangs in the United States.* Albany: State University of New York Press.

Cureton, S. R. (2009). Something wicked this way comes: A historical account of Black gangsterism offers wisdom and warning for African American leadership. *Journal of Black Studies, 40,* 347–361.

Curry, G. D., Ball, R. A., & Decker, S. H. (1996). Estimating the national scope of gang crime from law enforcement data. In C.R.Huff (Ed.), *Gangs in America* (pp. 21–36). Thousand Oaks, CA: Sage Publications.

Klein, M. W. (1995). *The American street gang.* New York: Oxford University Press.

Miller, W. B. (1982/1992). *Crime by youth gangs and groups in the United States.* Washington, DC: U.S. Department of Justice, Office of Juvenile Justice and Delinquency Prevention.

Miller, W. B. (2001). *The growth of youth gang problems in the United States: 1970–1998.* Washington, DC: Office of Juvenile Justice and Delinquency Prevention.

Perkins, U. E. (1987). *Explosion of Chicago's Black street gangs: 1900 to the present.* Chicago: Third World Press.

Spergel, I. A. (1995). *The youth gang problem.* New York: Oxford University Press.

Vigil, J. D. (1988). *Barrio gangs: Street life and identity in Southern California.* Austin: University of Texas Press.

Vigil, J. D. (2002). *A rainbow of gangs: Street cultures in the mega-city.* Austin: University of Texas Press.

Vigil, J. D. (2007). *The projects: Gang and non-gang families in East Los Angeles.* Thousand Oaks, CA: Sage.

Transnational Gangs: Mexico and Central America

Cruz, J. M. (2010). Central American maras: From youth street gangs to transnational protection rackets. *Global Crime, 11,* 379–398.

Danelo, D. J. (2009). Disorder on the border. *Proceedings* (October), 45–47.

Decker, S. H., & Pyrooz, D. C. (2010a). Gang violence worldwide: Context, culture, and country. *Small Arms Survey 2010.* Geneva, Switzerland: Small Arms Survey.

The Economist. (2010). Organized crime in Mexico: Under the volcano. *The Economist* (October 16), 29–31.

The Economist. (2011). The rot spreads: Organized crime in Central America. *The Economist* (January 22), 45–46.

Valdez, A. (2007). *Gangs: A guide to understanding street gangs* (5th ed.). San Clemente, CA: LawTech Publishing Co.

Gangs in Other Countries (Outside Europe)

Covey, H. C. (2010). *Street gangs throughout the world.* Springfield, IL: Charles C Thomas.

Decker, S. H., & Pyrooz, D. C. (2010a). Gang violence worldwide: Context, culture, and country. *Small Arms Survey 2010.* Geneva, Switzerland: Small Arms Survey.

Hagedorn, J. M. (2008). *A world of gangs: Armed young men and gangsta culture.* Minneapolis: University of Minnesota Press.

Prison Gangs

Camp, C. G., & Camp, G. M. (1988). *Management strategies for combating prison gang violence.* South Salem, NY: Criminal Justice Institute.

Camp, G. M, & Camp, C. G. (Eds.). (1985). *Prison gangs: Their extent, nature and impact on prisons.* Washington, DC: U.S. Department of Justice.

Daniels, S. (1987). Prison gangs: Confronting the threat. *Corrections Today, 66,* 126 and 162.

Decker, S. H., Bynum, T., & Weisel, D.L. (1998). Gangs as organized crime groups: A tale of two cities. *Justice Quarterly, 15,* 395–423.

Early, P. (1991). *The hot house: Life inside Leavenworth Prison.* New York: Bantam Books.

Eckhart, D. (2001). Civil cases related to prison gangs: A survey of federal cases. *Corrections Management Quarterly, 5*(1), 60–65.

Fleisher, M. S. (1989). *Warehousing violence.* Newbury Park, CA: Sage.

Fleisher, M. S. (1995). *Beggars and thieves: Lives of urban street criminals.* Madison: University of Wisconsin Press.

Fleisher, M. S., & Decker, S. (2001). An overview of the challenge of prison gangs. *Corrections Management Quarterly, 5*(1), 1–9.

Fong, R. S., & Fogel, R. E., (1994–95, Winter). A comparative analysis of prison gang members, security threat group inmates and general population prisoners in the Texas Department of Corrections. *Journal of Gang Research, 2,* 1–12.

Gaes, G., Wallace, S., Gilman, E., Klein-Saffran, J., & Suppa, S. (2002). The influence of prison gang affiliation on violence and other prison misconduct. *The Prison Journal, 82,* 359–385.

Griffin, M. L., & Hepburn, J. R. (2006). The effect of gang affiliation on violent misconduct among inmates during the early years of confinement. *Criminal Justice and Behavior, 33,* 419–448.

Irwin, J. (1980). *Prisons in turmoil.* Boston: Little, Brown.

Jacobs, J. B. (1974). Street gangs behind bars. *Social Problems, 21,* 395–409.

Jacobs, J. B. (1977). *Stateville: The penitentiary in mass society.* Chicago: University of Chicago.

Jacobs, J. B. (2001). Focusing on prison gangs. *Corrections Management Quarterly, 5*(1), vi–vii.

Johnson, R. (1996). *Hard time: Understanding and reforming the prison.* Belmont, CA: Wadsworth.

Knox, G. W. (1998). *An introduction to gangs* (4th ed.). Peotone, IL: New Chicago Press.

Olson, D. E., & Dooley, B. (2006). Gang membership and community corrections populations: Characteristics and recidivism rates relative to other offenders. In J. F. Short & L. A. Hughes (Eds.), *Studying youth gangs* (pp. 193–202). Lanham, MD: AltaMira Press.

Pyrooz, D. C., Decker, S. H., & Fleisher, M. (2011). From the street to the prison, from the prison to the street: Understanding and responding to prison gangs. *Journal of Aggression, Conflict and Peace Research, 3,* 12–24.

Ralph, P., Hunter, J., Marquart, W., Cuvelier, J., & Merianos, D. (1996). Exploring the differences between gang and non-gang prisoners. In C. R. Huff (Ed.), *Gangs in America* (2nd ed., pp. 241–256). Thousand Oaks, CA: Sage.

Reiner, I. (1992). *Gangs, crime, and violence in Los Angeles.* Los Angeles: Office of the District Attorney of the County of Los Angeles.

Sanchez-Jankowski, M. S. (2003). Gangs and social change. *Theoretical Criminology, 7,* 191–216.

Schlosser, E. (1998). The prison-industrial complex. *The Atlantic Monthly* (December), 51–77.

Scott, G. (2004). "It's a sucker's outfit": How urban gangs enable and impede the integration of ex-convicts. *Ethnography, 5,* 107–140.

Shelden, R. G. (1991). A comparison of gang members and non-gang members in a prison setting. *The Prison Journal, 81*(2), 50–60.

Silberman, M. (1995). *A world of violence.* Belmont, CA: Wadsworth.

Stevens, D. J. (1997, Summer). Origins and effects of prison drug gangs in North Carolina. *Journal of Gang Research, 4,* 23–35.

Toch, H. (2007). Sequestering gang members, burning witches, and subverting due process. *Criminal Justice and Behavior, 32,* 274–288.

Toch, H., & Adams, K. (1988). *Coping, maladaptation in prison.* New Brunswick, NJ: Transaction Publishing.

Notes

1. President's Commission on Law Enforcement and Administration of Justice (1967); National Commission on the Causes and Prevention of Violence (Mulvihill & Tumin, 1969); National Advisory Commission on Criminal Justice Standards and Goals (1973).

2. *Ethnicity* merely reflects cultural differences, whereas *race* overlaps with ethnicity and "refers to a group that is defined as culturally or physically distinct and, furthermore, ranked on a social hierarchy of worth and desirability" (Telles & Ortiz, 2008, p. 23). Mexican Americans have an ambiguous status. "Although Mexican Americans are often referred to as an ethnic group and not as a race, they were referred to as the latter in earlier times and arguably continue to be referred to and treated as such in societal interactions today" (p. 24).

3. This paragraph contains verbatim extracts (with minimal editing) from Stratfor Global Intelligence (Burton and West, 2009).

Urban Gangs and Violence

Introduction

Urban gang problems are examined in this chapter. The main focus here is on *serious violent gangs* in large cities. New research is presented which groups cities according to their experiences with serious gang violence over a 14-year period. Case studies of two cities' gang histories are reviewed that illustrate nationwide differences in serious violent gangs. Both the correlates (features of gangs themselves) and the context of serious gang violence are then examined. The chapter begins with research on the longevity of gang activity in urban areas.

Serious violent gangs have been characterized by a diverse set of criteria (see Serious Violent Gangs in the Further Reading section at the end of this chapter). Most notably, they are typically found in large cities and distinguished by structural features, first, the degree of organization (e.g., have rules, meetings, and leadership roles, a large membership—at least 15 gangs and more than 200 members but with wide variations—and a multiple-year history). Second, these gangs are involved in violent crimes that physically harm victims, including homicide, aggravated assault, robbery, and nonlethal firearm injuries. These violent crimes frequently occur out of doors, on the streets, and in very small set spaces in neighborhoods with high violence rates. Moreover, such violent crimes likely involve multiple victims and suspects and have a contagious feature. Third, drug trafficking and firearm offenses are also very common. Last, these gangs tend to have both female and male members.

Onset of Urban Gang Problems

The oldest and very large gang problem cities were noted in Chapter 1 as New York City, Chicago, and Los Angeles. But do any cities in which gang activity emerged much later also display serious gang problems? To gain insight into the timing of the spread of gang problems across U.S. cities, the 2000 National Youth Gang Survey (NYGS) asked respondents for the approximate year when their current youth gang problem began, more simply referred to as *year of onset*. Figure 8.1 shows the cumulative percentage of cities by year of onset for each of four population groups.

Cities with very large population sizes (100,000 or more persons) experienced a much higher rate and earlier onset of gang activity than all other cities. Approximately one-third of these cities reported gang problems before 1985, and an additional 50% reported an onset of gang problems in the following 10-year period (i.e., from 1985 to the mid-1990s). These patterns are increasingly less pronounced across the remaining population groups, suggesting a cascading pattern of gang proliferation from the larger to the smaller populated cities and rural areas. This is also reflected in the average year of onset across population groups, which was 1985 for cities with populations of 100,000 and greater, 1988 for cities with populations of 50,000 to 100,000, 1990 for cities with populations of 25,000 to 50,000, and 1992 for cities with populations of less than 25,000 persons.

Figure 8.1   Patterns of Gang Emergence

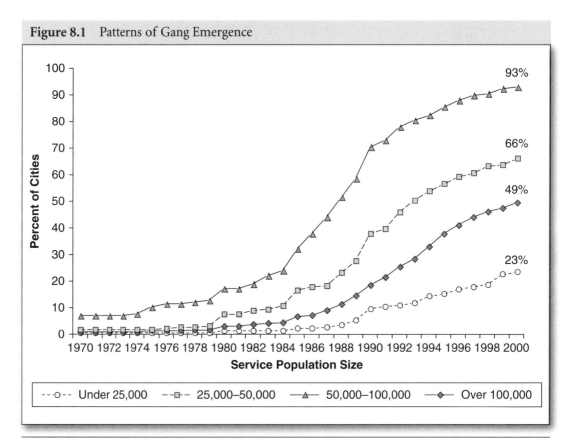

Source: Egley, Howell, & Major, 2006, p. 10

Gangs in newer gang problem jurisdictions are qualitatively different from traditional gangs in jurisdictions where gang problems began much earlier. Notably, gangs in the late-onset group commit fewer homicides (Egley, Howell, Curry, & O'Donnell, 2007). In localities that report onset of gang activity before 1990, 47% experienced gang homicides versus only 16% of localities that first reported gang activity in 2000 or later. In addition, 91% of localities in the former group reported gang-related firearm use, and 100% reported gang-related aggravated assaults, versus 65% and 74%, respectively, in the latter group. The before-1990 onset group also reported more gangs (7, on average) and gang members (178, on average) than localities in the latter group, 4 gangs and 34 gang members, respectively.

The consistency of gang problems also varies directly with city size. Very large cities tend to have persistent gang activity in comparison with small cities, which report gang activity sporadically. Uncertainty about gang presence is likely a main reason in these less populated areas.

In summary, NYGS data reveal a cascading pattern (of earlier to later onset) from the largest to the smallest localities and from urban to rural areas. A majority of cities, including those with populations greater than 100,000, reported the emergence of gang activity before 1996, which helps explain the peak in gang activity in the mid-1990s. The next section presents another view of cities' gang problem histories, grouping them according to similarities in their histories across more than a decade.

Consistency and Seriousness of Gang Problems in Large Cities

The analysis that follows is the first of its kind. *Trajectory modeling* has been widely used in the classification of individuals according to their pattern of offending over time (Lacourse et al., 2003; Piquero, 2008). In 2004, researchers began to apply this group-based trajectory method to model the criminal careers of geographic areas, like street segments and census tracts, to capture communities' trajectories across time and space (Griffiths & Chavez, 2004; Weisburd, Bushway, Lum, & Yang, 2004). This research, in turn, led Howell and colleagues (2011) to explore the possibility of modeling cities' gang problem histories. This revealed that some cities in which officials say they have gang activity tend to report a consistent presence of gangs, while others experience no gang activity over time, rapid increases over time, rapid decreases, fluctuating presence of gang activity, or other more complex trends between 1996 and 2009. Key results from the first modeling of cities' gang problem histories follow.

This exploratory research has used the 14 years of the NYGS (from which overall trend data were shown in Chapter 7). Table 8.1 shows the consistency of reported gang activity in cities with populations greater than 50,000 persons in the NYGS (referred to as larger cities). First, nearly 9 out of 10 (86.3%) cities reported gang activity in 2009. Second, almost 8 out of 10 (79.8%) of these cities consistently (each year) reported gang activity across the entire 6-year period, from 2005 to 2009. In sharp contrast, fewer than 3 out of 10 (23.8%) smaller cities, towns, and villages consistently reported gang activity during the same period. Stated another way, persistent or stable gang activity is almost four times more likely to be found in larger cities.

Table 8.1 Law Enforcement Reports of Gang Activity, 2005 to 2009

Area Type	Gang Activity Reported in 2009	Gang Activity Consistently Reported, 2005–2009
Larger cities	86.3	79.8
Suburban counties	51.8	38.9
Smaller cities	32.9	23.8
Rural counties	17.0	11.0

Source: Egley & Howell, 2011, p. 2

These data and other prior NYGS analyses clearly demonstrate that gang activity—in terms of size of gang membership and the occurrence of gang violence—remains largely concentrated in the most populated areas in the United States (Egley et al., 2004, 2006; Howell, 2006). Therefore, the next analysis focuses only on jurisdictions with populations greater than 50,000. This permits an examination of cities with more persistent gang activity—where gang activity is not only more prevalent, but also more serious, and thus of greater interest here with respect to common patterns.

Figure 8.2 displays six identifiable groups in the analysis of a subgroup of 598 larger cities (only cities with populations greater than 50,000 among the 1,517 localities in the trajectory analysis of gang activity prevalence shown in Chapter 7). The most predominant group is the 418 cities in the trajectory group T5 (69.9% of the total sample), which reported a persistent and chronic gang problem over the 14-year period. Three groups of cities (T1, T2, and T3) showed increasing gang activity in the second half of the measurement period (Figure 8.2). The T1 group (25 cities; 4.2%) exhibited some minor increases in gang activity between 1996 and 2002 that escalates thereafter, but not to a chronic level. In both the T2 (4.8%; 29 cities) and T3 (9.2%; 55 cities) groups, gang activity escalates very sharply in the second half of the measurement period even though they had different experiences from 1996 to 2002. The T2 group, comprised of 29 cities, showed virtually no presence of gang activity at the beginning of the survey period until 1998, and gang activity rose sharply thereafter to near complete persistence from 2004 onward. Gang activity initially decreased in the T3 group of cities, and then sharply increased after 2002. The remaining two groups of cities (T4; 6.5%; 39 cities; and T6; 5.4%; 32 cities) reported decreasing gang activity almost entirely throughout the 14-year period.

Overall then, three main patterns are seen in the six groups of cities: first, no change (T5 group), second, increasing gang activity (T1, T2, and T3 groups) and third, decreasing gang activity (T4 and T6 groups). Seeing these distinctive patterns raises the question of what factors might account for cities' various gang problem histories. Because of the novelty of this research, explanations are not available at this point for

Figure 8.2 Trajectory Model: Presence of Gang Activity in Cities With Populations Greater Than 50,000

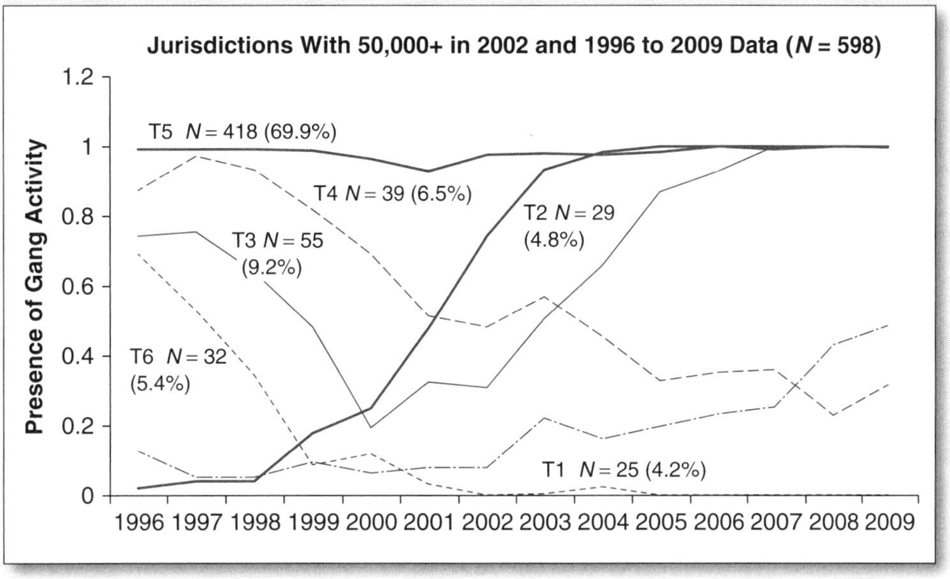

Source: Howell et al., 2011, p. 6

the trends observed here. Rather, the purpose in this initial application of trajectory analysis is to identify the main patterns of persistent gang activity; identifying factors associated with group patterns is a task that lies ahead. In the next section, Howell and colleagues apply trajectory modeling to see if serious gang problem cities could be grouped on another dimension: violent crime involvement over the 14-year period.

Serious Gang Problem Trends

The above analyses demonstrate that cities can be grouped in terms of their distinctively patterned gang problem histories. With this in mind, the next step is to assess the relative seriousness of gang activity among cities. For the purposes of this analysis, homicide is considered to be a primary indicator of serious gang activity.

Gang-Related Homicides and Serious Gang Activity

Homicides characterize serious gang problem cities more than any other factor, and these are heavily concentrated geographically in the United States. Most cities have no gang homicides, and those that do usually report very few of them from year-to-year in the NYGS (Egley et al., 2006). Rather, it is in a subset of larger cities (populations greater than 50,000) and adjacent suburban counties where the overwhelming majority (96%) of them occurred in 2009 (Table 8.2).

The trajectory analysis presented in Figure 8.3 examines trends in the proportion of all homicides that are gang related and, in contrast to the previous section, offers a more pointed investigation of *seriousness* of gang problems nationwide. Further, the analysis that follows is limited to cities with populations in excess of 100,000 persons and incorporates total annual homicide counts from the Federal Bureau of Investigation's Uniform Crime Report (UCR) data for all very large cities participating in the NYGS between 1996 and 2009. Overall, 247 cities met the criteria for inclusion. (Only two of the eligible cities were excluded from the analysis due to complete missing data for the entire time period.) Proportional homicide rates—along which cities' patterns are aligned here—were determined by dividing the total number of gang homicides reported in the NYGS annually by the total number of homicides reported for the city in the UCR, multiplied by 100.

The largest cluster of *very large cities* with populations greater than 100,000 is seen in trajectory group T2 ($N = 105$; 42.5% of the total sample of 247 cities). This group of 105 cities experienced a very high rate of gang-related homicides over the 14-year period. Approximately 20% of annual homicides—or 1 in 5—were gang related. Yet the next largest group of cities (T4; $N = 71$; 28.7%) experienced the highest rate of gang-related homicides, with approximately 40% of all homicides—or 4 in 10—gang-related over the same period. Another sizeable group, 55 very large cities

Figure 8.3 Trajectory Model: Percent of Gang-Related Homicides, Populations Greater Than 100,000

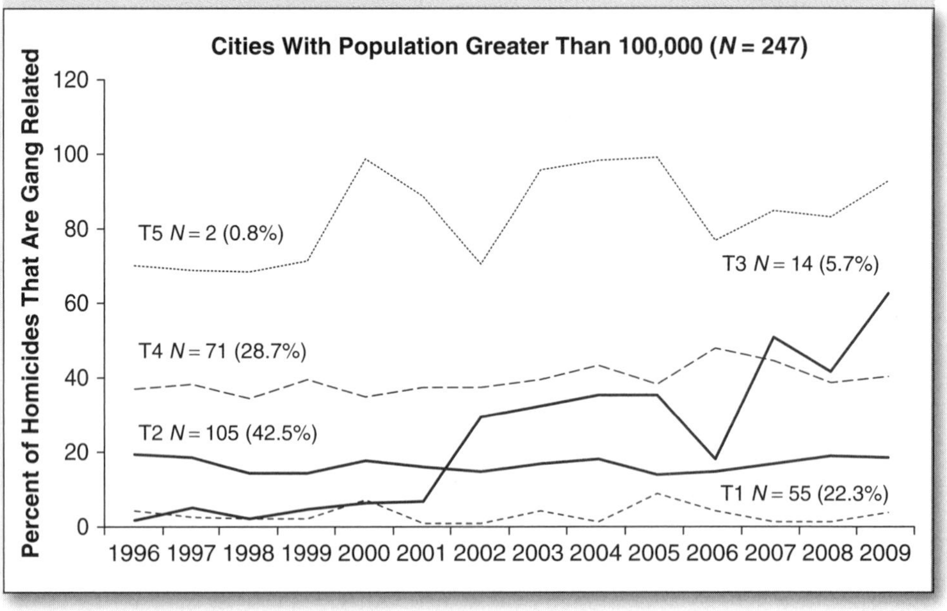

Source: Howell et al., 2011, p. 7

(T1; 22.3%) averaged between 2% and 8% gang-related homicides over the 14-year period. The two remaining trajectory groups (T3; $N = 14$; 5.7% and T5; $N = 2$; 0.8%) experienced increases in annual gang-related homicides over the entire period. A cluster of 14 cities (T3) experienced a very sharp increase from 2001 to 2009, from only around 4% to more than 60% over this 10-year period. Finally, two very large cities (T5; Inglewood, California, and Salinas, California) stand out for having started the 14-year period at nearly 70% of all lethal violence that was gang related. By 2009, however, more than 90% of all homicides in these cities are gang related.

Three important observations can be made from this analysis of gang-related homicides as a proportion of the total number of homicides reported in the UCR for very large cities. Overall, more than 7 out of 10 very large cities (72%, T2, T4, and T5 combined) reported a consistently high level or increasing proportion of gang-related homicides over the 14-year period. Second, a remarkable degree of consistency in the rate of gang-related homicides across trajectory groups is observed. Third, *none* of the trajectory groups found in these cities displayed a pattern consistent with a decline in the prevalence of gang homicide.

A Snapshot of Gang Homicide in the Largest Cities

This section reviews the current prevalence and concentration of high homicide levels in *very large* U.S. cities, with populations greater than 100,000 persons.[1] In the 167 NYGS cities that provided gang homicide data in 2009, Egley and Howell (2011) found 110 (66%) reported a total of 1,017 gang homicides in 2009. This total represents a 2% increase from 2002, a 7% increase from 2005, and an 11% increase from 2008 for these same cities (see Table 7.2 in Chapter 7).

Figure 8.4 shows the total percent of homicides that were gang related among cities with populations of 100,000 or more (for which homicide data were reported in 2009). Separate figures are presented for Chicago and Los Angeles because of their historically very large numbers of gang homicides. Overall, approximately one-quarter of all homicides in this category of cities were gang related in 2009. By comparison, one-half of the homicides in Los Angeles and one-third in Chicago were gang related.

Prior research has contrasted historical gang homicide patterns in Chicago and Los Angeles (Howell, 1999; Maxson, 1999). No other cities in the United States—nor worldwide for that matter—come close to producing the volume of gang homicides found in these two cities. As discussed in Chapter 1, in each city, unique histories produced several waves of gang development that further exacerbated gang problems.

Aside from the prominence of Chicago and Los Angeles as gang homicide capitals, many other cities—particularly very large ones with populations greater than 100,000—have extremely high gang homicide rates and should not be ignored. As shown in Figure 8.3, 71 U.S. cities in this population category experienced very high gang homicide rates from 1996 to 2009, accounting for approximately 40% of all homicides.

San Antonio researchers' analysis of 28 homicides that occurred in the course of their study yielded five distinct circumstances in which these occur and six different motives (Valdez, Cepeda, & Kaplan, 2009). Key circumstances include drug-related

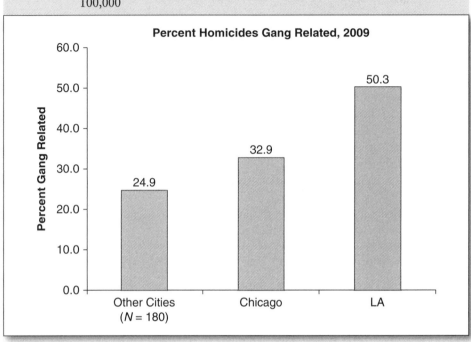

Figure 8.4 Gang Homicide Prevalence, Cities With Populations Greater Than 100,000

Source: Howell, J. C., Egley, A. Jr., Moore, J. P., Tita, G., & Griffiths, E. (2011)

disputes (typically an argument associated with drug transactions), gang disputes over a variety of gang-related issues, assaults in which victims are attacked without notice, and *rolling out* (a gang exit rite that entails a physical beating by several gang members). Key motives include personal vendettas (feuds between victims and offenders), gang revenge or retaliation, gang rivalries, territorial trespassing, gang solidarity (camaraderie), and spontaneous retaliation (spur-of-the-moment retribution or defense.

Case Studies of Gang Problems in Large Cities

The next sections feature two of these cities (Pittsburgh, Pennsylvania, and St. Louis, Missouri) to gain insights about urban settings in which gang problems are very serious. Pittsburgh is in trajectory group four above (averaging about 40 gang homicides per year) and St. Louis is in trajectory group two (averaging about 20 gang homicides). Pittsburgh gangs have a long history of involvement in drug trafficking and guns (Gordon, 2010), and several St. Louis gangs have a long history of violence and other criminal activity (Decker & Van Winkle, 1996; Rosenfeld et al., 1999).

Pittsburgh Case Study

Pittsburgh appears to be somewhat typical among very large cities (populations greater than 100,000) with respect to the long-standing seriousness of its gang problem. Before examining its history, we pause to reveal Pittsburgh's location in the above trajectory groups. Pittsburgh is among larger cities in trajectory group T5 (Figure 8.2) that consistently experienced gang activities throughout the 14-year survey period from 1996 to 2009. In Figure 8.3, Pittsburgh is among the very large cities in the trajectory group T4 that consistently reported a high level of gang homicides (approximately 40% of total annual homicides, on average) during the 1996 to 2009 period.

Having experienced onset of gang activity in 1991, Pittsburgh is considered a "late onset" city (Howell, Egley, & Gleason, 2002). However, unlike other cities in that onset group that did not report a large problem with violent crimes before the end of the decade, Pittsburgh quickly developed a serious gang problem, as we shall see in the next section. Its gang activity developed in two stages, which are characterized here as onset of gang activity and recent gang activity.

Onset of gang activity in Pittsburgh. According to Tita (1999) and Cohen and Tita (1999), a surge in drug-related arrests (apparently driven by crack cocaine offenses) preceded onset of gang activity in Pittsburgh. "Shots fired" calls (citizen-initiated emergency 911 calls to police) spontaneously increased in "a classic epidemic" during a pregang period (1990–1991) in census tracts[2] distributed widely throughout the city (p. 195). Tita and Cohen (2004) observed that it was precisely in the high-violence communities that gangs emerged, which "was followed by a contagious spread of shots fired activity in gang tracts or tracts adjoining them," fueling an epidemic of gun violence or "contagious diffusion" to other areas (p. 195).

Violent urban street gangs, including sets (subgroups) of Black Crip and Blood gangs, began to take hold in Pittsburgh during the latter half of 1991, according to Tita and Cohen (2004). Gang emergence continued through 1993, and stabilized in 1994–1995 with no new gangs forming and no gangs desisting (Tita, Cohen, & Engberg, 2005).

Perhaps largely attributable to the violent community context within which they formed, Tita and Ridgeway assert that "all of the gangs included in this [Pittsburgh] study share one thing in common: They are known to be violent" (p. 217), and Tita (1999) contends they have earned "respect" and fear from the community. Given the territorial and retaliatory natures of urban youth gang violence, it is reasonable to expect that gang-related violence would follow predictable spatial and temporal patterns. In short, Tita and Ridgeway (2007) state, "One might expect set space to serve as a sort of lightning rod for intergang violence" (p. 217). From the onset of gang activity in Pittsburgh, about two-thirds of all gang homicides were gang motivated (intergang disputes, initiation activities, or spontaneous drive-by killings).

Recent gang activity in Pittsburgh. Tita and colleagues' on-site study of Pittsburgh ended in 1995. This section summarizes the Pittsburgh Police Department's responses to the NYGS from 1996 onward. Beginning in 1996, with a well-publicized Federal Racketeer Influenced and Corrupt Organization Act indictment of a local street gang, which "had an enormous impact on all Pittsburgh gangs. Gangs and gang violence virtually disappeared from the streets of Pittsburgh" (Tita et al., 2005, p. 281). From 1996 to 1999, the number of gangs reported by the Pittsburgh Police Department dropped 77%, from 86 to just 20.

For several years thereafter, Pittsburgh police responses to the NYGS characterized the city's gang problem as somewhat stabilized but at a serious level, particularly in drug trafficking, aggravated assault, and firearm use. Law enforcement considered a majority of the gangs to be "drug gangs." The proportion of gang members that is Black has remained virtually unchanged over the 14-year period, averaging almost 86%. Gangs still were well established in certain areas of the city as at the beginning of the new millennium, with subgroups based on age, gender and geographical area. Then the gang problem turned noticeably larger and more serious. In 6 out of 9 years (2001 through 2009), Pittsburgh police said the city's gang problem was "getting worse" in three main respects.

First, law enforcement reported more gangs from the mid-2000s onward, which likely contributed to increased intergang conflicts and gang violence. While only 20 gangs were counted in 1999, an average of 36 was reported each year during 2001 to 2009, and this is a conservative estimate because Pittsburgh police count multiple sets as one gang. Although police had more difficulty estimating the number of gang members during the early part of this decade, during 2005 to 2008 an average of 869 gang members were reported each year. The few very large gangs were estimated to have between 95 and 200 members.

Second, gang members apparently remained in the gangs for multiple years. In the 1990s, police had estimated that 7 out of 10 gang members were juveniles. By 2008, this proportion had dropped to one-half, reflecting either more involvement of older gang members or aging of earlier members.

Third, inmates returning from prison may have reconnected with some of the gangs or joined outright as a result of relationships they formed in prison. Survey respondents said the returning inmates influenced local gang activity in important ways in the new decade, including drug trafficking, access to weapons, and violence itself. Three-fourths of these former inmates were estimated to be adults. From 2003 to 2006, the Pittsburgh Police Department reported more than 20 gang homicides each year along with increases in gang-aggravated assaults.

In sum, Pittsburgh's gang problem developed quickly and worsened measurably over time. Widespread drug dealing and gunplay preceded early gang emergence. Gangs emerged in the high-violence communities, followed by a contagious spread of shots fired activity in gang tracts or tracts adjoining them. Once the gangs developed a reputation for violence and earned respect for this, gang violence stabilized, but at a high level. Now, inmates returning from prison may well be refueling existing gangs.

St. Louis Case Study

Like Pittsburgh, St. Louis appears to be typical of very large cities with respect to the long-standing seriousness of its gang problem. In Figure 8.3, St. Louis is among the very large cities in the trajectory group T2 that consistently reported a high level of gang homicides (approximately 20% of total annual homicides, on average) during the 1996 to 2009 period.

In the following description of St. Louis gangs, Monti (1993) illustrates how gangs vary in their key characteristics depending on the degree to which their surrounding communities were well settled. In describing St. Louis communities as "less settled," "relatively settled," and "more settled," Monti refers to degrees of social disorganization, as indicated by racial/ethnic heterogeneity, rapid population movement, poverty, and unstable families and schools. His study of St. Louis gangs covered nearly a decade, beginning in 1984, corresponding with a period of accelerated gang growth across the city. These are based on his interviews of hundreds of male and female gang and non-gang youths and other knowledgeable persons, including police officers (see Monti, 1991). The gangs Monti (1993) describes include only the "better-established groups composed largely, but not exclusively, of teenagers or young men in their twenties [and also] slightly older gangs that had established reputations for fierceness and, in most cases, a merchandising operation for illegal drugs" (p. 228). In particular, numerous nascent gangs comprised of younger children in north St. Louis and teenage gangs in south St. Louis are excluded.

Gang activity in north St. Louis. Known gangs located in the "less-settled" north St. Louis area numbered 19 in 1988—most of which had not been in existence long, typically only since 1984 (Monti, 1993). Most of the gangs in this area were still in the process of expanding and contracting, and only eight gangs had stable territories. Moreover, few of them had ties with other gangs. This introduces "brittleness" to gangs that makes them more vulnerable to outside attacks and seductions. Only Black males belonged to these gangs, but seven had female auxiliaries. Gang size varied enormously—but most had more than 30 members. These gangs were not particularly well organized. More than one leader was common and at least one person typically served as an "enforcer." Moreover, "the gangs had no strict hierarchy, or officer corps or formal rules for developing a 'gang policy,' and no set custom for initiating or removing members" (p. 233). Virtually all of the gangs in this area were described as "fighting gangs."

Gang activity in near north suburbs of St. Louis. Gangs in "relatively settled" communities (i.e., near north suburbs) of St. Louis numbered 14 in 1990 (Monti, 1993). The two largest gangs had more than 100 members and both maintained up to four age-graded sets that corresponded with the members' school status. Of these gangs, 12 had a fixed territory (3 of which were female) and another gang, the Gangster Disciples, claimed a connection to the Chicago supergang of the same name. The remaining group was actually a drug gang. Criminal activities involving members of these gangs varied widely, although some patterned behavior was observed. Some junior high members began with burglaries and graduated into drug dealing. Others stole cars in auto theft rings. Some gangs were actively involved in drug distribution.

Gangs in this area were generally well established, such that children took their gang loyalties with themselves to schools. "There were children at every elementary school who were active gang members or serious about claiming some gang affiliation. Youngsters from all the municipalities were brought together at the junior and senior high schools" as a result of school desegregation plans (Monti, 1993, p. 242). Not surprising, "gang fights in city schools were more routine [in this area] and more likely to involve substantial violence" (p. 242). The "unsettled character" [poor school climate] of schools contributed to the seriousness of gang-related crime in and around the school, including drug and weapon use, intimidation, extortion, disruptiveness, and encroachment of nonstudents into the schools.

The early involvement of children in gangs in this area of St. Louis was noteworthy. "Children grew up knowing where they lived and the gang they eventually would join, if they chose to become a member" (Monti, 1993, p. 243). Even elementary school–aged children played at being a gang member. "While much of this play and imitative behavior had a game-like quality, the children took it seriously and appreciated the significance of what it meant to be a gang member" (p. 243). It was through role play such as "the wearing of colors, and flashing of signs that children explored the role of 'gang member' and identified themselves as potential recruits. The self-selection process continued at the junior and senior high schools" (p. 243).

Gang activity in central and south St. Louis. Gangs in "more-settled" areas (i.e., central and south St. Louis) were not as numerous as in the less-settled and relatively settled areas (Monti, 1993). This is attributable to two factors. First, part of this area, the central corridor, was very well settled and had only two gangs. Second, the gangs in the south St. Louis area were populated by Black youth from families that either lived in public housing projects or in relatively poor zones into which they had been relocated from housing developments downtown.

The principal established gangs in this area—the Thunder Katz, Peabody Boyz, Southside Posse, and South Side Gangsters—had fewer potential recruits, but the family and friendship ties among them were strong. Hence, these gangs were less vulnerable to outside attacks and seductions. Gangs in these areas were well integrated into the daily routines of the institutions and residents that shared the area with them; therefore, they seldom engaged in serious crimes or sustained criminal activities over a long period of time. In addition, these gangs did not exploit the vulnerability of the community; instead, they actually defended its integrity.

Features of Gangs That Contribute to Serious Gang Problems

The Pittsburgh and St. Louis case studies illustrate an important point. Even though each of these cities has experienced similar seriousness of gang problem histories, their gang problems differ in some key aspects that may have important implications for intervention. The main difference between them is that St. Louis has an older, more established gang problem with considerable diversity in various sectors of the city. In contrast, Pittsburgh has more homogeneous gangs with very extensive involvement in drug trafficking.

The number of gang members, number of gangs, and other structural characteristics are important indicators of serious gangs in large cities. A brief summary of important studies on each of these indicators follows. Table 8.2 shows variations in the number of gang members and gangs among population categories.

Number of Gang Members

Three findings about the number of gang members are important. First, gang members are far more prevalent in gangs within large cities. Approximately 8 out of 10 gang members are located in cities with populations above 50,000 and adjacent suburban counties in 2009 (Table 8.2). Second, Egley and colleagues (2006) conclude gang members are far more numerous in densely populated cities that report a persistent gang problem. Third, Decker and Pyrooz (2010b) found that in the largest cities both the number of gang members and number of gangs "are strong, positive, and significant" correlates of gang homicides (p. 369), but that the number of gang members is a much stronger correlate than the number of gangs, although both measures are statistically significant. This is understandable given that some law enforcement agencies count multiple sets as one gang while others enumerate only larger gangs.

Table 8.2 Distribution of Gangs, Gang Members, and Gang Homicides by Area Type

	Gangs (%)	Gang Members (%)	Gang-Related Homicides (%)*
Larger cities	44.1	55.6	74.4
Suburban counties	21.4	23.3	22.1
Smaller cities	29.1	18.3	2.3
Rural counties	5.4	2.7	1.2

*Note: Total reflects only homicides reported to NYGS, and is not a national total due to missing data and sampling design

Source: Egley & Howell, 2010

Number of Gangs

The number of gangs is a potent correlate of overall gang crime, particularly where multiple gangs are active in a restricted area. Approximately two-thirds of all gangs are located in cities with populations above 50,000 and adjacent suburban counties in 2009 (Table 8.2). Generally speaking, studies in Chicago (Block, 2000; Block & Block, 1993) and Los Angeles (Hutson et al., 1995) show that the number of gangs present in an area significantly relates to the area's overall level of violence and other crimes.[3] In the southeast area of Chicago, where three or four different named gangs were sometimes active in the same area, Block (2000) illustrates this point:

> The conflict between the Black Disciples and BGDN has resulted in many deaths in District 7, where all four major black gangs were active, and in the housing projects in District 2. The southeast area includes the former turf of the El Rukins (now renamed the Black P. Stone Nation) and the oldest Mexican neighborhood in Chicago, where the Latin Kings are active. (p. 378)

Block (2000) examined gang homicides in very small *grid squares,* consisting of 150 meters on a side, revealing that "the relationship between the number of gangs that are active in an area and the levels of assaults and drug-related incidents is remarkably high" (p. 379).[4] This research found that gangs do not necessarily need to be large to carry out a large number of homicides. In fact, street gangs other than the largest four Chicago gangs were responsible for more police-recorded offenses of any type than any one of the top four. Many of these smaller street gangs were relatively new in Chicago, predominantly Latino, and continuously fighting among themselves over limited turfs.

Gang Structure

Previous research has not always been clear on the correlation between gang structure and gang violence, largely because much of the research viewed gangs as either highly organized (resembling adult organized-crime groups) or unorganized. Thrasher is widely recognized for his *natural history of the gang,* which accounts for how gangs develop and evolve from a play group to a full-fledged criminal gang. For instance, Thrasher (1927/2000) observed that gangs in the embryonic stage may evolve into a more highly organized unit, from a "diffuse" gang to a "solidified" criminal gang, and eventually, possibly become a "conventionalized" (formalized) organization if it, for example, becomes an athletic club or a pleasure club of some sort. But as pointed out in Chapter 2, few highly formalized gangs have been found elsewhere. Taylor (1990a) asserts that some gangs evolve into criminal organizations.

In Focus 8.1
Gang Transformations in Detroit

Taylor (1990b) identified three different motivational gang categories in Detroit: scavenger, territorial, and corporate. *Scavenger* gangs mainly consist of lower and underclass youths. They are urban survivors who prey on the weak and have no common bond except their impulsive behavior and need to belong. They engage in senseless, spontaneous crimes for fun. Leadership constantly changes (p. 105). Scavengers are also the social and violent gangs that Yablonsky (1959) identified, and the "42 Gang" of Chicago that Shaw and McKay (1942) studied.

Territorial gangs, Taylor (1990b) contends, evolve from scavenger gangs, when they define a territory and become more organized for a specific purpose. Someone assumes a leadership role, which is part of the process of organization. They actively defend their territory, the area of their "business." They become "rulers" of their turf, generally defined as neighborhood and ethnic boundaries, using physical violence as their only enforcement tool against invaders (pp. 107–108).

Corporate gangs are highly organized for the purpose of engaging in illegal money-making ventures. In Chicago, for a time, they served as "farm teams" for mafia organizations. They took on many of the characteristics of a business organization, motivated only by profit. They operated by rules and military-like discipline. Their membership came from all social classes (Taylor, 1990b).

Taylor (1990b) illustrates the transformation of a scavenger gang to a territorial gang, then to a corporate gang with the case of the 42 Gang in Chicago. Even as scavengers, they were considered one of the most vicious gangs in the United States. Then they staked a territory, grew, and became more formal in their organization. Some of them graduated into the lower ranks of the Capone mob, although some of the mobsters thought the 42 Gang "too crazy" for organized crime work. Most of them eventually were maimed, killed, or imprisoned for murder, armed robbery, or rape.

The growth of the lucrative drug trade in the mid- to late 1980s redirected scavenger gangs into independent organizations in their own right. Taylor (1990b) contends that the advent of the new era of drug use and trafficking gave new life and purpose to scavenger gangs. Without the Job Corps and other programs, they saw the drug business as a way out of poverty. These conditions combined to move Black youths for the very first time into "mainstream major crime." The Detroit scavenger gangs became part of organized crime in America. They became "imperialists" (p. 113). As Taylor put it, "The fact is that drugs have taken street gangs and given them the capability and power to become social institutions" (p. 114).

Source: Taylor, 1990a, 1990b, pp. 105–114. Reprinted with permission from Sage Publications, Inc..

As research became more sophisticated, findings have become clearer. For example, Esbensen and colleagues (2001) show that when gangs are even "somewhat organized" (i.e., have initiation rites and so on), members self-report involvement in more serious delinquent acts than other youths. In other research, a study of detained youths in three

Arizona sites, Decker, Katz, and Webb (2008) found that "the more organized the gang, even at low levels of organization, the more likely it is that members will be involved in violent offenses, drug sales, and violent victimizations" (p. 169).

Egley and Howell (2010) associate other structural characteristics of gangs with gang violence. Using NYGS data, gangs that have subgroups based on age, gender, and geographic area or territory are associated with higher homicide levels. Among 2009 survey respondents, asked directly whether gangs in their jurisdiction have such subgroups, only 33% of localities that acknowledged gang activity reported none of these three subgroupings. Almost one-third (32%) reported one of the three subgroups, 18% reported any two of them, and 17% reported having all three subgroups. While the existence of any one of these subgroups predicted gang violence, odds ratios reached statistical significance with any two of the subgroups, and the odds ratio more than doubled from a condition in which only one subgroup was present to the presence of all three subgroups. This national research strongly suggests that the greater the organization of gangs into subgroups of some sort, the greater the likelihood of gang-related violence.

In sum, evidence is growing "that measures of gang membership and the number of gangs are . . . robust, whether reported by individual gang members or official agencies" (Decker & Pyrooz, 2010b, p. 371). And both the number of gangs and gang members predict gang homicide in very large cities and also when gang subgroupings exist.

Contextual Conditions That Contribute to Serious Gang Problems

Contextual conditions contribute to serious gang problems (Hughes, 2006; Hughes & Short, 2005). The important contribution of the general diffusion of gang culture to ongoing gang problems was discussed in Chapter 7. Broadcast media, public officials, law enforcement, and others typically associate street gangs with drug involvement. Law enforcement also views gang-involved members returning from confinement as a very important factor, along with intergang conflict (Figure 8.5). Other research discussed here emphasizes the importance of firearm possession and use, particularly in the intersection of violent hot spots and gang set space. Last, the group process itself (interactions among gang members) is discussed as a vital context for gang violence.

Drug-Related Factors

Law enforcement views drug-related activity as the strongest factor influencing local gang violence (Figure 8.5). The Pittsburgh case study illustrates two ways in which gang involvement in illicit drug trafficking contributes to violence, by co-occurrence in communities, and the second mechanism is direct involvement. Among proposed explanations for the surge in youth violence and gang homicides in the late 1980s and early 1990s, Blumstein (1995a, 1995b, 1996) and Blumstein and Rosenfeld insist that the so-called crack cocaine epidemic is more valid than any other single explanation: "As the crack epidemic spread in the mid- and late-1980s, so did the danger around inner city drug markets, driving up the incentive for more kids to arm themselves in an increasingly threatening environment. That environment also became a prime recruiting

Figure 8.5 Factors Influencing Local Gang Violence, National Youth Gang Survey, 2009

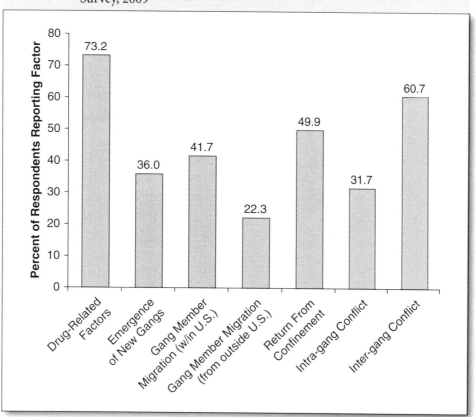

ground for urban street gangs" (1999, p. 162). In two separate studies, Cork (1999) and Grogger and Willis (1998) attempted to show a causal connection between youth violence and a presumed crack cocaine epidemic, but actual crack use could not be distinguished from other more widespread means of ingesting cocaine (Golub & Johnson, 1997), and none of these studies empirically established the expected connection directly to gangs. Blumstein and Rosenfeld's claim is unsubstantiated. The crack cocaine phenomenon was not as widespread as Blumstein and Rosenfeld presumed; it was limited to but a few cities. Moreover, Ousey and Lee (2004) note "different drug 'epidemics' have hit different cities at different points in time"; hence, Blumstein and Rosenfeld's hypothesis does not universally apply. It is interesting to note that in Canada, Hagan and Foster (2000) found juvenile homicide rates increased sharply in the mid- to late 1980s in concert with the U.S. increase without the presence of any crack cocaine epidemic. Also, drug distribution and related drug wars are overwhelmingly the province of adult criminal organizations and cartels (Eddy et al., 1988; Gugliotta & Leen, 1989; Klein, 2004), not typically street gangs. There are exceptions, of course.

Another contributing factor to the sharp rise in drug-related arrests observed in the United States in the late 1980s and early 1990s was the "war on drugs" initiated by the

Reagan administration (Howell, 2003a). To be sure, Howell and Decker (1999) conclude gang participation, drug trafficking, and violence occur together, and more so, according to Coughlin and Venkatesh (2003), in certain cities with more organized gangs, and where gangs are intermixed with drug trafficking groups. The Pittsburgh case study is an excellent illustration of this latter condition. In other words, these problems overlap considerably but, as Decker (2007) notes, "Conflict between gangs accounts for more gang violence, including homicide, than does involvement in the drug trade" (p. 392).

Law enforcement officers themselves recognize infrequent gang involvement in drug distribution. From the 1996 NYGS onward, Howell and colleagues (2002) found only a minority of gang-problem jurisdictions report that gangs controlled a majority of the drug distribution in their jurisdiction. The bulk of the evidence from law enforcement, field studies, and youth surveys finds that most gangs lack key organizational characteristics to effectively manage drug distribution operations. Decker (2007) outlines specific criteria as follows, and few street gangs meet these criteria:

> First, gangs must have an organizational structure with a hierarchy of leaders, role, and rules. Second, gangs must have group goals that are widely shared by members. Third, gangs must promote stronger allegiance to the larger organization than to subgroups within it. Finally, gangs must possess the means to control and discipline their members to produce compliance with group goals. (p. 392)

In general, research supports a connection between drug-related offenses and lethal violence rates in multiple cities, but racial differences are apparent along with what Ousey and Lee (2004) consider "preexisting social conditions." Future explanatory models must therefore be more complex than Blumstein's hypothesis of an impact of illicit drug *markets* on rates of crime. For example, Martinez, Rosenfeld, and Mares (2008) claim indicators of social disorganization (socioeconomic disadvantage and residential instability) predict drug activity, which, in turn, leads to higher levels of criminal violence (aggravated assault and robbery). Violence rates may be higher where gang problems are especially serious. Several studies of individuals' increased violence rates while involved in gangs suggest this possibility.

At the individual level, Bjerregaard's (2010) research to date supports each of three potential connections among involvement in gangs, drugs, and violence. The first possibility is that gang membership promotes involvement in drugs and that both gang membership and drug involvement in turn increase violence. Second, drug involvement or violent behavior may precede gang membership. Third, gang membership, drug involvement, and violent behaviors all occur simultaneously and are simply manifestations of the same antecedent factors.

In a nationwide test of the relationships between individual gang membership, drug involvement, and violence in the National Longitudinal Survey of Youth, Bjerregaard (2010) found that gang membership influences drug involvement in teenagers, though less than anticipated. Misleading impressions of greater gang–drug involvement are reported when the sequencing of gang, drug, and violence involvement is not separated in the analysis.[5] Across several longitudinal studies with high-risk, large-city samples, Krohn and Thornberry (2008) found drug use and involvement in drug sales increase with gang membership and decrease when youth leave gangs, but drug sales remain elevated in some sites. This excellent review of longitudinal studies also finds that "the weight of the evidence suggests that street gangs do facilitate or elicit increased involvement in delinquency, violence and drugs" (p. 147).

For example, in the Pittsburgh study, Loeber and colleagues (2008) concluded that in the youngest cohort, "drug dealing and gang membership in late childhood significantly increased the risk of violence in late adolescence and dealing and gun carrying in early adolescence significantly increased the risk of violence in late adolescence" (p. 163). In this study, "drug dealing, gang membership, and gun carrying were strongly related to serious violent offending over time. . . . Gang members who carried guns had about three to five times the risk of later violence even with controls for violence" (p. 163). The most plausible explanation is that "criminal offending, as well as drug use, drug dealing, gangs, and gun carrying, are often concentrated in neighborhoods that are poor, densely populated, and racially segregated and have a transient population" (p. 160).

Other studies show that youth substance use, drug dealing, and gang homicide co-occur (Loeber et al., 2008), as discussed in Chapter 5. A long-term and very comprehensive Pittsburgh study of child, adolescent, and young adult crime reveals that violence is most likely to occur where drug trafficking, gun ownership and use, and gang activity intersect (Loeber et al., 2008). These are commonplace components of gang activity in large cities where gangs have a longer history, are more entrenched in communities, and tend to be heavily armed (Howell, 1999). It is in this setting that intergroup rivalries are more intense, and violent, producing what Block and Block (1993) identify as "peaks and valleys" in homicides and other violence. Indeed, "much gang violence appears to have a contagious character, spreading from one neighborhood to another and outliving the initial source of the problem" (Decker, 2007, p. 398). Notwithstanding the gravity of this form of gang violence, most violent incidents may well occur within gangs, between members and factions—involving status considerations within interpersonal and group dynamics that are discussed subsequently.

In sum, at the individual level, gang involvement typically promotes participation in violence, drug use, and drug trafficking and perhaps prolongs gang member involvement in drug sales, and these crimes can co-occur. But the sequencing can vary from one neighborhood or community to another, from one city to another, and from one cohort to another in the same city.

Returning Gang Inmates

Current knowledge of prison gangs is scant, as indicated previously. Moreover, it is widely recognized that national prison data seriously underestimate the proportion of inmates who are gang involved (because of inmates' reluctance to divulge their gang affiliations). However, in recent years, the issue of gang members returning from secure confinement has received greater attention, in part because of the growing numbers of inmates who are released annually. Sabol, Minton, and Harrison (2007) estimate that nearly 700,000 prison inmates arrive in communities throughout the

United States each year. The proportion of these that are gang involved is unknown. A pioneering Illinois study has shed some light on this matter.

In Olson and Dooley's (2006) statewide study of more than 2,500 adult inmates released from Illinois prisons during 2000, nearly one-quarter of them were identified as gang members. Three-fourths (75%) of the gang members had a new arrest versus 63% of the nongang inmates in the follow-up period of almost 2 years. Gang members also were rearrested more quickly following release from prison, and were more likely to be arrested for violent and drug offenses than were nongang members. More than half (55%) of the gang members were readmitted to Illinois prisons within the 2-year follow-up period, compared to 46% of the nongang members. Of those cited for a violation of the term of parole, gang members were more likely than nongang members to receive a violation for an arrest for a more serious offense (52% versus 42%, respectively). Similar findings emerged from Huebner, Varano, and Bynum's (2007) study of a sample of 322 young men aged 17 to 24 years released from prison in a midwestern state: "As hypothesized, men who were involved in a gang or were drug dependent before entering prison had higher reconviction rates and recidivated more quickly than men who did not report involvement in gangs or drug use" (p. 208).

The return of gang members from prison to communities is a noticeable problem for approximately two-thirds of the gang problem jurisdictions nationwide (Egley et al., 2006). This is not a new issue, however. Of the agencies reporting the return of gang members from confinement in 2001, nearly two-thirds (63%) reported that returning members "somewhat" or "very much" contributed to an increase in violent crime among local gangs; 69% reported the same for drug trafficking. The impact of returning members is much greater on localities with more long-standing gang problems (Howell, 2006). In the view of law enforcement in the 2008 NYGS, returning gang inmates contribute rather evenly in the following ways, in descending order of importance: violent crime, drug trafficking, property crime, access to weapons, and dress and demeanor.

Intergang Conflict

Numerous studies show that most intergang violence, including homicides, is commonly related to several circumstances (maintenance of set space, interpersonal "beefs," normative or order violations, surprise assaults, and drug-turf disputes) and a variety of motives (defending one's identity as a gang member, defense of the gang's honor and reputation, set space trespassing, gang revenge or retaliation, personal spontaneous retaliation) (Braga, 2004; Hughes & Short, 2005; Papachristos, 2009; Tita et al., 2005). Interestingly, the bulk of intergang violence is directed against others of the same racial/ethnic background, often in battles for prominence or turf disputes.

The most detailed information on dispute pretexts among gangs and gang members comes from two studies in which research teams observed thousands of gang members' daily behaviors over several years. In the first one, a Chicago study, Hughes and Short's (2005) analysis of Short's observations in the 1960s of more than 2,600 incidents involving members of 20 lower or working-class gangs over 3 years revealed valuable insights on typical dispute pretexts.[6] When outcomes were examined, those that precipitated violence per se were largely retaliations (55%), followed by identity attacks (36%), and normative or order violations (31%).

In Focus 8.2
Gang Dispute Pretexts

1. *Norm violations* include "annoying" behavior, failure to fulfill an obligation, ignoring, causing or contributing to another's loss, boasting, cheating, unacceptable demeanor, and taking and/or violating another person's property.

2. *Noncompliance with an order* is begrudging acquiescence or (implicit or explicit) refusal to comply with demands to engage in or to not engage in a certain behavior.

3. *Actions that suggest a need to defend others* are behaviors directed toward one party that elicit a verbal or physical sanction from one or more observers.

4. *Money or debts* involve failure to repay a financial loan upon collection by the lender.

5. *Unfair or rough play* includes actions that are inconsistent with the explicit or implicit rules governing participation in an organized athletic contest.

6. *Identity attack* is any direct attack on personal or gang identity—including accusations, insults, challenges, and physical violations that do not result in bodily harm—and degrading rejections and yelling.

7. *Concern regarding opposite sex relations* involves behaviors that are upsetting because of the challenge they represent to another party's claim to a romantic interest.

8. *Territory or neighborhood honor* include actions undertaken by one party that directly threaten another's claim to a particular territory.

9. *Racial concerns* involve physical or verbal actions undertaken by one party against another solely because of race.

10. *Fun or recreational incidents* include physical or verbal actions that lead to hostilities during leisure activities such as "playing the dozens" and less organized body punching and horsing around.

11. *Retaliation* is action undertaken by one party to avenge a previous attack or any other perceived or actual wrongdoing by the targeted party or a close associate.

12. *Rumors* include physical or verbal actions undertaken by one party to seek redress for negative information allegedly disseminated by the targeted party.

13. *Robbery* involves intimidating or violent actions undertaken by one party to acquire money or property from another.

14. *Misunderstanding* includes intended or actual actions directed at one or more persons who are mistakenly identified or accused of a wrongdoing that did not occur.

15. *General troublemaking* is physical or verbal harassment for no apparent reason.

16. *Other pretexts* include physical or verbal actions undertaken for other reasons, such as doing what someone else asked, indirectly harassing one person or group of persons by harming others, and preempting an anticipated attack.

Source: Hughes & Short, 2005, pp. 48–49

Earlier, W. B. Miller's (1966) pioneering study of Boston gangs revealed the nature of gang-based retaliations. Over a 2-year period of field observation of 200 members of seven violent gangs by field workers, approximately 54,000 behaviors and interpersonal sentiments were recorded. Although violent intergang conflicts were extremely rare, Miller found that "gang members fight to secure and defend their honor as males; to secure and defend the reputation of their local area and the honor of their women; and to show that an affront to their pride and dignity demands retaliation" (p. 112).

Firearm Possession and Use

Many gang studies show that the felt need for protection is the primary reason gang members own or carry an illegal gun (Lizotte et al., 1994). "It is therefore not surprising," explain Lizotte and colleagues, "that when boys join gangs, the probability of carrying a hidden gun is substantially increased" (p. 829). Beginning in the 1970s, youth gangs were reported to have more weapons of greater lethality (W. B. Miller, 1982/1992). Stretesky and Pogrebin (2007) assert that gang-related gun violence can be understood in terms of self and identity that are heavily rooted in self-concepts of masculinity and reinforced by other gang members. Bjerregaard and Lizotte (1995) add that gangs are more likely to recruit adolescents who own a gun, and gang members are more than twice as likely as nongang members to own a gun for protection. "Gang members carry guns to protect themselves and their turf from rival gangs, who, in turn, must arm" (Lizotte et al., 2000, p. 830). Furthermore, Block and Block note that gun availability fuels gang violence. Lizotte and colleagues (2000) explain, "If one travels in a dangerous world of youth armed illegally and defensively with firearms, it only makes sense to carry a gun. Therefore, peer gun ownership for protection increases the probability of gun violence" (p. 830).

In the Rochester study cited here, during early adolescence, younger boys who are members of a gang have a higher probability of gun carrying. "In late adolescence, involvement in serious drug trafficking, independent of gang involvement, is a much stronger factor explaining hidden gun carrying"—typically after gang involvement has ended (Lizotte et al., 2000, p. 829). In this study and in the Pittsburgh longitudinal delinquency study referenced earlier in this chapter, extensive involvement in drug sales and gun carrying typically come later in the life course, as demonstrated in Pittsburgh and discussed in Chapter 5.

Violent Hot Spots and Gang Set Space

Violent hot spots and gang turf often co-occur, and these neighborhood places are said to have criminal *trajectories* in their own right (Griffiths & Chavez, 2004; Tita & Cohen, 2004). In Chicago in the mid-1990s—at the height of gang violence, Block (2000), explains,

Even the largest street gangs were active only in limited areas. Police, community residents, and gang members perceived these activity areas, "turfs," differently. Some gangs are ephemeral. They may occupy a fixed area for only a few months. For other gangs, a core turf may exist for generations, [but overall, gang] boundaries are constantly shifting, are extremely tedious to map. (p. 369)

In the course of interviews with gang members, Tita, Cohen, and Endberg (2005) asked them to map places where they came together as a sociological group; that is, their set space. This study indicates that the location of gang set space is usually a very small geographic area, much smaller than neighborhoods or even census tracts. Block's (2000) study showed that multiple-gang activity was observed in just 4.5% of the total grid squares citywide ($N = 25,744$), however, "these squares accounted for 23.0% of the assaults and 44.3% of all drug-related incidents " (p. 379). Papachristos's (2009) Chicago study demonstrates that these homicidal patterns of gang conflict are extremely stable over time. "Gang members come and go, but their patterns of behavior create a network structure that persists and may very well provide the conduit through which gang values, norms, and culture are transmitted to future generations" (p. 119).

Group Process

As discussed in Chapter 5, the typical gang is constantly in flux. As long as the gang is in an inactive state, that is, not attacked by a rival, even fringe members can keep it alive. Jankowski (1991) observed that the Los Angeles Mexican American gangs became violently active when the hard-core element grew in size or when an age cohort with a large hard-core membership moved into a central leadership position. "If the hard core is large, then the gang will be violently active, but if the hard core is small, then the gang will be violently inactive, regardless of the static structure" (137). Cohesive gangs challenge more rival gangs and, in turn, are more frequently challenged themselves.

In prior research, Klein (1969) lists six key connections between group process and gang violence:

1. Gang process selectively crystallizes factors leading to violent behavior.

2. The gang legitimates displays of aggression.

3. The gang inadvertently leads to violence on behalf of the gang because this mistakenly appears to be expected.

4. The gang permits or reinforces the commission of violent acts.

5. The gang helps to draw attention to immediate as opposed to long-range consequences of activity.

6. The gang is a medium for the status-deprived member to seek peer group status. (pp. 1143–1144)

This is not to say that gangs routinely encourage their members to "get out and be violent" (Klein, 1969, p. 1144). Indeed, some gangs are not involved in violence at all, and in other cases the gang acts as a constraining force on its members, prohibiting or discouraging behavior which otherwise would result in discomfort for the group as a whole. "The gang more commonly does nothing so much as wonder what to do" (p. 1144). Spergel (1966) described this situation as follows: "A dozen youngsters suddenly find themselves walking down the street to a fight, and eight or ten or even all of them, individually, may be wondering why he is there" (p. 115).

The importance of group process in gangs and the implications of this for the violent character of many gangs, according to Decker (2007), cannot be overstated. Decker (1996) delineated a seven-step process to accounts for observed gang violence spurts—episodic gang conflicts that wax and wane and sometimes extend over a number of years (Block and Block, 1993). The process Decker (1996) observed in St. Louis begins with a loosely organized gang:

1. Loose bonds to the gang

2. Collective identification of threat from a rival gang (through rumors, symbolic shows of force, cruising, and mythic violence), reinforcing the centrality of violence that expands the number of participants and increases cohesion

3. A mobilizing event, possibly, but not necessarily, violence

4. Escalation of activity

5. Violent event

6. Rapid de-escalation

7. Retaliation (p. 262)

As Horowitz (1983) explains, "In seeking to protect and promote their reputation, gangs often engage in prolonged 'wars,' which are kept alive between larger fights by many small incidents and threats of violence" (p. 94). One gang may claim *precedence,* which means that the other group must challenge that gang if they want to retain their honor and reassert their reputation. "Whatever the 'purpose' of violence, it often leads to retaliation and revenge creating a feedback loop where each killing requires a new killing" (Decker & Van Winkle 1996, p. 186), often "spreading from one neighborhood to another and outliving the initial source of the problem" (Decker, 2007, p. 398).

Similarly, Papachristos (2009) found that murders spread in the neighborhood through a process of social contagion as gangs respond to threats from other gangs. Specific murders, particularly those in public view, have the intended effect of threatening the social status and ranking of groups within the neighborhood. Thus, in order to avoid subjugation themselves, other gangs "must constantly (re)establish their social status through displays of solidarity—in this case, acts of violence—which, in turn, merely strengthen these murder networks" (p. 76). Papachristos's (2009) findings "suggest that gangs are not groups of murderers per se, but rather embedded social networks in which violence ricochets back and forth . . . [and] what begins as a single murder soon generates a dozen more as it diffuses through these murder networks" (p. 76). These events become, in effect, "dominance contests" in which "violence spreads through a process of social contagion that is fueled by normative and behavioral precepts of the code of the street" (p. 81).

Concluding Observations

Three main patterns are seen in the trajectory modeling of larger cities' gang problem histories over the 14-year period: (1) no change, (2) increasing gang activity, and (3) decreasing gang activity. This finding raised the question of whether even larger cities might have more consistent serious gang problem histories, particularly with gang violence. Further examination of very large cities' (populations in excess of 100,000 persons) percent of homicides that are gang related, revealed that more than 7 out of 10 reported a consistently high level or increasing proportion of gang-related homicides over the 14-year period. Surprisingly, none of the groups of very large cities evidenced a decreasing pattern. Gang problems are more persistent in very large cities than smaller ones, particularly in cities with populations of 250,000 or more that report gang activity every year. But nearly 8 out of 10 cities with populations of 100,000 or more report gang activity with the same consistency.

Therefore, gang activity and its associated violence remains an important and significant component of the U.S. crime problem. While it has been reasonably assumed that gang-related violence would be following the overall dramatic declines in violent crime nationally, analyses reported in this chapter and by Howell and associates (2011) is overwhelming evidence to the contrary—that is, gang violence rates have continued at exceptional levels over the past decade *despite* the remarkable overall crime drop. Gang violence that is rather commonplace in very large cities seems largely unaffected by, if not independent from, other crime trends—with the possible exceptions of drug trafficking and firearm availability.

The number of gang members, number of gangs, and other structural characteristics are important indicators of serious gang activity in large cities. Both the number of gangs and gang members predict gang homicide in very large cities, particularly where gang subgroupings exist, based on age, gender, and geographical area.

Several contextual factors inflame gang set space, including drug trafficking, inmates returning from prison, intergang conflict, and firearm possession and use. Set space (as originally noted in Chapter 3, where gang members most frequently hang out) serves as a sort of "lightning rod" for intergang violence (Tita & Ridgeway, 2007). Group processes, especially contagion, can enliven gang members and unify them in collective violence. At the individual level, gang involvement promotes or facilitates individual participation in violence, drug use, and drug trafficking and perhaps prolongs gang member involvement in drug sales, and these crimes can co-occur.

The influence of each of the indicators of serious gang activity and contextual factors associated with gang violence can vary from one neighborhood or community to another, from one city to another, and from one gang to another in the same city. Serious gaps remain with respect to prison gangs and the impacts of returning gang-involved inmates.

DISCUSSION TOPICS

1. Why are gang problems in small cities, towns, villages, and rural counties seldom persistent?

2. What might account for disappearance of gang activity?

3. Why do gang problems become increasingly more serious in some cities and not in others?

4. What are some distinctive features of gang activity in the two case studies (Pittsburgh and St. Louis)?

5. What are the implications of the case studies' unique features for intervention?

RECOMMENDATIONS FOR FURTHER READING

Serious Violent Gangs

Braga, A. A., Kennedy, D. M., & Tita, G. E. (2002). New approaches to the strategic prevention of gang and group-involved violence. In C. R. Huff (Ed.), *Gangs in America III* (pp. 271–285). Thousand Oaks, CA: Sage.

Decker, S. H. (2007). Youth gangs and violent behavior. In D. J. Flannery, A. T. Vazsonyi, & I. D. Waldman (Eds.), *The Cambridge handbook of violent behavior and aggression* (pp. 388–402). Cambridge, MA: Cambridge University Press.

Decker, S. H., Katz, C. M., & Webb, V. J. (2008). Understanding the black box of gang organization: Implications for involvement in violent crime, drug sales, and violent victimization. *Crime and Delinquency, 54,* 153–172.

Decker, S. H., & Pyrooz, D. C. (2010b). On the validity and reliability of gang homicide: A comparison of disparate sources. *Homicide Studies, 14,* 359–376.

Griffiths, E., & Chavez, J. M. (2004). Communities, street guns and homicide trajectories in Chicago, 1980–1995: Merging methods for examining homicide trends across space and time. *Criminology, 42,* 941–975.

Howell, J. C. (1999). Youth gang homicides: A literature review. *Crime and Delinquency, 45,* 208–241.

Howell, J. C. (2006). *The impact of gangs on communities* (NYGC Bulletin No. 2). Tallahassee, FL: National Youth Gang Center.

Hutson, H. R., Anglin, D., Kyriacou, D. N., Hart, J., & Spears, K. (1995). The epidemic of gang-related homicides in Los Angeles County from 1979 through 1994. *Journal of the American Medical Association, 274,* 1031–1036.

Klein, M. W., & Maxson, C. L. (1989). Street gang violence. In M. E. Wolfgang, & N. A. Weiner (Eds.), *Violent crime, violent criminals* (pp. 198–234). Newbury Park, CA: Sage.

Lien, I.-J. (2005a). Criminal gangs and their connections: Metaphors, definitions, and structures. In S. H. Decker & F. M. Weerman (Eds.), *European street gangs and troublesome youth groups* (pp. 31–50). Lanham, MD: AltaMira Press.

Lien, I.-J. (2005b). The role of crime acts in constituting the gang's mentality. In S. H. Decker & F. M. Weerman (Eds.), *European street gangs and troublesome youth groups* (pp. 105–125). Lanham, MD: AltaMira Press.

Maxson, C. L. (1999). Gang homicide: A review and extension of the literature. In D. Smith & M. Zahn (Eds.), *Homicide: A sourcebook of social research* (pp. 197–220). Thousand Oaks, CA: Sage.

Maxson, C. L., Gordon, M. A., & Klein, M. W. (1985). Differences between gang and nongang homicides. *Criminology, 23,* 209–222.

Papachristos, A. V. (2009). Murder by structure: Dominance relations and the social structure of gang homicide. *American Journal of Sociology, 115,* 74–128.

Tita, G. E., & Abrahamse, A. (2004). Gang homicide in LA, 1981–2001. *Perspectives on Violence Prevention, 3,* 1–18.

Tita, G. E., & Abrahamse, A. (2010). *Homicide in California, 1981–2008: Measuring the impact of Los Angeles and gangs on overall homicide patterns.* Sacramento, CA: Governor's Office of Gang and Youth Violence Policy.

Tita, G. E., Cohen, J., & Engberg, J. (2005). An ecological study of the location of gang "set space." *Social Problems, 52,* 272–299.

Tita, G. E., & Ridgeway, G. (2007). The impact of gang formation on local patterns of crime. *Journal of Research in Crime and Delinquency, 44,* 208–237.

Crack Cocaine "Epidemic"

Brownstein, H. (1996). *The rise and fall of a violent crime wave: Crack cocaine and the social construction of a crime problem.* Guilderland, NY: Harrow and Heston.

Cockburn, A., & St. Clair, J. (1998). *Whiteout: The CIA, drugs and the press.* London: Verso.

Cork, D. (1999). Examining space-time interaction in city-level homicide data: Crack markets and the diffusion of guns among youth. *Journal of Quantitative Criminology, 15,* 379–406.

Golub, A., & Johnson, B. D. (1997). Crack's decline: Some surprises among U.S. cities. *Research in Brief.* Washington, DC: National Institute of Justice.

Grogger, J., & Willis, M. (1998). *The introduction of crack cocaine and the rise in urban crime rates.* National Bureau of Economic Research Working Paper No. W6353. Cambridge, MA: National Bureau of Economic Research.

Hartman, D. A., & Golub, A. (1999). The social construction of the crack epidemic in the print media. *Journal of Psychoactive Drugs, 31,* 423–433.

Reeves, J. L., & Campbell, R. (1994). *Cracked coverage: Television news, the anti-cocaine crusade, and the Reagan legacy.* Durham, NC: Duke University.

Sampson, R. J. (2008). Rethinking crime and immigration. *Contexts, 7,* 28–33.

Pittsburgh Case Study

Cohen, J., Cork, D., Engberg, J., & Tita, G. E. (1998). The role of drug markets and gangs in local homicide rates. *Homicide Studies, 2,* 241–262.

Cohen, J., & Tita, G. E. (1999). Spatial diffusion in homicide: Exploring a general method of detecting spatial diffusion processes. *Journal of Quantitative Criminology, 15,* 451–493.

Tita, G. E. (1999). *An ecological study of violent urban street gangs and their crime.* Unpublished dissertation. Pittsburgh, PA: Carnegie Mellon University.

Tita, G. E., & Cohen, J. (2004). Measuring spatial diffusion of shots fired activity across city neighborhoods. In M. F. Goodchild & D. G. Janelle (Eds.), *Spatially integrated social science* (pp. 171–204). New York: Oxford University Press.

Tita, G. E., Cohen, J., & Endberg, J. (2005). An ecological study of the location of gang "set space." *Social Problems, 52,* 272–299.

Tita, G. E., & Griffiths, E. (2005). Traveling to violence: The case for a mobility-based spatial typology of homicide. *Journal of Research in Crime and Delinquency, 42,* 275–308.

Tita, G. E., & Ridgeway, G. (2007). The impact of gang formation on local patterns of crime. *Journal of Research in Crime and Delinquency, 44,* 208–237.

Studies on Gang Retaliations

Block, C. R., & Block, R. (1993). Street gang crime in Chicago. *Research in Brief.* Washington, DC: U.S. Department of Justice, National Institute of Justice.

Decker, S. H. (2007). Youth gangs and violent behavior. In D. J. Flannery, A. T. Vazsonyi, & I. D. Waldman (Eds.), *The Cambridge handbook of violent behavior and aggression* (pp. 388–402). Cambridge, MA: Cambridge University Press.

Hughes, L. A., & Short, J. F. (2005). Disputes involving gang members: Micro-social contexts. *Criminology, 43,* 43–76.

Papachristos, A. V. (2009). Murder by structure: Dominance relations and the social structure of gang homicide. *American Journal of Sociology, 115,* 74–128.

Tita, G. E., Riley, K. J., & Greenwood, P. (2005). *Reducing gun violence: Operation Ceasefire in Los Angeles.* Washington, DC: National Institute of Justice.

Valdez, A., Cepeda, A., & Kaplan, C. (2009). Homicidal events among Mexican American street gangs: A situational analysis. *Homicide Studies, 13,* 288–306.

Gang Features and Structure

Decker, S. H., & Pyrooz, D. C. (2010b). On the validity and reliability of gang homicide: A comparison of disparate sources. *Homicide Studies, 14,* 359–376.

Esbensen, F., Winfree, L. T., He, N., & Taylor, T. J. (2001). Youth gangs and definitional issues: When is a gang a gang, and why does it matter? *Crime and Delinquency, 47,* 105–130.

Katz, C. M., Webb, V. J., & Schaefer, D. (2000). The validity of police gang intelligence lists: Examining differences in delinquency between documented gang members and non-documented delinquent youth. *Police Quarterly, 3,* 413–437.

Pyrooz, D. C., Fox, A. M., & Decker, S. H. (2010). Racial and ethnic heterogeneity, economic disadvantage, and gangs: A macro-level study of gang membership in urban America. *Justice Quarterly, 14,* 1–26.

Vigil, J. D. (1993). The established gang. In S. Cummings & D. J. Monti (Eds.), *Gangs: The origins and impact of contemporary youth gangs in the United States* (pp. 95–112). Albany: State University of New York Press.

Notes

1. This analysis was performed by Arlen Egley, Jr., Administrator and Co-Principal Investigator of the NYGS, reported in Egley and Howell (2011) and Howell, Egley et al. (2010).

2. A census tract is considered to be a reasonable approximation of a "neighborhood" or a "community" (Griffiths & Chavez, 2004, p. 942).

3. Studies in several established gang cities show that when multiple gangs are active within specific geographic areas, violent, drug, and property crime rates are higher—especially violent crimes, in Chicago (Block, 2000; Papachristos, 2009), St. Louis (Decker & Van Winkle, 1996; Monti, 1993), Boston (Braga, Papachristos, & Hureau, 2010), and Pittsburgh (Tita & Cohen, 2004; Tita, Cohen, & Endberg, 2005; Tita & Ridgeway, 2007).

4. Squares with no gang activity had an average of 2.88 assaults. Those with one active gang averaged 13.5 assaults. Those squares with four gangs averaged 42.7 assaults in 1996 (Block, 2000, p. 379).

5. For example, see Bellair & McNulty, 2009.

6. These data come from Hughes and Short's reanalysis of disputes that Short & Strodtbeck's (1965/1974) team of detached workers observed over a 3-year period between 1959 and 1962.

What Works

Gang Prevention

Introduction

This chapter provides information on effective strategies and programs for preventing gang involvement. Chapter 5 explained how delinquent behavior and gang involvement unfold over time as risk factors accumulate. The next three chapters provided information on girls in gangs (Chapter 6), the continuing presence of gangs across the United States and other countries (Chapter 7), and the greater seriousness of gang problems in very large cities (Chapter 8). Taken together, these four chapters strongly suggest that the prevention of gang activity is of paramount importance. Gang problems exist to some extent in many countries—albeit to a less serious degree (with the exception of Mexico and Central American countries)—thus each jurisdiction that experiences gang activity should first assess its gang problem before launching prevention initiatives. Chapter 10 focuses on integrating intervention and suppression strategies with prevention, with particular emphasis on comprehensive frameworks for addressing gang problems.

The purposes for devoting an entire chapter to gang prevention are threefold. First, juvenile delinquency precedes gang involvement, which necessitates more expanded systemwide strategies. Second, communities should integrate delinquency prevention and gang prevention strategies and programs, forming a continuum. Third, there are some complicated contexts that prevention programs and strategies must address, particularly the ambiguity that typically is associated with very young gangs and gang members, and bullying behaviors in schools.

How to systematically assess gang activity in schools and in the community is described in the first section of this chapter. The next section presents a framework for organizing delinquency prevention and early intervention programs and strategies. The final section provides information on promising and effective prevention programs.

Gang Intervention: Risk-Focused, Data-Driven, and Research-Based Gang Prevention

The likelihood of successfully preventing gang activity is greater if community initiatives are focused on risk factors and are data driven and research based. This approach involves first, determining the presence of research-based risk factors for gang membership (provided in Chapter 5) along with an assessment of the scope and nature of gang activity, and second, providing a continuum of evidence-based programs.

Community Assessments

Before starting a program for delinquency and gang prevention, a community should conduct a gang-problem assessment to identify elevated risk factors that lead to child delinquency and gang involvement. Several studies strongly indicate that communities must define youth gangs, locate them, and identify and target the youth who are at greatest risk of joining (Howell, 2010a). Because every community has its own characteristics, each must agree on a unique definition that will guide its data collection and strategic planning. Chapter 3 presented a practical definition that should be considered as a guide for the assessment:

- The group has five or more members.
- Members share an identity, typically linked to a name and often other symbols.
- Members view themselves as a gang, and are recognized by others as a gang.
- The group associates continuously, evidences some organization, and has some permanence.
- The group is involved in an elevated level of criminal activity.

Stakeholders in every community that conducts an assessment should agree upon the definition that will guide its data collection and strategic planning. This definition will serve as a useful point of departure.

As part of its Comprehensive Gang Model, OJJDP has published *A Guide to Assessing Your Community's Youth Gang Problem,* a user-friendly resource to assist communities that are conducting a gang-problem assessment (Office of Juvenile Justice and Delinquency Prevention, 2009a). This guide simplifies the data-collection process, helping communities determine types and levels of gang activity, gang crime patterns, community perceptions of local gangs and gang activity, and gaps in community services for gang prevention. In Vance County, North Carolina, a community organization, the Juvenile Crime Prevention Council (JCPC), conducted its first countywide assessment of gang activity using the OJJDP guide (Vance County Gang Assessment Project, 2010). The JCPC has begun using this excellent self-study (which was conducted entirely by county residents) as a basis for developing a continuum of prevention and intervention programs and strategies with guidance provided in the OJJDP (2009b) *Comprehensive Gang Model: Planning for Implementation.* This initiative is part of the statewide North Carolina Department of Juvenile Justice and Delinquency Prevention implementation of the Comprehensive Gang Model.

Ideally, the assessment should provide an understanding of the "evolution of gangs in time and space" within the city, community, or neighborhood (Hughes, 2006). To help communities understand their unique gang situation, an assessment should answer these questions:

- Who is involved in gang-related activity and what is the history of these gangs?
- What crimes are these individuals committing?
- When are these crimes committed?
- Where is gang-related activity primarily occurring?
- Why is the criminal activity happening (e.g., individual conflicts, gang feuds, gang members acting on their own)?

In addition to helping communities answer questions about gang emergence, OJJDP's Comprehensive Gang Model promotes a problem-solving approach to gang-related crime, asking communities to identify:

- Neighborhoods with many risk factors for gang involvement
- Schools and other community settings in which gangs are active
- Hot spots of gang crime
- High-rate gang offenders
- Violent gangs

Although this cannot be done quickly, it would be helpful if the assessment could provide at least preliminary information on the structure of the most active gangs. For prevention purposes, it is important to know age and gender composition, and the relative size of gangs. In less violent crime territories, social gangs or school-based gangs are also likely to be found, on which prevention and intervention strategies should be focused.

To assist with program planning and development, the OJJDP Strategic Planning Tool (www.nationalgangcenter.gov/SPT/Programs/110) provides the following:

- A list of risk factors for delinquency and gang membership organized by age
- Data indicators (i.e., measures of risk factors)
- Data sources (from which relevant data can be retrieved)
- Hyperlinks connecting risk factors with effective programs that address them
- A "Community Resource Inventory," for community planning groups to record information on existing programs
- Information on promising and effective juvenile delinquency and gang programs
- Strategies that address specific risk factors for various age groups

The next step is to identify program gaps and develop and coordinate a continuum of prevention and intervention program services and sanctions, in concert with a targeted strategy of community and government agency responses to serious and violent gang activity. Prevention and intervention services should be directed to the neighborhoods, schools, and families from which gangs emanate. Before presenting the prevention and early intervention framework, it is important to consider two complicated contexts within the framework that prevention programs and strategies must address—starter gangs and bullying.

Starter Gangs

Rather than immediately joining serious, violent street gangs, most youth become involved in less criminal groups referred to here and earlier as *starter gangs*. The term intentionally underscores this important point: prevention work needs to focus in particular on newly formed gangs along with at-risk youngsters in both neighborhood and school settings.

But how do the starter gangs form? As noted in Chapter 3, Huff (1989) specified three ways: (1) from conflict between adolescent groups at regularly scheduled competitive events that festers and grows; (2) from similar conflict that develops among young groups in public gathering places such as malls and on the streets, and other social gatherings; and (3) when a previously gang-involved youth moves into town. In addition, the youngest cliques in larger gangs might be considered starter gangs. Examples include the "pee wee" and "wannabe" cliques in established gangs.

In Focus 9.1
How the San Diego Del Sol Gang Formed

One of the original members of Del Sol explained how he and a number of other boys formed their gang. In their teenage years, they began going to parties. Other youths at the parties would ask them, "Where are you from?" Such queries can be challenges or simple questions. However, when asked by members of other gangs, the questions are typically challenges. To avoid conflict, the reply "I'm not from anywhere" means the respondent does not claim a gang—at least not at the moment. It also is a way of losing some face and may result in being attacked anyway for being a punk (a weakling).

After being roughed up at several encounters, a group of Del Sol boys (named after a low-income housing development) decided that the next time they went to a party together, they would "carry" (bring weapons). If anyone asked them where they were from, they would reply "*Del Sol. Y que?*" The use of "Y que?" in the context of Mexican American gangs means, "And what are you going to do about it?" It is an unequivocal challenge. If they were attacked, they would bring out their weapons and fight back. After a number of encounters where challenges were made and answered, Del Sol became recognized as a gang.

Source: Sanders, 1994, p. 43

Sanders's (1994) account of how a group came to be recognized as a gang and also to view themselves as a gang as a result of conflict is a common example. Other avenues to gang formation are common, of course. Starter gangs described in this chapter often form as a result of experimenting with gang culture and symbols. Others are formed in barrios in which belonging to a gang in one's neighborhood is expected (Vigil, 1988).

Child delinquents are more likely to participate in starter gangs than other children— or to initiate group formation themselves. Prior involvement in both delinquency and aggression increases this likelihood. Indeed, as Craig and colleagues (2002) suggest, the problem behaviors associated with becoming a gang member appear to emerge before age 10. The neighborhood context (e.g., lack of adult supervision and associating with older delinquents) elevates risk of gang exposure. Klein and Crawford (1967) first explained this in a pioneering study: "When a number of boys in a neighborhood withdraw from similar sets of environmental frustrations and interact with one another enough to recognize, and perhaps generate common attitudes, the seeds of the group are sown" (p. 67). Vigil (1993) delineates seven steps that are often involved in the gang joining process. These steps apply to children in the school setting, beginning in elementary school and unfolding as the child enters middle school. Vigil's steps presume that gangs already exist in and around the school. Vigil's seven steps to gang joining are as follows:

1. In elementary school, children may have heard about gangs and are thus not surprised by them when they see these barrio groups in middle school. They already have heard about conflicts involving them in the barrio and at the schools that they soon will enter.

2. Seeing these gangs at middle school for the first time validates their importance, and provides impetus for incoming children's own gang affirmation.

3. At middle school, children commonly are coping "with the various groupings and cliques there that have carved out their social niches. Gangs are already evident at this level, and the school yard has its separate barrio gang hangouts where the youngsters gravitate" (p. 102).

4. The most vulnerable children enter middle school with poor academic achievement, and their family strains and street exposure make matters worse. "As they become more and more involved in the oppositional subculture, they become increasingly disdainful of teachers and school officials—and in the process become budding dropouts" (p. 103).

5. As intergroup conflicts increase in intensity and frequency, an individual's reputation spreads as a known and committed gang member. At middle school, ready-made friends and protectors await the uneasy child, some of whom are known to him, and he begins to gravitate to those that appear to provide some protection.

6. The child or adolescent who joins the gang may feel compelled to do so, for to be isolated is to invite disaster. Faced with the prospect of belonging to nothing and no one, youngsters in this situation may feel that they *must* join the gang, "even though the requisites for membership are quite demanding and life threatening" (p. 104).

7. Youths may first casually associate with a gang, perhaps a chance bonding. Walking home from school with friends, a child might stop along the way with older friends and perhaps engage in delinquency in the company of gang members with whom he had been hanging out. Mutual acceptance is followed by an initiation—a "baptism" into the gang—he is now committed to it.

Gang activity is not confined to the school grounds themselves, of course. It begins to escalate very early on school days, once students begin to gather in the neighborhood, at bus stops, or on the way to school, and gang-related violence peaks earlier in the day (Figure 9.1) on school days than on nonschool days and also earlier than other violent crimes involving juveniles (Wiebe, Meeker, & Vila, 1999). Modern-day school environments may give rise to gangs and more gang activity when student bonds to schools and teachers are broken by poor school climates characterized by zero tolerance policies, less support for students in overcoming individual risk factors, harsh one-size-fits-all punishments, and elevated school suspension and expulsion rates.

The gang-forming process likely begins with a small group of rejected-aggressive children. Cairns and Cairns (1994) reason that aggressive and antisocial youths begin to affiliate with one another in childhood, while Kupersmidt, Coie, and Howell (2003), as well as Warr (2002) find this pattern of aggressive friendships continues through adolescence. Fleisher (1995) notes an internal process characterizing rejected children: "The group behavior of rejected children follows the internal social processes that typify children's school playgroups. Boys and girls tacitly classify themselves into those

Figure 9.1 Gang Offenses per Time of Day

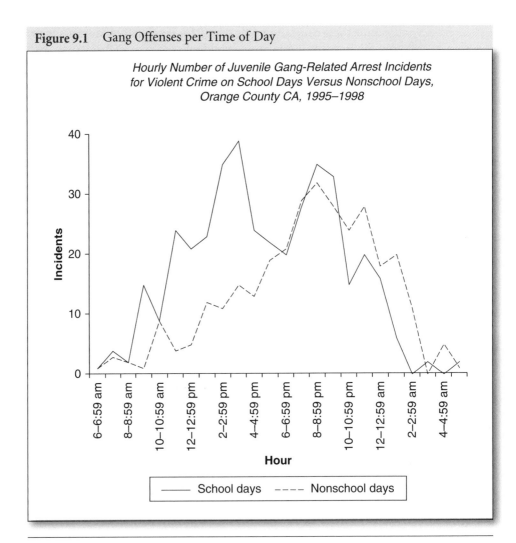

Source: Wiebe, Meeker, & Vila, 1999. Reprinted with permission.

who are rejected, neglected, accepted, aggressive, unpopular, disliked and liked, and so on" (p. 119). Social rejection by prosocial youths, according to Coie and Dodge (1998), may serve to channel rejected-aggressive children toward deviant peer groups.

Based on a French study, Debarbieux and Baya (2008) suggest that some starter gangs may emerge in "difficult" schools. This research identified current or future gang members among a small subgroup of "highly rebellious" students "who consider that everything is wrong with school and that teachers are awful, who commit aggression more often than others, who are punished repeatedly and more frequently than the others, and who have developed a feeling of hatred and rejection of everything that represents order" (p. 214). These students (ages 12 and older) most often attended the most "difficult" schools—those characterized by greater levels of student victimization, self-reported violence, poor student-teacher relations, and systems of punishment that pupils did not accept well. This core group (4% to 5% of all students) was responsible for most disorder and violence in 16 schools that were studied. This subgroup contained a significant proportion of gang members—up to 11% in the most "difficult" schools.

> This does not mean that they constitute a "group," or the beginnings of a gang. They could be isolated individuals although [many do] identify with each other and form a real group. If, initially, they were not considered gang members, their being on the margins of the school norms slowly leads them to identify themselves as a group. (p. 215)

Ongoing association away from school is the next phase of gang development. A major consequence of estrangement in school (feelings of powerlessness, exclusion, and fear for one's safety) is what Hutchison and Kyle (1993) believe is a strong identification with the immediate peer group in the barrio who share the same feeling and experience. Fleisher (1998) adds that incipient members of the newly formed groups begin to display signs of involvement:

> Like the groups they left behind at school, rejected boys and girls adopt the equivalent of distinctive school clothing and colors, the insignia of membership. Members learn group cheers, rhymes, and folklore, wear group clothing, engage in rites of passage and intensification, uphold communal values (like school children's loyalty to their school) and they give themselves a name. (p. 119)

In Focus 9.2
Behaviors Associated With Joining a Gang

- Negative changes in behavior, such as
  o withdrawing from family
  o declining school attendance, performance, or behavior
  o staying out late without reason

(Continued)

(Continued)

- o unusual desire for secrecy
- o confrontational behavior, such as talking back, verbal abuse, name calling, and disrespect for parental authority
- o sudden negative opinions about law enforcement or other adults in positions of authority (e.g., school officials or teachers)
- o changes in attitude about school, church, or other normal activities or change in behavior in these activities

- Unusual interest in one or two particular colors of clothing or a particular logo
- Interest in gang-influenced music, videos, and movies
- Use and practice of hand signals to communicate with friends
- Peculiar drawings or gang symbols on schoolbooks, clothing, notebooks, or even walls
- Drastic changes in hair or dress style or having a group of friends who have the same hair or dress style
- Withdrawal from longtime friends and forming bonds with an entirely new group of friends
- Suspected drug use, such as alcohol, inhalants, and narcotics
- Presence of firearms, ammunition, or other weapons
- Nonaccidental physical injuries such as being beaten or injuries to hands and knuckles from fighting or a gang initiation
- Unexplained cash or goods, such as clothing or jewelry

Source: National Gang Center (*Parents' Guide to Gangs*)

Vigil (1993) asserts that regular group involvement in illegal activity is the next step toward gang formation. Exclusion from school may further facilitate crossing the tipping point toward gang formation, but as Thrasher (1927/2000) noted, the group does not become a gang until it begins to excite disapproval and opposition. Parents or neighbors may look upon it with suspicion or hostility; the storekeepers or the cops may begin to chase it; contact is made with a rival or an enemy in a nearby gang; or some representative of the community steps in and tries to break it up. "This is the real beginning of the gang, for now it starts to draw itself more closely together" (p. 10). Klein (1995) identifies two "signposts" that indicate actual "tipping." The first one is a commitment to a criminal orientation or willingness to use violence. The second signpost is when the gang-to-be takes on a collective criminal orientation as a group, "a gang," that is set apart from other groups in the community. "Now the delinquent group has become a gang" (Fleisher, 1995, p. 119).

If gang activity is not curbed in its earliest stages, more serious gang activity can develop. Gangs are present in many schools in the United States. The *National Survey of American Attitudes on Substance Abuse XV: Teens and Parents* (National Center on Addiction and Substance Abuse, 2010) found that 45% of high school students and 35% of middle school students say that there are gangs or students who consider themselves to be part of a gang in their schools. During the 2007–2008 school year, 43% of public high schools and 35% of middle schools experienced discipline problems related to gangs at least once (Robers, Zhang, Truman, & Snyder, 2010).

Figure 9.2 illustrates the clustering of gangs in and around schools as indicated by the location of gang-related crimes in Guilford County, North Carolina. The spatial distribution of gang-related crimes in this figure illustrates the dual presence of young gangs inside schools and within a 1-mile radius around schools in the community. Interestingly, traditional school security measures (security guards, metal detectors, locker checks, and the like) do not appear to be solutions, in and of themselves, to gang and other disruptive problems at school (Howell & Lynch, 2000). Other interventions need to be implemented in schools and in the surrounding community to address gang activity.

Figure 9.2 Guilford County Schools and Gang-Related Incidents, January–December 2009

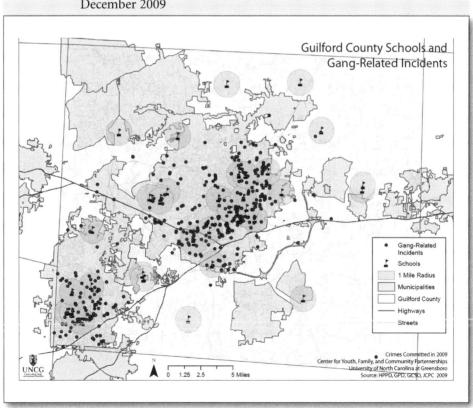

Source: Graves, K.N., Ireland, A., Benson, J., DiLuca, K., Chiu, K., Johnston, K., Varner, L., McCoy, S., & Sechrist, S. (2010, September). *Guilford County Gang Assessment: OJJDP Comprehensive Gang Assessment Model.* Greensboro, NC: Center for Youth, Family, and Community Partnerships, University of North Carolina at Greensboro. Reprinted by permission of the illustrator, Kelly N. Graves.

Bullying

Student bullying, as Lassiter and Perry (2009) suggest, is a potential contributor to gang joining in and around schools, and particularly when it makes children feel unsafe. Bullying should not be disregarded; this can be a serious problem in which children and adolescents use power through frequent acts of aggression to intimidate and control others, and make them feel powerless in their relationships (Pepler, Madsen, Levene, & Webster, 2004). During the 2007–2008 school year, Robers and fellow researchers (2010) found 44% of public high schools, 22% of middle schools, and 21% of primary schools experienced discipline problems related to student bullying at least once a week or daily. Larochette, Murphy, and Craig (2010) assert that most bullying at school appears to be racially motivated, and Mexican Americans and African Americans are the most common victims, depending on the location.

Interestingly, bullying behavior causes aggressive and violent behavior, rather than the other way around (Kim et al., 2006), although aggressive youth are actively involved, to be sure (White & Loeber, 2008). In this way, bullying may lead to more serious offenses (Figure 9.3), may encourage possible gang involvement, or is symptomatic of starter gangs. Early intervention in problem behaviors such as bullying are more likely to be successful than delaying action until physical violence occurs.

In a study involving 15 known gang-problem schools across the United States, Carbone-Lopez, Esbensen, and Brick (2010) found that girls and boys are equally affected and that boys were more likely to experience direct or physical forms of bullying while girls were more likely to report being teased or joked about. The research team recommended that "prevention and intervention efforts can be gender-neutral in their approach. In particular, efforts to reduce students' involvement in delinquent and criminal activity should also reduce their risk of being bullied at or en route to school" (p. 343).

In another revealing study, Baldry and Farrington (2007) employed *meta-analysis* of all types of evaluated bullying prevention programs. Using statistical procedures to synthesize and compare clusters of studies on a given topic, this enabled the researchers to examine a wider range of program evaluations in a more systematic manner than is possible in a conventional program-by-program review. Baldry and Farrington concluded that "the effectiveness of anti-bullying programs is not proven" (p. 201; Tofi, Farrington & Baldry, 2008). Furthermore, Ferguson, Miguel, Kilburn, and Sanchez (2007) concluded that the positive effects are too small to have practical benefits. However, Olweus, Limber, and Mihalic (1999) note that the highly regarded Bullying Prevention Program has demonstrated additional benefits including significant improvements in the "social climate" of the class, as reflected in students' reports of improved order and discipline, and reduced reports of general antisocial behavior, such as vandalism and fighting—outcomes which greatly enhance its practical benefits.

Osher, Dwyer, and Jackson (2004) recommend that threat assessment be considered a component of a comprehensive approach toward maintaining a safe school. Such an assessment can identify students who may need additional services as well as more general problems in the school environment, such as bullying, that merit broader

Figure 9.3   The Youth Violence Continuum

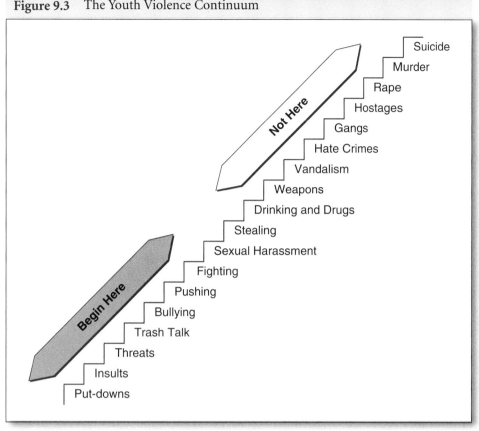

Source: Lassiter, W. L., & Perry, D. C. (2009). Reprinted with permission.

attention. The Virginia model for student threat assessment (Cornell & Sheras, 2006) proves most successful to date for this purpose.

A Framework for Prevention and Early Intervention

Windows of opportunity for delinquency prevention and early intervention are shown in Figure 9.4. Because gang membership is presented as a pathway to serious and violent delinquency, delinquency prevention programs must work to target gang involvement. As explained in Chapter 5, the top section of the figure shows the major risk factor domains that influence youth: family, school, peer group, individual characteristics, and community. At birth—or beginning in the prenatal period for some infants—the biological family is the central influence on infants and children. During preschool, and especially in elementary school and onward, the array of risk factors expands as some children are exposed to negative influences outside the home (particularly school problems and delinquent peers). Family, school, and peer influences continue from childhood to young adulthood, although family influences gradually fade as friends become more important. In addition, individual characteristics and community factors can come into play at any point during childhood and adolescence. It is important to provide an array of prevention and intervention programs across the entire continuum that address each of the risk factor domains.

Strategies for Delinquency and Gang Intervention and Prevention

This section focuses on broad intervention strategies or methods of addressing gang involvement and the big picture of continuum building—in a manner that reduces risk across the multiple contexts. This is a useful framework for organizing community programs because it helps engage agency and community representatives in the collective enterprise of continuum building. The next section contains recommended programs that address the specific problem behaviors in Figure 9.3. Readers are reminded that protective factors are not addressed in this chapter. As explained in Chapter 5, the research foundation on these is inadequate at this time.

Primary prevention programs and strategies are needed to target the entire youth population in high-crime and high-risk communities, particularly in reducing risk factors for delinquency.[1] *Secondary prevention programs and strategies* are needed to target youths at risk of gang involvement. Each of these components helps to reduce the number of youths who join gangs. *Intervention programs and strategies* are needed to provide sanctions and services for younger youths who are actively involved in gangs

Figure 9.4 Windows of Opportunity for Delinquency Prevention and Early Intervention

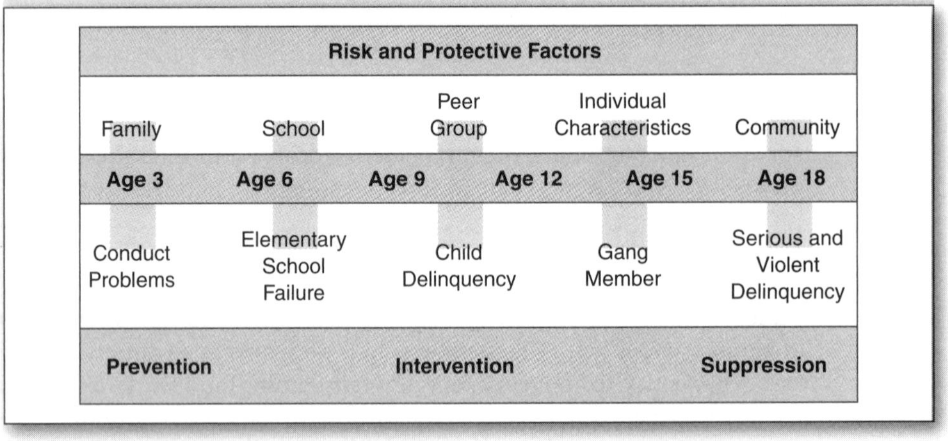

Risk and Protective Factors				
Family	School	Peer Group	Individual Characteristics	Community
Age 3	**Age 6**	**Age 9**	**Age 12**	**Age 15**	**Age 18**
Conduct Problems	Elementary School Failure	Child Delinquency	Gang Member	Serious and Violent Delinquency
Prevention		**Intervention**		**Suppression**

Source: Howell, 2009, p. 151

to separate them from gangs. If the intervention strategies address risk and protective factors at or slightly before the developmental points at which they begin to predict later gang involvement and other problem behaviors, they are more likely to be effective (Institute of Medicine, 2008). These components are integrated in the Office of Juvenile Justice and Delinquency Prevention's *Comprehensive Gang Prevention, Intervention, and Suppression Model* (National Gang Center, 2010a). (This model and both intervention and suppression programs are described in Chapter 10.)

Several large-scale initiatives have been undertaken in America's communities aimed at radically modifying communitywide economic conditions or at least improving the lot of residents most adversely affected while reducing crime. The latest massive governmental initiative to rescue residents of inner cities from the clutches of poverty is the Moving to Opportunity for Fair Housing Demonstration (MTO) described in Chapter 6. Unlike previous grand schemes that aimed to help residents moderate the effects of poverty, the MTO helped them leave the disadvantaged public housing environments. This experiment proved to be a remarkable success for girls, whose lives were buffered from crime and gangs, but generally a failure for boys, who maintained gang connections and criminal activity. The MTO experiment suggests that high-quality mother-daughter relationships can provide valuable insulation for ordinary at-risk girls. But can such relationships buffer girls who are "beyond risk" for multiple negative outcomes such as the San Antonio girls identified in the upcoming discussion? What can be done to promote positive youth development for girls in this circumstance when moving out of the neighborhood to better opportunities is not a realistic option?

Building a Continuum of Effective Delinquency Prevention and Early Intervention Programs

This section illustrates the process of building countywide, citywide, or statewide continuums of delinquency prevention and early intervention programs. Implementing effective programs is not as simple as pulling a model program off the shelf (Lipsey, Howell, Kelly, Chapman, & Carver, 2010). Three main approaches can be taken to translate the body of research evidence about effective programs into practice for everyday use by practitioners and policy makers, what commonly is called *evidence-based practice* (EBP) (Lipsey et al., 2010). In each of these approaches, studies must meet very high standards of methodological quality. The first approach is to conduct an evaluation of each program and, if it is found ineffective, to use that evidence to improve or terminate it. This approach is cost-prohibitive in statewide initiatives. The second approach is to draw on lists of model programs, with evidence of effectiveness, as certified by an authoritative source. "A third approach is to implement a type of program that has been shown to be effective on average by a meta-analysis of many studies of that program type, but to do so in the manner that the research indicates will yield that average effect or better" (p. 17). The Campbell Collaboration is at the forefront of systematic reviews in criminology and criminal justice. Several of these reviews provide guidelines for juvenile justice programs.

In Focus 9.3
Repositories of Research-Based Delinquency Prevention Programs

- OJJDP Strategic Planning Tool: www.nationalgangcenter.gov/SPT/
- Model Programs Guide (Office of Juvenile Justice and Delinquency Prevention): www.dsgonline .com/mpg2.5/mpg_index.htm
- Blueprints for Violence Prevention: www.colorado.edu/cspv/index.html
- National Registry of Effective Programs and Practices (Substance Abuse and Mental Health Services Administration, U.S. Department of Health and Human Services): http://nrepp.samhsa.gov
- Exemplary and Promising Safe, Disciplined and Drug-Free Schools Programs (U.S. Department of Education): www.ed.gov/admins/lead/safety/exemplary01/index.html
- What Works Clearinghouse, U.S. Department of Education (on educational interventions, some of which address youth violence and substance abuse prevention): http://ies.ed.gov/ncee/wwc/
- Campbell Collaboration (systematic reviews in criminology and criminal justice): www .campbellcollaboration.org/library.php

---

A major caveat is in order. "At present, we know relatively little about the effects of taking evidence-based programs to scale in public health and related areas of mental health, education, welfare, and criminal justice" (Lipsey et al., 2010, p. 13). For example, major shortcomings exist in achieving high fidelity with evidence-based substance abuse and violence prevention programs in community settings (Fagan, Hanson, Hawkins, & Arthur, 2008). In addition, in two national assessments, Gottfredson and Gottfredson (2001) and Hallfors and Godette (2002) found poor implementation for many delinquency and violence prevention programs that schools attempted to adopt. Transporting *exemplary, blueprint,* or otherwise *model programs* into everyday practice settings also has proved to be more difficult than originally assumed. Lipsey and colleagues (2010) explain,

> There are many challenging issues associated with translating an evidence-based program into routine practice in a way that closely replicates the relevant circumstances of the original research. As a result, the desirable program effects on delinquency and subsequent offending found in the research studies often are attenuated when those programs are scaled up for general application. (p. 19)

Numerous research-based program options are available that communities should consider for inclusion in their continuum of delinquency prevention and early intervention programs. In Focus 9.3 contains links to key repositories of research-based delinquency prevention programs, including those mentioned in the subsequent sections. Because of the availability of these electronic databases, and the comprehensive information they contain on programs, individual programs need not be described in detail here. Readers are reminded that our main focus is on programs that prevent early development of juvenile delinquency and other problem behaviors. By addressing a variety of risk factors for gang involvement, these programs can make a contribution to preventing gang membership. Among the many effective delinquency prevention programs, attention is drawn here to those that stand out as holding considerable promise in the gang field as a result of having successfully targeted high-risk youth and families in high-risk settings. In addition, the reviewed programs often serve racial/ethnic minorities and girls.

The remainder of this chapter illustrates how effective delinquency prevention programs can be integrated with gang programs to form a continuum in accord with the developmental model presented in Chapter 5. Figure 5.3 illustrates a more specific way of viewing programs that should help program developers and managers grasp the interconnection of programs at these more targeted levels of intervention, which are organized by age and setting. The four levels are preschool programs, school entry, later childhood, and early adolescence.

Preschool Programs

Preschool intellectual enrichment and child skill training programs generally target children who have low intelligence and attainment. The main goals of these programs are improved cognitive skills, school readiness, and social and emotional development.

For very early intervention, the combination of home visiting and parent training is the most effective approach with high-risk families. The most widely recognized program, Nurse-Family Partnership (NFP), provides first-time, low-income mothers of any age with home visitation services from public health nurses. Nurse home visitors work with families in their homes during pregnancy and the first 2 years of the child's life. Importantly, involvement of other family members and people in the mother's social network is emphasized. (For additional information, see the Blueprints for Violence Prevention, In Focus 9.3.) For very young abused children, trauma-focused cognitive behavioral therapy (TF-CBT) is designed to help 3- to 18-year-olds and their parents overcome the negative effects of traumatic life events such as child sexual or physical abuse (accessible at the National Registry of Effective Programs and Practices, In Focus 9.3).

The Perry Preschool is an excellent preschool intellectual enrichment program. It has even proved effective in reducing serious and chronic delinquency in long-term follow-ups. This program provides high-quality education for disadvantaged children ages 3 to 4 to improve their capacity for future success in school and in life. The HighScope curriculum emphasizes an open approach to learning; children are active participants. The follow-up of experimental groups at age 27 showed that significantly fewer program group members than nonprogram group members were chronic offenders (Schweinhart et al., 2005). In addition, study subjects had fewer arrests at age 19, significantly fewer arrests for drug manufacturing and drug distribution offenses and less involvement in serious fights, gang fights, causing injuries, and police contact. (For additional information, see the OJJDP Model Programs Guide, In Focus 9.3.)

School Entry

Family settings are particularly difficult to penetrate for service delivery. Lerman and Pottick (1995) contend that youth and family service agencies rarely make parents in troubled families aware of available services and how to access them. Child welfare, social services, mental health, education, and juvenile justice agencies must serve as portals for information access and assistance. Preventing gang involvement among children who are alienated from their very own families and schools—particularly in communities characterized by concentrated disadvantage—is a formidable challenge. A single program for families cannot be expected to have large impacts. Numerous programs tied together in a continuum, serving multiple children and families simultaneously with integrated services are required to achieve noticeable success under these conditions. Several strategies for school entry youth and their parents are recommended.

- Identify local family-related factors that may be related to gang involvement, including family structure, native language, immigration status, economic issues, and educational attainment levels.
- Provide parental gang awareness training.
- Educate youth to modify their perception that gang membership is beneficial.
- Increase parental supervision and monitoring of children.
- Improve parents' involvement in and support for their children's academic progress.
- Provide family-strengthening or effectiveness training
- Support first-time, low-income mothers with education, nutrition, health, safety, and human service resources.

Schools And Families Educating Children (SAFEChildren) is a family-focused preventive intervention designed to increase academic achievement and decrease risk for later drug abuse and associated problems such as aggression, school failure, and low social competence. Initially targeting first-grade children and their families living in inner-city Chicago neighborhoods, SAFEChildren has two components. The first component is a multiple-family group approach that focuses on parenting skills, family relationships, understanding and managing developmental and situational challenges, increasing parental support, skills and issues in engaging as a parent with the school, and managing issues such as neighborhood problems (e.g., violence). Families participate in 20 weekly sessions (2 to 2.5 hours each) led by a trained, professional family group leader. The second component is a reading tutoring program for the child. Among families designated as high risk, Tolan, Gorman-Smith, and Henry (2004) found improved parenting skills, and among children also assessed as high risk, there was a decrease over time in child aggression and hyperactivity in participants compared to controls. (For additional information, see the National Registry of Effective Programs and Practices, In Focus 9.3.)

The Incredible Years Series is a set of three comprehensive, multifaceted, and developmentally based curricula for parents, teachers, and children designed to promote emotional and social competence and to prevent, reduce, and treat behavior and emotion problems in young children ages 2 to 10, at risk for, or presenting, conduct problems (defined as high rates of aggression, defiance, oppositional, and impulsive behaviors). Multiple randomized control group evaluations of the child training series indicate significant increases in children's appropriate cognitive problem-solving strategies and more prosocial conflict management strategies with peers, and reductions in conduct problems at home and school. (For additional information, see the Blueprints for Violence Prevention, In Focus 9.3.)

The Toronto-based Child Development Institute's Under 12 Outreach Project (ORP) was developed as an intervention for child delinquents. Augimeri and colleagues (2006) explain how ORP employs a multisystemic approach, combining interventions that target the child, the family, and the child-in-the-community. The program uses a variety of established interventions: skills training; training in cognitive problem solving, self-control strategies, cognitive self-instruction; family management skills training; and parent training. Both the ORP and its parallel gender-sensitive program for girls, Earlscourt Girls Connection (EGC), are fully manualized and are in various stages of replication in the United States, and in other countries. (For additional information, see the OJJDP Strategic Planning Tool, In Focus 9.3.)

PeaceBuilders is a schoolwide violence prevention program for elementary and middle schools (K–8). A high school program is also being piloted in several locations. The program incorporates a strategy to change the school climate created by staff and students and is designed to promote prosocial behavior among students and adults. Children learn six simple principles: (1) praise people, (2) avoid put-downs, (3) seek wise people as advisers and friends, (4) notice and correct hurts that you cause, (5) right wrongs, and (6) help others. Adults reinforce and model behaviors at school, at home, and in public places. Flannery and colleagues' (2003) evaluation found that at follow-up, there was an 89% decrease in physical aggression. Further, an 82% decrease in verbal aggression was observed. (For additional information, see the OJJDP Model Programs Guide, In Focus 9.3.)

The Montreal Preventive Treatment Program is an early intervention program that has demonstrated an impact on gang involvement even though it was not developed with this purpose in mind. It was designed to prevent antisocial behavior among boys, ages 7 to 9, of low socioeconomic status who had previously displayed disruptive problem behavior in kindergarten. This program demonstrated that a combination of parent training and childhood skill development can steer children away from gangs. Tremblay, Masse, Pagani, and Vitaro's (1996) evaluation of the program showed both short- and long-term gains, including less delinquency, less substance use, and less gang involvement at age 15.

Striving Together to Achieve Rewarding Tomorrows (CASASTART) is an effective family- and school-centered program that was designed to keep high-risk 8- to 13-year-old youth free of substance abuse, delinquency, and gang involvement. CASASTART works through a partnership between the lead agency (that drives the collaborative process), schools, and the police. The CASASTART partnership centers around three main agencies: (1) the lead agency (that drives the collaborative process), (2) schools, and (3) the police. Each case manager serves 15 to 18 children and their families for a 2-year period. There are eight program components including mentoring, family services, education services, and case management. Compared with control group youth, CASASTART clients were less likely to report at follow-up the use of any drugs or gateway drugs, involvement in drug trafficking, and violent acts. (For additional information, see the OJJDP Model Programs Guide, In Focus 9.3.)

School systems across the United States are implementing the federal Office of Special Education Program's Positive Behavioral Interventions and Supports (PBIS) for improving and responding more constructively to problem behavior. This is a three-level framework, intended for all school levels, that is structured from low to high intensity. Dwyer and Osher (2000) describe the program as follows:

- *Level 1—Build a schoolwide foundation* that targets all students. The main goal is to create a positive social culture in schools in which prosocial behaviors are explicitly taught and reinforced for all students. Research at the University of Oregon's Institute on Violent and Destructive Behavior suggests that most schools with effective schoolwide systems that focus on learning and behavior can prevent at least 80% of problematic student behaviors.
- *Level II—Provide early intervention strategies* that target students who may be at risk for violence or disruption—this includes approximately 10% to 15% of students who continue to experience behavioral problems even when schoolwide interventions are in place. Youth served at this level should meet several of the early warning criteria. Of course, immediate attention would be given to youth who have made serious threats of violence.
- *Level III—Provide intensive intervention* that targets students who have already engaged in disruptive or violent behavior. This is the remaining 5% to 10% of children who experience significant emotional and behavioral problems. Specific interventions and their intensity depend on the nature, severity, and frequency of each child's emotional and behavioral problems. Students who meet any of the imminent warning signs should receive help or control measures immediately, although cautions should be exercised to avoid an overreaction.

This is a promising framework with the potential of improving school climate and possibly linking those students with serious behavioral problems to needed services, although Kutash, Duchnowski, and Lynn (2006) note that results to date do not qualify PBIS as an evidence-based practice. Most experts in the field agree that PBIS is in its infancy, and the most promising results have been found when PBIS was implemented in conjunction with functional behavioral assessments of troubled students for serious behavioral problems.

Osgood and Anderson (2004) identify the afterschool period as a critical opportunity for delinquency prevention because the lack of adult supervision of youngsters is conducive to delinquency. A meta-analysis of after-school programs (Gottfredson et al., 2004) found that middle school programs reduced delinquency but elementary school programs did not. The most effective programs were structured ones that focused on social competency (e.g., developing self-control) or interpersonal skill development.

A RAND review of research on afterschool programs (Beckett, Hawken, & Jacknowitz, 2001) identified 17 good practice (research-based) standards for successful programs (e.g., educational attainment, emotional development, and health). More recently, Gottfredson, Cross, and Soule's (2007) statewide evaluation of afterschool programs in Maryland produced some important findings. Smaller well-structured programs (i.e., a published curriculum), and programs that employed more highly educated staff produced reductions in one or more problem behaviors. (For additional information on this curriculum, see the Blueprints for Violence Prevention, In Focus 9.3.)

Later Childhood and Early Adolescence

Second to protection, what both boys and girls at greatest risk of gang involvement need most are services. A wide variety of substance abuse prevention programs is available in evidence-based program repositories listed in the In Focus 9.3 box. (See especially the National Registry of Effective Programs and Practices.) LifeSkills Training (LST) is a program that seeks to influence major social and psychological factors that promote the initiation and early use of substances. LifeSkills has distinct elementary (8- to 11-years-old) and middle school (11- to 14-years-old) curricula that are delivered in a series of classroom sessions over 3 years. The sessions use lecture, discussion, coaching, and practice to enhance students' self-esteem, feelings of self-efficacy, ability to make decisions, and ability to resist peer and media pressures by teaching drug resistance skills, personal self-management skills, and general social skills. (See the University of Colorado Blueprints, In Focus 9.3.)

Numerous effective family strengthening programs benefit girls as well as boys (Kumpfer & Alvarado, 1998). Brief strategic family therapy (BSFT) is a short-term, problem-focused intervention with an emphasis on modifying maladaptive patterns of interactions that has proved effective with Mexican American families and those of other races and ethnicities (Robbins & Szapocznik, 2000), and also with aggressive girls (Nickel, Luley, Nickel, & Widermann, 2006). A major BSFT goal is to improve family functioning, including effective parental leadership and management, positive parenting, and parental involvement with the child and his or her peers and school. Sessions are conducted at locations that are convenient to the family, including the family's home in some cases. The target population in general is children and adolescents between 8 and 17 years of age displaying or at risk for developing behavior problems, including substance abuse, conduct problems, delinquency, sexually risky behavior, aggressive or violent behavior, and association with antisocial peers (accessible at the National Registry of Effective Programs and Practices, In Focus 9.3).

Close parent-child bonds reduce the likelihood of running away. Dedel (2006) relays most runaways are older teenagers, ages 15 to 17, with only about one-quarter ages 14 and younger. Juveniles of different racial/ethnic origins run away at about the same rates, and boys and girls run away in equal proportions. Although juveniles from all socioeconomic statuses run away, the majority are from working-class and lower-income homes, possibly because of the additional family stress created by a lack of income and resources. The family strengthening programs reviewed previously can help prevent running away. The Incredible Years Series is another highly regarded program for this age-group. It consists of a set of three comprehensive, multifaceted, and developmentally based curricula for parents, teachers, and children, designed to promote emotional and social competence and to prevent, reduce, and treat behavioral and emotional problems in young children. Early conduct problems, which this program prevents, predict running away. (See the University of Colorado Blueprints, In Focus 9.3.)

The Gang Resistance Education and Training (GREAT) Program is a school-based gang prevention curriculum for girls and boys. In uniform, law enforcement officers teach the curriculum in entire classrooms of mainly middle school students during a 13-week course. In addition to educating students about the dangers of gang involvement, the lesson content places considerable emphasis on cognitive-behavioral training, social skills development, refusal skills training, and conflict resolution. Along with the middle school curriculum, the GREAT Program includes three other components: (1) an elementary school curriculum, (2) a summer program, and (3) families training. Esbensen and colleagues (2011) preliminary results from a 7-city experimental evaluation of the revised G.R.E.A.T. Program (only one year following treatment) are positive overall. The program appears to have short-term effects on the intended goals of reducing gang involvement and improving youth-police relations (more positive attitudes about police), as well as on interim risk or skills. Specifically, compared to non-G.R.E.A.T. students, the G.R.E.A.T. students were more likely to report frequent use of refusal skills, greater resistance to peer pressure, and less positive attitudes about gangs.

Gottfredson, Gottfredson, Payne, and Gottfredson (2005) state the first priority for bonding students to schools is a safe school. Safety and security derive from two conditions: (1) an orderly, predictable environment where school staff provide consistent, reliable supervision and discipline; and (2) a school climate where students feel connected to the school and supported by their teachers and other school staff. A balance of structure and support requires, according to Lassiter and Perry (2009), an organized, schoolwide approach that is practiced by all school personnel. Misguided school policies such as zero tolerance break the bond between students and their school and teachers because such blanket rules cannot be enforced evenly, especially when harsh punishments are attached to violations. When adolescents feel cared for by people at their school and feel like a part of their school, they are less likely to use substances, engage in violence, or initiate sexual activity at an early age. Students who feel connected to school also report higher levels of emotional well-being.

Responding in Peaceful and Positive Ways (RiPP) is a school-based violence prevention program designed to provide students ages 10 to 14 in middle and junior high schools with conflict resolution strategies and skills. It combines a classroom curriculum of social/cognitive problem solving with real-life skill-building opportunities such as peer mediation. Program outcomes include fewer disciplinary violations for violent offenses and school suspensions, and fewer fight-related injuries (Meyer, Farrell, Northup, Kung, & Plybon 2000). Overall, RiPP helps improve school climate because it has been proven to decrease school disciplinary code violations for violent behaviors and to decrease student-reported frequency of drug use, violent behavior, and fight-related injuries. This program has demonstrated some impact on involvement in gang activity by addressing the relationship between self-image and gang-related behaviors. (For additional information, see the National Registry of Effective Programs and Practices, In Focus 9.3.)

Disruptive behaviors are damaging to healthy school climates. But these can be addressed effectively with a large variety of evidence-based delinquency programs. Wilson, Lipsey, and Derzon (2003) reviewed 221 studies of school-based interventions for aggressive or disruptive behavior by students and found that well-implemented demonstration programs can be highly effective. This systematic review and synthesis provides new information on the types of programs that have proved effective and also common features of them and key factors in their implementation. The most effective programs, in descending order of effectiveness, are

- Behavioral, classroom management
- Therapy, counseling
- Schools within schools (separate schools or classrooms)
- Academic services
- Social competence, cognitive–behavioral component
- Social competence, no cognitive–behavioral component

Some of these programs were implemented in preschool, but most of them were placed in elementary schools, followed by middle schools and high schools. Interestingly, programs that targeted very young children (5 years and under in pre-school and kindergarten) and adolescents (14 years and up in high schools) tended to show greater reductions in aggression—more than programs for elementary and middle school children. Programs that targeted high-risk students tended to show greater reductions in aggression. In addition, programs were most effective in settings where the base rates of aggressive behavior were high. Otherwise, the effects of school-based programs did not vary greatly with the age, gender, or ethnic mix of the research samples.

While Teske and Huff (2011) suggest that communities commonly lack *alternative schools* and suspended or expelled students often are not placed in them, Wilson and Lipsey (2007) find that *schools within schools* that successfully address aggressive and disruptive behavior employ universal programs and targeted programs for selected or indicated children have proven effective. This category includes alternative schools that may award General Education Development (GED) diplomas. However, priority should be given to serving students with histories of truancy, suspensions, and dropout because of their elevated risk for gang involvement.

School systems are increasingly difficult to access with delinquency and gang services, in large part because of the resistance to acknowledging a gang problem, and the plethora of surrounding myths. The National Gang Center recommends the following school intervention strategies (Arciaga et al., 2010):

- Build a comprehensive framework for the integration of child and adolescent services that links the juvenile justice system with human service and other related agencies, including schools, child welfare services, mental health agencies, and social services.

- Create an infrastructure consisting of client information exchange, cross-agency client referrals, a networking protocol, interagency councils, and service integration.
- Target potential and current serious, violent, chronic gang-involved juvenile offenders for resource priority.
- Provide case management by a particular agency for case conferencing and to coordinate services for offenders and the families of gang youth.
- Provide mentoring of at-risk and gang youths, counseling, referral services, gang conflict mediation, and anti-gang programs at schools in the community.
- Provide close supervision and monitoring of gang-involved youth by agencies of the juvenile or criminal justice system and also by community-based agencies, schools, and grassroots groups.
- Provide intensive probation supervision linked with more structured behavioral or skill-building and multimodal interventions.
- Provide direct placement and referral of youth for employment, training, education, and supervision.
- Provide alternatives to gang involvement, including remedial and enriched educational programs for gang youth with academic problems and vocational and apprentice training. (p. 4)

Community Safety

Community safety is the responsibility of adults on behalf of children and adolescents of all ages. Several promising community-centered programs should be considered. Every community should have violence prevention programs in its continuum, and specific services for children who are exposed to violence—and victims themselves—at a very young age in their homes, neighborhoods, and at school and going to and from school (Flannery et al., 2007). More attention needs to be given to within-gang victimization (Melde, Taylor, & Esbensen, 2009) and victimization of nongang youth by gang members (Rennison & Melde, 2009) in each of these contexts. Flannery and colleagues (2007) identify several program priorities:

- Mental health professionals should be placed in schools to immediately identify children who need services and deliver or coordinate them.
- Interventions must include families and peer groups.
- Prevention services must address risk and protection at multiple levels and across multiple systems.
- Prevention services must also give priority to development of positive coping skills, competencies, and problem-solving skills that will help children and adolescents deal effectively with high levels of exposure to violence and victimization. (p. 315)

The National Gang Center has recognized the Child Development-Community Policing Program as providing an effective structure for linking youth exposed to violence with needed services. This program, developed by the Yale Child Study Center in New Haven, Connecticut, in 1992, has been replicated in many communities across the United States. It teams police with mental health clinicians in providing interdisciplinary intervention to children and families who are victims, witnesses, or perpetrators of violent crimes (Office of Juvenile Justice and Delinquency Prevention, 1997). Interventions should also be based in hospital emergency rooms to help break the cycle of retaliatory gang violence. CeaseFire Chicago includes this component (described in Chapter 10).

A key component of the OJJDP Gang Reduction Program is one-stop resource centers that make services accessible and visible to members of the community (National Gang Center, 2010a). These services include prenatal and infant care, after-school activities, truancy and dropout prevention, and job programs. In Richmond, Virginia, the One-Stop Resource Center is located in the middle of an apartment complex that houses more than 4,000 residents. Many of Richmond's programs are housed in the center, including a free health clinic and computer lab for area youth. The *Movimiento Ascendencia* (Upward Movement), a promising program for Mexican American girls, operated in Pueblo, Colorado, in the early 1990s for the purposes of preventing them from joining gangs and reducing their gang involvement. Most of the girls served in this program were in need of prevention and intervention services. Williams, Curry, and Cohen (2002) report the program successfully provided a safe haven for girls in the target area, and a variety of other needed services.

The PanZOu Project in North Miami Beach serves at-risk and gang-involved or high-risk (mainly Haitian) youth and their families, providing a safe place for these youth to congregate and engage in healthy pro-social activities (National Gang Center, 2010a). Provided services include intensive case management, strengthening families classes, job skills development and placements, life skills and conflict resolution, links to individual or family counseling, academic support, and other services as needed. The major intervention services clients utilize are intensive case management, conflict resolution, job skills development, and placement and community service programs. Prevention clients are primarily engaged in the Strengthening Families Program (a best-practice model program implemented for the first time with Haitian families), the Early Literacy Program (targeted to serve high-risk children in first grade through third grade in an afterschool program), Developing Intelligent Voices of America (DIVA) Program (social, emotional, and behavioral competence development for at-risk young women ages 8 to 18), and Man Up (a similar program for boys ages 8 to 18). The program evaluation concluded that "the implementation of the OJJDP GRP comprehensive model was largely a success in North Miami Beach" (Hayeslip & Cahill, 2009, p. 268).

Both Fleisher (1998) and Vigil (2010) recommend supervised residential centers for the purpose of insulating girls from violent community contexts. These centers, as Fleisher envisions them, would have three specific objectives: "(1) to shelter and protect girls; (2) to provide job training, and job placement; and (3) to ensure a healthy start for gang girls' children" (p. 219). In short, these centers would serve as a one-stop resource for a variety of services and sources of assistance. Victimized and "beyond risk" girls should receive mental health screening and assessment, and are likely to benefit from cognitive-behavioral therapy, treatment for traumatic victimization and posttraumatic stress disorder, and other mental health services; health screening, health education, and basic health services; substance abuse services, and a full spectrum of on-grounds experiential learning and educational services (A. Valdez, 2007).[2]

To draw attention to services, Adler and colleagues (1984) strongly recommend family-level interventions with Mexican American families. For access to families, they suggest *block therapy*, a technique for engaging families that are in most need of assistance. A local priest or clergyman could initially convene these groups. This would be a very practical starting point for families of all racial/ethnic backgrounds.

Boys and Girls Clubs Gang Prevention Through Targeted Outreach (GPTTO) is another promising program for preventing gang involvement. The overall philosophy of the program is to give at-risk youths ages 6 to 18 what they seek through gang membership (e.g., supportive adults, challenging activities, and a place to belong) in an alternative, socially positive setting. Arbreton and McClanahan's (2002) evaluation concluded that more frequent GPTTO Club attendance is associated with the following positive outcomes: (1) delayed onset of one gang behavior (less likely to start wearing gang colors); (2) less contact with the juvenile justice system (less likely to be sentenced to a residential placement by the court); (3) fewer delinquent behaviors (less likely to steal and less likely to start smoking marijuana); (4) improved school outcomes (higher grades and greater valuing of doing well in school); and (5) more positive social relationships and productive use of out-of-school time (engaging in more positive afterschool activities and increased levels of positive peer and family relationships). (For additional information, see the OJJDP Model Programs Guide, In Focus 9.3.)

A Framework for Continuum Integration

Because considerable research indicates that gang involvement overlaps substantially with serious, violent, and chronic offending (Howell, 2003b, pp. 83–84), communities should integrate their gang programs with their continuum of delinquency programs. The recommended framework for this endeavor is the *Comprehensive Strategy for Serious, Violent, and Chronic Juvenile Offenders* (Figure 9.5) (Wilson & Howell, 1993, 1994; Howell, 2003a, 2003b, 2009). The Comprehensive Strategy is a framework for putting research into practice in the juvenile justice and human service field in order to improve outcomes. Thus, implementing it involves integrating evidence-based services derived from sound research, into state and local prevention and intervention systems, ideally in statewide approaches (Lipsey et al., 2010).

From a broader perspective, the Comprehensive Strategy is a two-tiered system for responding proactively to juvenile delinquency. In the first tier, delinquency prevention, youth development, and early intervention programs are relied on to prevent delinquency and reduce the likelihood that at-risk youth will appear in the juvenile justice system. If those efforts fail, then the juvenile justice system, the second tier, must make proactive responses by addressing the risk factors for recidivism and associated treatment needs of the offenders, particularly those with a high likelihood of becoming serious, violent, and chronic offenders.

More specifically, the Comprehensive Strategy framework is structured around seven levels of parallel program interventions and sanctions, moving from least to most restrictive, plus aftercare for youth released from secure facilities:

1. Community primary prevention programs oriented toward reducing risk and enhancing strengths for all youth

2. Focused secondary prevention programs for youth in the community at greatest risk but not involved with the juvenile justice system or, perhaps, diverted from the juvenile justice system

3. Intervention programs tailored to identified risk and need factors, if appropriate, for first-time minor delinquent offenders provided under minimal sanctions (e.g., diversion or administrative probation)

4. Intervention programs tailored to identified risk and need factors for nonserious repeat offenders and moderately serious first-time offenders provided under intermediate sanctions (e.g., regular probation)

5. Intensive intervention programs tailored to identified risk and need factors for first-time serious or violent offenders provided under stringent sanctions (e.g., intensive probation supervision or residential facilities)

6. Multicomponent intensive intervention programs in secure correctional facilities for the most serious, violent, and chronic offenders

7. Postrelease supervision and transitional aftercare programs for offenders released from residential and correctional facilities

A key feature of the Comprehensive Strategy is that it is a forward-looking administrative model, organized around risk management. Achieving this requires use of offender management tools (particularly risk assessment, needs assessment, and a disposition matrix) to ensure that resources are targeted on higher risk offenders. This will save money and improve program effectiveness, provided that offenders are carefully matched with needed services in case management plans that are well implemented. Quality assurance measures are needed to ensure that comprehensive treatment plans are executed in accordance with established timetables and service protocols.

The Comprehensive Strategy has been implemented successfully in several states (Howell, 2003b, 2009) including North Carolina, which also is the first state that has undertaken the integration of gang and delinquency programs statewide.

In 2006, the North Carolina Department of Juvenile Justice and Delinquency Prevention (DJJDP) provided 11 grants to counties for the purpose of implementing the prevention and intervention components of the Comprehensive Gang Model in a Gang Violence Prevention Program. Although the total project period was only 2 years, several sites made noteworthy progress in targeting gang activity on an ongoing basis, particularly the Charlotte-based Gang of One program, with an impressive array of prevention and intervention programs and services (www.gangofonecharlotte.org/).

In late 2009, North Carolina became the first state to begin implementing a statewide gang prevention and intervention initiative, the Community-Based Youth Gang Violence Prevention Project (North Carolina DJJDP, 2009). It was based in part on a statewide assessment that revealed an elevated level of gang activity in schools across the state (North Carolina Department of Juvenile Justice and Delinquency Prevention

Figure 9.5 The Comprehensive Strategy for Serious, Violent, and Chronic Juvenile Offenders

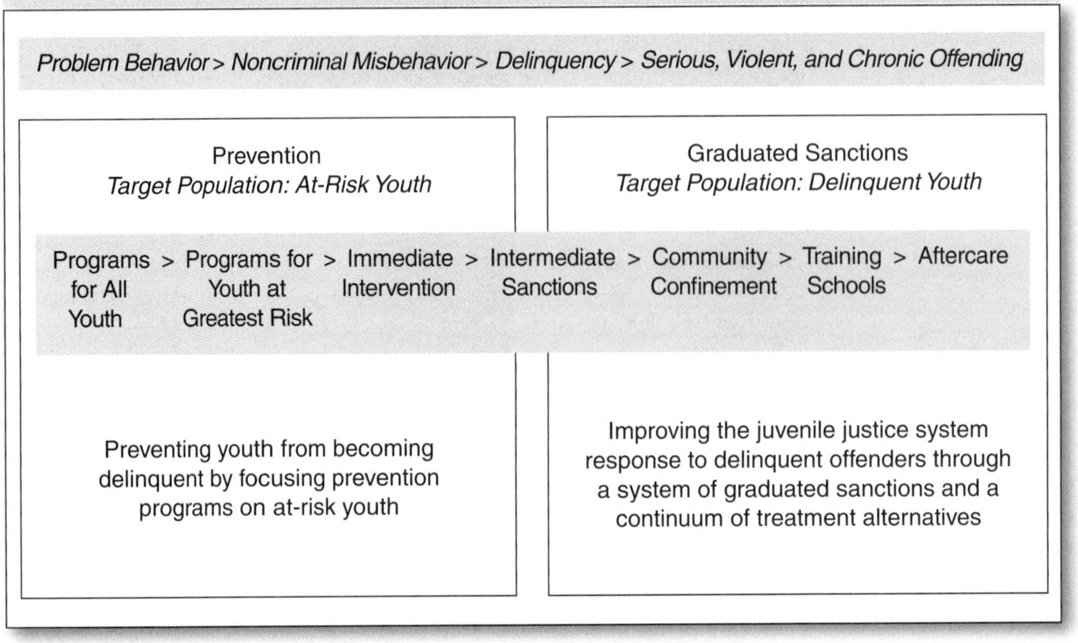

Problem Behavior > Noncriminal Misbehavior > Delinquency > Serious, Violent, and Chronic Offending

Prevention *Target Population: At-Risk Youth*	Graduated Sanctions *Target Population: Delinquent Youth*

Programs > Programs for > Immediate > Intermediate > Community > Training > Aftercare
for All   Youth at   Intervention   Sanctions   Confinement   Schools
Youth   Greatest Risk

Preventing youth from becoming delinquent by focusing prevention programs on at-risk youth	Improving the juvenile justice system response to delinquent offenders through a system of graduated sanctions and a continuum of treatment alternatives

and Department of Public Instruction, 2008). Awards were made to more than half of the state's counties (most of which are rural areas), those that expressed an interest in addressing potential gang problems. Eligible applicants for awards under this $5 million program were restricted to the DJJDP Juvenile Crime Prevention Councils (JCPCs). These councils already existed, established by the 1998 NC Juvenile Justice Reform Act (S.L. 1998–202) in all 100 North Carolina counties, with statutory authority to implement the prevention component of the DJJDP Comprehensive Strategy framework. The JCPCs are charged with preparing comprehensive delinquency prevention plans, and developing a continuum of programs based on risk and needs assessments. Hence the JCPCs are well positioned to carry out gang assessments and program implementation, although both activities proved challenging in the state's rural counties.

The Community-Based Youth Gang Violence Prevention Project was created with the goal of preventing and reducing high-risk behavior and gang involvement statewide (North Carolina DJJDP, 2009). A standard definition was provided for the purpose of guiding the assessment, that closely resembles the practical definition in Chapter 3, and applicants were directed to the OJJDP (2009a) assessment guide, and OJJDP (2009b) implementation manual, as well as the OJJDP Strategic Planning Tool as resources. A condition of funding was that "assessment, program development, and implementation must reflect the Comprehensive Gang Model components" (p. 1).

Funding was also restricted to prevention and intervention programs and activities (North Carolina DJJDP, 2009). Priority A required participating counties to conduct a gang problem assessment, with emphasis placed on generating a clear and concise understanding of the local risk factors for delinquency and gang involvement. Once the assessment was completed (or if one already existed from the 2006 DJJDP program), for which up to a 15-month time frame was allowed, counties could move to Priority B, continuation funding to build programming around the gang assessment findings. At this time, many of the counties have completed a gang problem assessment and begun funding gang programs. This initiative bears close watch as the first U.S experiment to fund statewide Comprehensive Gang Model programming.

Using programs described in this chapter, both youth at risk of delinquency and gang involvement can be mainstreamed into programs. This is a matter of targeting geographical areas for programming in which gangs are active—which also are very likely to have elevated delinquency rates. It also is important to accumulate knowledge of the gang context in which youth are involved. For example, based on his multiyear study of the epidemiology of violence and drugs among 26 Mexican American gangs on the south and west sides of San Antonio, Valdez (2003) constructed a useful typology of four distinct gang types, the main features of which are sketched here from Valdez's detailed descriptions and illustrations with actual gangs observed in his long-term study. Valdez's (2003) typology has excellent practical value for guiding intervention strategies.

- *Criminal adult dependent* (4 gangs). "In these gangs, adults provided access to illegal drugs, weapons, drug-dealing networks, and national and international (Mexican) markets for stolen merchandise" (p. 21). Adults also provide protection against rival gangs and adult criminals. Drug dealing (particularly heroin) is the primary source of illegal income for these gangs. Violence was not uncommon among them, and it typically revolved around business transactions.
- *Criminal non-adult dependent* (5 gangs). Although this type of gang is similar to the criminal adult dependent gangs in its centralized organizational structure, it differs from those gangs in being more loosely knit with a flexible leadership structure, and it is less influenced by adults. These gangs also tend to be more territorially based than the first type, and members are involved more often in personal fights within the gang and with rival gang members.
- *Barrio-territorial* (12 gangs). These gangs, which constitute the majority of the gangs in Valdez's study, are younger than others and tend to operate independently of any adult gang influence and devoid of any centralized organizational structure. These are located in a variety of neighborhoods on the south and west sides of San Antonio, ranging from those with single-family homes to public housing units. Criminal activities tend to be less gang directed and more individual based, and include drug dealing auto theft, burglary, robbery, and other less serious crimes.
- *Transitional* (5 gangs). The remaining five gangs fit into this category. They are smaller than others and semi-organized with a loose leadership structure, often centered around a charismatic leader. These gangs are often formed along residential lines, such as living in the same building, in subareas of public housing projects, or in particular

neighborhoods. Partying with drugs and alcohol are common activities. Criminal activities usually are individually based and include drug dealing, auto theft, and burglaries. School-based gangs are a subset of these smaller, less structured gangs. These gangs are formed and maintained in the junior high and high schools.

Valdez's classification is very practical because of the readily apparent implications for intervention. Other gang classification schemes do not have this advantage, such as Klein and Maxson's (2006) typology of five gang types: traditional, neotraditional, compressed, collective, and specialty. In contrast, in Valdez's typology, the first two types clearly distinguish adult dominated (street gangs) from adolescent dominated (youth gangs). This helps specify both the location and age level of target groups for prevention and intervention activities. The barrio-territorial gangs are largely neighborhood based. This gang class is very common in inner-city Mexican American and African American neighborhoods, again providing guidance to prevention and intervention activities. The transitional gangs are similar to starter gangs. School and neighborhood social settings should be the location of prevention and early intervention activities to thwart gang formation and mount early intervention measures.

Concluding Observations

Preventing youth from joining gangs is challenging, and most programs have not shown noteworthy results. Several factors contribute to this challenge. Many alienated youth seek a place where they are accepted socially and often find it in the streets. Most youth who join gangs experience many risk factors and family, school, and community problems. Joining a gang can be a natural process for many youth in socially and economically deprived areas of large cities. The gang may already be there, in their neighborhood, and some of their friends and relatives may belong to it. Under these conditions, gangs' promises of protection may gradually envelop these vulnerable youth.

A major prevention problem is the lack of gang awareness in schools, among community leaders, and among parents. Gottfredson and Gottfredson's (2001) national study showed that, in the 10% of schools with the greatest student gang participation rates, only 18% of principals recognized that gangs were a problem in their schools. Fortunately, school resource officers and safe and drug-free school coordinators recognize gang activity more frequently. All school systems that suspect gang activity should immediately undertake a gang assessment as described earlier.

It is possible to prevent youth from joining gangs. The first level of prevention involves strengthening the core social institutions that let them down from the outset. School climate environments need to be improved and student "connectedness" to schools must be strengthened. Students must remain "tethered" to the school, their families, and community institutions such as faith-based organizations to buffer them from gang attractions. This means the school suspension, expulsion, and dropout rates must be reduced. Zero tolerance policies must be relaxed, and tempered by risk assessments. Interventions with youth at high risk for delinquency and gang involvement must be applied early in life. Effective afterschool and alternative education programs are imperative. Community violence prevention programs are necessary to reduce the level of community violence and personal victimization of vulnerable youth. Programs should target girls and boys and both White and minority youth. Poor implementation of gang-related programs in schools is a significant problem (Gottfredson & Gottfredson, 2001). For effective implementation, delinquency and crime reduction programs that target children and adolescents must adhere with high fidelity to the requirements of the original model and target high-risk offenders (Lipsey, 2009; Lipsey et al., 2010).

Because youth who join gangs typically experience multiple problems in multiple developmental domains that change over time, a continuum of age-graded strategies and programs is needed to prevent gang involvement. These strategies and programs need to target risk factors in family, school, and community contexts. The next chapter describes how to link prevention and early intervention programs with later interventions for gang-involved youth and gang suppression strategies.

DISCUSSION TOPICS

1. Class exercise: Pick one or more risk factors for gang involvement in Table 5.2 in each developmental domain, select a target age range, review the programs that are linked to it in the OJJDP Strategic Planning Tool, and build an ideal continuum of prevention and early intervention programs: www.nationalgangcenter.gov/SPT

2. What strategies and programs would you propose to reduce female gang involvement in San Antonio's public housing sector? Take into account the results of the Moving to Opportunity experiment.

3. What about males? Would you use the same strategies and programs for them?

4. Would your strategies and programs work as well elsewhere, and for White and African American children and adolescents?

5. After a community builds and implements a continuum of programs and structures, how can effectiveness of the entire continuum be determined?

RECOMMENDATIONS FOR FURTHER READING

OJJDP Comprehensive Gang Model

National Gang Center. (2010a). *Best practices to address community gang problems: OJJDP's Comprehensive Gang Model*. Washington, DC: Author. Retrieved from www.ncjrs.gov/pdffiles1/ojjdp/231200.pdf

Difficulties Associated With Moving Model Programs Into Everyday Practice

Barnoski, R. (2004). Outcome evaluation of Washington State's research-based programs for juvenile offenders. Document No. 04–01–1201. Olympia: Washington State Institute for Public Policy. Retrieved from http://www.wsipp.wa.gov/

Dodge, K. A. (2001). The science of youth violence prevention: Progressing from developmental epidemiology to efficacy to effectiveness to public policy. *American Journal of Preventive Medicine, 20*(1S), 63–70.

Fagan, A. A., Hanson, K., Hawkins, J. D., & Arthur, M. W. (2008). Implementing effective community-based prevention programs in the community youth development study. *Youth Violence and Juvenile Justice, 6,* 256–278.

Gottfredson, G. D., & Gottfredson, D. C. (2001). *Gang problems and gang programs in a national sample of schools.* Ellicott City, MD: Gottfredson Associates.

Karoly, L. A., Greenwood, P. W., Everingham, S. S., Houbé, J., Kilburn, M. R., & Rydell, C. P. (1998). *Investing in our children.* Santa Monica, CA: RAND.

Lipsey, M. W., Howell, J. C., Kelly, M. R., Chapman, G., & Carver, D. (2010). *Improving the effectiveness of juvenile justice programs: A new perspective on evidence-based practice.* Washington, DC: Georgetown University, Center for Juvenile Justice Reform.

Welsh, B. C., Sullivan, C. J., & Olds, D. L. (2010). When early crime prevention goes to scale: A new look at the evidence. *Prevention Science, 11,* 115–125.

Strategies for Prevention and Early Intervention

Dodge, K. A. (2001). The science of youth violence prevention: Progressing from developmental epidemiology to efficacy to effectiveness to public policy. *American Journal of Preventive Medicine, 20*(1S), 63–70.

Howell, J. C. (2000). *Youth gang programs and strategies.* Washington, DC: U.S. Department of Justice, Office of Juvenile Justice and Delinquency Prevention.

Howell, J. C. (2010a). Gang prevention: An overview of current research and programs. *Juvenile Justice Bulletin.* Washington, DC: U.S. Department of Justice, Office of Juvenile Justice and Delinquency Prevention.

Howell, J. C. (2010b). Lessons learned from gang program evaluations: Prevention, intervention, suppression, and comprehensive community approaches. In R. J. Chaskin (Ed.), *Youth gangs and community intervention: Research, practice, and evidence* (pp. 51–75). New York: Columbia University Press.

Karoly, L. A., Greenwood, P. W., Everingham, S. S., Houbé, J., Kilburn, M. R., & Rydell, C. P. (1998). *Investing in our children.* Santa Monica, CA: RAND.

Moore, J. W., & Hagedorn, J. M. (2001). Female gangs. *Juvenile Justice Bulletin. Youth Gang Series.* Washington, DC: U.S. Department of Justice, Office of Juvenile Justice and Delinquency Prevention.

Welsh, B. C., & Farrington, D. P. (2006). *Preventing crime: What works for children, offenders, victims, and places.* Dordrecht, The Netherlands: Springer.

Welsh, B. C., & Farrington, D. P. (2007). Save children from a life of crime. *Criminology and Public Policy, 6,* 871–879.

Welsh, B. C., Sullivan, C. J., & Olds, D. L. (2010). When early crime prevention goes to scale: A new look at the evidence. *Prevention Science, 11,* 115–125.

Preschool Programs

For additional information, see the Model Programs Guide: www.dsgonline.com/mpg2.5/mpg_index.htm

School Entry Programs

For additional programs, see the OJJDP Strategic Planning Tool at www.national gangcenter .gov/SPT/Planning-Implementation/Strategies/0. (Note that the database includes program structures.)

School-Based Intervention

Wilson, S. J., & Lipsey, M. W. (2007). School-based interventions for aggressive and disruptive behavior: Update of a meta-analysis. *American Journal of Preventive Medicine, 33* (Supplement), S130–S143.

Notes

1. Prevention methods can be classified as universal, selected, or indicated. Universal programs are applied to a complete population; selective programs are applied to a high-risk subgroup of the population; and indicated programs are applied to identified cases such as offenders.

2. Stephanie Rapp made excellent suggestions for the mix of services, particularly for girls.

10

What Works

A Comprehensive Gang Program Model

Introduction

The history of gang intervention is littered with programs that failed outright or were of questionable effectiveness, from Thrasher's first evaluation in the 1930s into the 1990s. Owing to progress that has been made in anti-gang programming over the past 15 years, at least a dozen programs have demonstrated evidence of effectiveness. Research-based gang intervention, suppression, and comprehensive programs are reviewed in this chapter.

Three caveats are important for applying program information that is presented here. First, there is more than one way of achieving evidence-based practice (EBP), as noted in Chapter 9. Unfortunately, the U.S. federal government has funded very few gang program evaluations, and only one gang research and development program, hence the literature in this area is very thin. In March 2011, OJJDP announced a new program of support for gang program evaluations that should help build a more generalizable body of knowledge of what works.[1] At the present time, only 12 evidence-based programs with a demonstrated gang impact have been identified. Two of these were reviewed in Chapter 9 and the remainder is reviewed in this chapter.

Like other social service disciplines, to a considerable degree, the gang field is fixated on replicating *model* or *exemplary* programs. The model programs approach entails selecting a recommended program from a list of research-supported programs and implementing it locally, but fidelity to the program developer's specifications for how the program is to be delivered is required to expect comparable results (Lipsey et al., 2010). In this approach, the recommended programs are supported in program-by-program reviews of the research. The programs typically considered for such reviews are specific brand-name programs that can be separately identified in the research literature, such as Functional Family Therapy, Multisystemic Therapy, and Multidimensional Treatment Foster Care. Well-known examples of model program repositories relevant to juvenile justice include the University of Colorado Blueprints for Violence Prevention project (Mihalic et al., 2001) and the Office of Juvenile Justice and Delinquency Prevention's Model Programs Guide (www2.dsgonline.com/mpg/).

As noted in Chapter 9, local agencies face many challenges when translating an evidence-based program into routine practice with high fidelity. Lipsey and colleagues (2010) identify a main drawback to the model program approach in the difficulties associated with achieving strict adherence to the prescribed protocol. First, it may not be possible to restrict the scaled-up program to the same population represented in the research. In everyday settings, the program is likely to serve a more heterogeneous population than was used in the research studies. Second, the service infrastructure within which the model program must operate is likely to be weaker than that organized by the program developer when conducting the evaluation research. Third, administrators frequently change, and replacements often introduce their own perception of what works. This can undermine support for existing evidence-based programs.

Another drawback to the model program approach is EBP programs in everyday practice can be overlooked. For example, Goldstein and Glick (1994) report that Aggression Replacement Training (ART) was developed in a New York State Division for Youth facility in central New York State, without the opportunity to randomly assign youth. Although ART has not met the criteria for a blueprint classification, it compared favorably with a program so classified when implemented in 21 juvenile courts in Washington State (working with general delinquents). Functional Family Therapy (FFT), a blueprint program, was implemented simultaneously in 14 of Washington's juvenile courts. When both programs were competently delivered, ART reduced 18-month felony recidivism rates by 24%, in comparison with a 38% reduction for FFT. But ART is much less expensive, and thus yielded a more favorable cost-benefits ratio of $11.66 in savings (avoided crime costs) for each taxpayer dollar spent versus $2.77 in savings for FFT (Barnoski, 2004). The lesson in this comparison is that some programs without the benefits of large-scale research and development support sometimes show worthwhile crime reductions and cost benefits that should not be ignored.

The second caveat is that a program need not be designed specifically to address gang crime to have positive gang crime reduction impacts. A reentry program, Operation New Hope, a case in point, is reviewed in this chapter and another one was reviewed in Chapter 9, the Montreal Preventive Treatment Program, which reduced gang involvement as a result of addressing precursor disruptive and delinquent behaviors, and another prevention program, Perry Preschool, reduced gang fights. Moreover, Howell and Egley (2005b) confirm gang and nongang offenders share common risk factors (see also Table 5.2), and as noted in Chapter 9, gang involvement overlaps substantially with serious, violent, and chronic offending. Hence, interventions potentially can reduce multiple problem behaviors.

Third, it is important to make a distinction between *program structures* and *program services*. Mark Lipsey is the first scholar to emphasize the importance of this distinction in evidence-based practice. In researching juvenile delinquency programs, he saw that communities typically have program structures that may not provide therapeutic services, and often serve other purposes in the course of controlling and rehabilitating juvenile offenders. This distinction is an important one; we should not expect program structures that do not typically provide a therapeutic service to reduce either delinquency or gang-related crime, including detention centers, correctional institutions, shelter care facilities, group homes, foster homes, and the like.

Homeboy Industries is an excellent example of the type of gang *program structure* that is needed to provide an alternative life course for gang members in inner cities where traditional services and supports are not available. This program structure is located in Boyle Heights, which Vigil (2007) describes as "an East L.A. neighborhood arguably with the highest concentration of gang activity in all of Los Angeles" (p. 93). Father Gregory J. Boyle created Homeboy Industries and Jobs for a Future in 1992, grassroots projects supported by the Dolores Mission, as an escape route from gangs, undergirded by the mantra, "Nothing stops a bullet like a job" (Fremon, 1995, 2004). Although it has not been evaluated as a program structure, Homeboy Industries successfully provided intervention services for the Los Angeles Gang Reduction Program, which has demonstrated evidence of effectiveness (reviewed subsequently). An intervention team is another example of a *program structure*. Among other things, intervention teams increase supervision of offenders, but as Bonta (1996) and Lipsey (2009) note, in and of themselves cannot be expected to have a significant impact unless effective services are provided, because it is the therapeutic services within such structured settings that have the actual power to produce change in offenders. *Program structures* make an important contribution in stabilizing gang offenders' behavior and everyday lives sufficiently to give treatment a chance to work.

The Comprehensive, Communitywide Gang Program Model

This chapter promotes a communitywide structure for organizing and integrating communities' gang programs. The original version of this model, called the Comprehensive, Community-Wide Gang Prevention, Intervention, and Suppression Program, was based on nearly a decade of research in a nationwide assessment of youth gang problems and programs led by the late Irving Spergel. This model synthesized both research on programs and best practices for successful intervention in gang activity.

Several observers fail to recognize that the Comprehensive Gang Model is a *program structure,* an organizational framework that guides users through a systematic data gathering, analysis, and strategic planning process. Thus, successful implementation depends upon community stakeholders' acknowledgment that they possibly or actually have a gang problem and also demonstrate a willingness to work together in solving it. Implementation failures are therefore shortcomings of community stake-holders, not inherent flaws in the framework. Others say the Comprehensive Gang Model is too complex. Granted, it is difficult to intervene effectively when gangs are entrenched in the cracks of society and require institutional and organizational modifications to reduce their impact. As of this writing, no shortcuts have been found in solving "the gang problem," which, in its most serious forms, is directly linked to fractures in social, cultural, educational, legal, and economic systems, many of which are referenced in Chapters 1 and 4. Organizations such as state and local governments (e.g., government immigration and housing policies) as well as core social institutions including schools (zero tolerance policies) and families (broken homes) contribute to gang problems. The impact of a single program in these multifaceted contexts is minuscule. Citywide and countywide initiatives are needed because gang members and other violent offenders often commit crimes outside their own neighborhoods. In addition, this model promotes organizational change, particularly in governmental agencies, and integration of multiple services provided by various agencies. The key assumption is that if human beings who preceded us created the conditions that permit gangs to thrive, then modern-day stakeholders have the capacity (collective efficacy) to change them (see Figure 4.1 in Chapter 4).

The most important contribution of the Comprehensive Gang Model structure in the aforementioned contexts may well be what Spergel, Wa, and Sosa (2006) suggest is the collective integration of specific intervention strategies and services. Short and Hughes (2010) generally characterize this function as "social capital and collective efficacy building" (p. 144). Indeed, Short and Hughes suggest that "a good argument can be made that this was what [Spergel and colleagues] were doing [in the first effective implementation of the model, the Little Village Gang Violence Reduction Program, reviewed subsequently] and that it may account for such success as they report" (p. 144). Along the same lines, Spergel (2010) suggests that "commitment to organizational change and development is at the core of the implementation" of the Comprehensive Gang Model strategies (pp. 236–238).

First Steps

The Comprehensive Gang Model incorporates five strategies: (1) organizational change and development, (2) community mobilization, (3) opportunities provision, (4) social intervention, and (5) suppression. Figure 10.1 shows how these strategies address program targets in ongoing activities, and each of these was generated from research on program initiatives in the national assessment by Spergel and colleagues. Brief descriptions of each of the five strategies follow.

1. *Community mobilization:* Involvement of local citizens, including former gang-involved youth, community groups, agencies, and coordination of programs and staff functions within and across agencies.

2. *Opportunities provision:* Development of a variety of specific education, training, and employment programs targeting gang-involved youth.

3. *Social intervention:* Involving youth-serving agencies, schools, grassroots groups, faith-based organizations, police, and other juvenile/criminal justice organizations in "reaching out" to gang-involved youth and their families, and linking them with the conventional world and needed services.

4. *Suppression:* Formal and informal social control procedures, including close supervision and monitoring of gang-involved youth by agencies of the juvenile/criminal justice system and also by community-based agencies, schools, and grassroots groups.

5. *Organizational change and development:* Development and implementation of policies and procedures that result in the most effective use of available and potential resources, within and across agencies, to better address the gang problem. (Spergel, 1995, pp. 171–296)

Communities that adopt the Comprehensive Gang Model will benefit from the simplified implementation process that OJJDP (2009b) has created. OJJDP synthesized the elements of the Comprehensive Gang Model into five steps:

1. The community and its leaders acknowledge the youth gang problem.

2. The community conducts an assessment of the nature and scope of the youth gang problem, leading to the identification of a target community or communities and population(s).

3. Through a steering committee, the community and its leaders set goals and objectives to address the identified problem(s).

4. The steering committee makes available relevant programs, strategies, services, tactics, and procedures consistent with the Comprehensive Gang Model's five core strategies.

5. The steering committee evaluates the effectiveness of the response to the gang problem, reassesses the problem, and modifies approaches, as needed. (National Gang Center, 2010a, p. 3)

The National Gang Center (2010a) has tested these steps in several settings.

A key to successfully combating youth gangs is to first mobilize the community. The reality is that communities respond to problems that their key leaders or stakeholders perceive to be important. The range of problems is broad, including natural disasters, public health problems, external threats, drugs, adult crime, juvenile delinquency, and gangs. In their literature review, Howell and Curry (2009) suggest several essential elements of successful community mobilization. These are organized in the preferred sequence as follows, but it is important to recognize that the sequencing of steps may vary in different communities. For example, if a community already is well organized to deal with a variety of youth problems—often by an existing coalition of some sort—it would be advisable to take advantage of this existing infrastructure to address a gang problem. This situation would preclude the need to take several of the preliminary steps outlined in what follows. The sequence of steps suggested here assumes that the community has a low degree of organization to address such social problems.

Figure 10.1   Comprehensive Gang Model

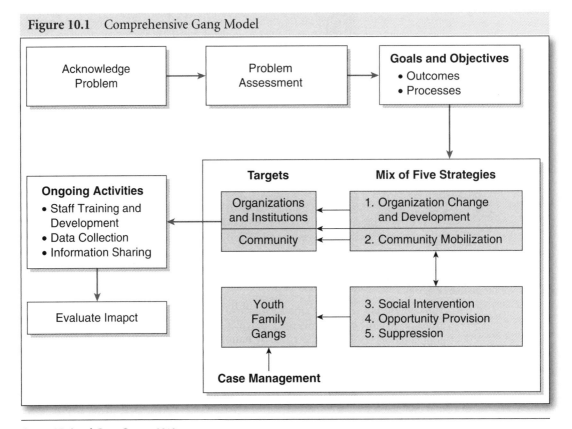

Source: National Gang Center, 2010a

Recognize the current or potential gang problem as a major threat to community safety and security.

Identify key neighborhood leaders in the community.

Contact key neighborhood leaders, local agencies, and community groups to discuss their concerns and to inquire about their interest in a collective effort.

Cultivate a natural leader, a community organizer, within the community. He or she must express deep feelings and impress on others that a problem exists and that something must be done about it.

Ensure that the community leader or organizer urges the broadcast media to become involved in all aspects of the mobilization process as early as possible.

Convene a meeting of community representatives and discuss emerging concerns. Think of these as community network meetings to formulate a cohesive plan.

Hold numerous individual meetings with many stakeholders to seek their support and refine priority problems.

Convene a second communitywide meeting. Think of this as a neighborhood involvement activity to heighten awareness of how community entities can interact to bring about a gang-free environment.

Direct the community or agency leaders, with the aid of the community organizer, to begin to involve and solicit the support of a variety of local agencies or community groups, former gang influentials, and even selected gang youth to alert the community to the gang problem.

Create or designate a formal community organization to act on behalf of the community group.

Conduct an objective assessment of the potential gang problem.

Set specific goals and objectives together with a timeline for their accomplishment. Seek to achieve small wins at the outset.

A community should not even think about programs and strategies for responding to gangs before it has assessed its own gang problem. Without the benefit of an empirical assessment, McCorkle and Miethe (2002) caution that community stakeholders run a high risk of being seriously mistaken about the nature of their gang problem. A common misstep is the assumption that local gangs, that in appearance may resemble distant Chicago or Los Angeles gangs (e.g., by hand signs, tattoos, or color of clothing), are actually far less dangerous. Every effort must be made in the assessment stage to discard preconceived notions because many of them are based on gang myths, as noted in Chapter 2. Guidelines for such an assessment were featured in Chapter 9.

The National Gang Center does not recommend naming gangs in the products of this assessment. This can backfire in five ways, by (1) increasing the notoriety of gangs and individuals, further emboldening groups; (2) helping gangs recruit by giving youth the mistaken impression that the gang can provide protection; (3) fueling conflict between groups, sparking retaliation; (4) increasing cohesion of gangs by publicizing their criminal activity, enhancing their image as "baddest" in the community; and (5) provoking fear among citizens, further reducing public confidence in the justice system and creating a panic that can misdirect policy makers.

An important point to underscore here is that communities should not wait for a tragic catalytic event to address gang activity. Sole reliance on law enforcement suppression strategies is typically the first response. Over the course of several years, communities then gradually embrace a more collaborative effort with community organizations and agencies. Success will come much more quickly in taking a balanced approach from the outset, one that is grounded in a careful assessment of gang activity.

Comprehensive Gang Model Administrative Structure

The recommended administrative structure for implementing the Comprehensive Gang Model is shown in Figure 10.2, consisting of a steering committee, the lead agency, a project coordinator, a research partner, and an intervention team (National Gang Center, 2010a, pp. 5–24).

A steering committee oversees the gang project. These individuals are policy or decision makers from agencies and organizations that have an interest in or responsibility for addressing the community's gang problem. These representatives should not only set policy and oversee the overall direction of the gang project, but they should take responsibility for spearheading efforts in their own organizations to remove barriers to services and fill service gaps. The steering committee should also develop effective criminal justice, school, and social agency procedures; and promote policies that will further the goals of the gang strategy. Ideally, this group oversees an assessment of the local gang problem and, using data obtained through the assessment, develops strategies to combat it.

Figure 10.2   OJJDP Comprehensive Gang Model Administrative Structure

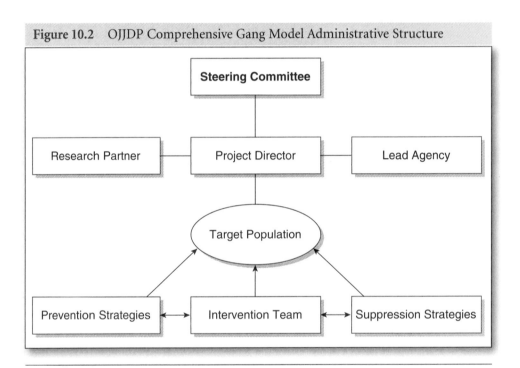

Source: National Gang Center, 2010a

The project coordinator or director is responsible for the day-to-day management of the project and reports directly to the steering committee.

Those with responsibility for addressing the problem—representatives of police, schools, probation, youth agencies, grassroots organizations, government, and others—form the assessment team and participate in identifying the gang problem's scope, severity, and unique features.

An intervention team is an essential component of the Comprehensive Gang Model. At a minimum, the following key agencies that are crucial to an effective intervention team's success should be represented on the team (National Gang Center, 2010a): law enforcement representatives involved in gang investigation and enforcement; juvenile and adult probation or parole officers who will have frequent contact with program clients; school officials who can access student educational data for program clients and leverage educational services; appropriate social service or mental health providers who can leverage services and provide outcome information to the team; a representative who can assist in preparing program clients for employment and find jobs for them; and outreach workers who can directly connect to program clients on the street, in their homes, or at school. Arciaga (2007) explains how the intervention team members work together to determine whether referred individuals are appropriate for their services in accordance with agreed-upon selection criteria, and then work as a team to serve these clients.

Because of the intensive work involved in helping youth negotiate their way around gangs in high-risk environments, and assisting others in extracting themselves from gangs, it goes without saying that "the outreach component of this model is critical to program success" (National Gang Center, 2010a, p. 20). The outreach workers' primary role is to build relationships with program clients and with other gang-involved and at-risk youth in the community. These workers generally are the primary source of program referrals and they often play an active role in delivering services, working closely with community service providers. In addition, "outreach workers are the intervention team's eyes and ears on the street, giving the team perspective on the personal aspects of gang conflicts and violence and how these affect the team's clients" (p. 20). The outreach workers also should play an active role in the development and execution of comprehensive treatment plans in intervention teams. According to Pyrooz, Decker, and Webb (forthcoming), accelerating the process of leaving a gang, or reducing the length of gang membership, is quite likely to reduce victimization and subsequent criminality. Research indicates that supportive peer and family networks play an important part in the desistance process. In addition, street evangelism conducted by Wilkerson (1962) produced a remarkable success story in redirecting gang members to alternative lifestyles.

The NGC has developed a Strategic Planning Tool (SPT) (www.nationalgang-center.gov/), introduced in Chapter 9, that communities can use to assess their youth violence and gang problem and use in conjunction with the Comprehensive Gang Model to develop a continuum of prevention, intervention, and suppression programs and strategies. For prevention planning, this tool includes research-based risk factors and indicators to assist communities in conducting a risk and program needs assessment. The SPT contains a section (the Community Resource Inventory) within which community planning groups can record and maintain information on existing programs, which would help enormously in identifying and filling program gaps. It supports a problem-solving approach to gang-related crime by engaging communities in the identification of elevated risk factors for gang involvement in communities and schools, locations of gang activity, hotspots of gang crime, and high-rate gang offenders and violent gangs. The SPT also contains information on promising and effective juvenile delinquency and gang programs and strategies that address specific risk factors among various age groups. Resource materials that assist communities in developing an action plan to implement the Comprehensive Gang Model are available online at www.nationalgangcenter.gov/Comprehensive-Gang-Model/About.

Implementing the Comprehensive Gang Model

The Comprehensive Gang Model aims to reduce gang involvement and gang crime at three levels: communities, gangs, and individuals. These are the specific priority goals:

1. Create a community environment that helps reduce youth gang crime and violence in targeted neighborhoods.

2. Prevent and reduce gang involvement and thereby reduce related gang crime.

3. Reduce gang-related crimes committed by particular gangs.

Stated in terms of a continuum, *prevention* programs are needed to target youths at risk of gang involvement, to reduce the number of youths who join gangs; *intervention* programs and strategies are needed to provide sanctions and services for younger youths who are actively involved in gangs to separate them from gangs; and law enforcement *suppression* strategies are needed to target the most violent gangs and older, most criminally active gang members. Therefore, a balanced and integrated approach employing multiple strategies is most likely to be effective, and the research supports this approach (Spergel, 2007; Spergel et al., 2006).

Wyrick (2006) further breaks down this continuum into five parts—with the addition of another strategy (reentry)—to facilitate the strategic planning process. *Primary prevention* targets the entire population in high-crime and high-risk communities. The key component is a one-stop resource center that makes services accessible and visible to the entire community, including prenatal and infant care, afterschool activities, truancy and dropout prevention, and job programs. *Secondary prevention* identifies young children (ages 7 to 14) at high risk and—drawing on the resources of schools, community-based organizations, and faith-based groups—intervenes with appropriate services before early problem behaviors turn into serious delinquency and gang involvement. *Intervention* targets active gang members and close associates, and involves aggressive outreach and program recruitment activity. Support services for gang-involved youth and their families help youth make positive choices. *Suppression* focuses on identifying the most dangerous gangs and weakens them by removing the most criminally active and influential gang members from the community. *Reentry* targets serious offenders who are returning to the community after confinement and provides appropriate services and monitoring. Of particular interest are displaced gang members who may cause conflict by attempting to reassert themselves in their former gang roles.

Figure 10.3 Focusing Anti-Gang Strategies

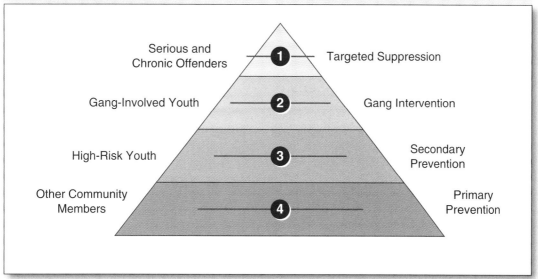

Source: Adapted from Wyrick, 2006, p. 56

Figure 10.3 presents a model that is useful for planning a continuum of programs and strategies within the above continuum segments.

- Group 1 (at the top of the triangle) represents serious, chronic, and violent gang and nongang offenders. These offenders make up a relatively small portion of the population, but commit a disproportionately large share of illegal activity.
- Group 2 consists of gang-involved youth and their associates, who make up a relatively larger share of the population. These youth are involved in significant levels of illegal activity but are not necessarily in the highest offending category. They typically range in age from 12 to 24 years old.
- Group 3 is made up of high-risk youth— ages 7 to 14 who have already displayed early signs of delinquency and an elevated risk for gang membership but are not yet gang involved. Most of these youth will not join gangs, but they represent a pool of candidates for future gang membership.
- Group 4 represents all youth living in a community where gangs are present.

These four groups should be targeted with the four basic strategies for combating gangs:

1. Members of Group 1 are candidates for targeted enforcement and prosecution because of their high level of involvement in crime and violent gangs and the small probability that other strategies will reduce their criminal behavior.

2. Members of Group 2 are candidates for intensive treatment services and supervision. Lipsey and Chapman (2011) recommend that such services include group therapy,

family therapy, mentoring, and cognitive-behavioral therapy—consisting of, for example, 48 hours of group therapy over a 24-week period.

3. Members of Group 3 are candidates for secondary prevention services, which are less intensive than those provided to group 2 but more intensive than those provided to youth in the community at large.

4. Members of Group 4 receive primary prevention services such as GREAT (discussed in Chapter 9).

OJJDP has developed an excellent tool to support the assessment process, a web-based Socioeconomic Mapping and Resource Topography (SMART) system, which provides a substantial amount of community-level data from the U.S. Census, Uniform Crime Reports (UCR), and a Community Disadvantage Index (a research and data derived index that provides a way to compare the risk level of one community versus another). The SMART system is free of charge and available online (go to the OJJDP website, www.ojp.usdoj.gov/ojjdp, and select "Tools"). Localities need only enter census tract numbers to download maps that depict high-risk areas for violent victimization. These areas often overlap with gang presence. This information should help assessment teams narrow the search for target areas.

Table 10.1 includes more specific information on the activities in each phase and, based on experiences of other comprehensive projects, the approximate length of time it takes to complete the activities in each phase (National Gang Center, 2010a). Each community's administrative structure and practices, community politics, and community readiness will dictate the actual length of each phase. Existing gang legislation also needs to be taken into account.

Table 10.1 Timeline for Implementing the Comprehensive Gang Model

Assessment and Planning (6–12 months)	Capacity Building (3–6 months)	Full Implementation (12–18 months)
• Identify key stakeholders.	• Complete the contract procurement process for program services.	• Initiate program services.
• Form a steering committee.	• Develop program policies and procedures.	• Initiate client referral process.
• Establish gang definitions.	• Advertise, hire, and train new staff to provide services.	• Begin client intake process for prevention and intervention clients.
• Hire a project coordinator.	• Develop a client referral and recruitment process.	• Begin conducting client reviews during intervention team meetings.
• Solicit a research partner.	• Select intervention team members.	• Begin providing focused suppression efforts.

Assessment and Planning (6–12 months)	Capacity Building (3–6 months)	Full Implementation (12–18 months)
• Conduct a comprehensive community assessment (including the Community Resource Inventory).	• Conduct intervention team training.	• Begin conducting community prevention activities.
	• Provide gang awareness training to project service providers.	• Reach program caseload capacity.
	• Develop a community resource referral and feedback process.	
	• Develop a program referral process.	
	• Train project partners on the project referral process.	
	• Develop a data collection and analysis system.	

Source: National Gang Center (2010a)

In Focus 10.1
Highlights of Gang-Related Legislation

- Of the 50 states and the District of Columbia (DC), all but three have enacted some form of legislation relating to gangs (Vermont, West Virginia, and Wyoming).
- 21 states and DC have passed gang prevention laws.
- 14 states and DC have enacted anti-carjacking statutes.
- More than half the states have laws that provide for enhanced penalties for gang-related criminal acts.
- 21 states' public nuisance laws count gang activity among the factors in determining a nuisance. Indiana has defined *real estate/dwellings* as "psychologically affected property" if they are the locations of criminal gang activity. By law, this factor must be disclosed in real estate transactions.
- 39 states and DC have legislation that defines *gang*.
- More than half of the states' legislatures have laws against graffiti.
- 22 states define *gang crime/activity*.
- 22 states have legislation on gangs and schools.
- Nearly 20% of the states have enacted laws that deal with gang-related databases.

Source: National Gang Center (2010b)

Evidence-Based Gang Programs

This section reviews effective gang-related intervention and suppression programs that have been identified since systematic reviews of gang programs among nine federal agencies began in 2005 (Howell, 2009). Programs are scored on the following widely accepted scientific standards for judging program effectiveness:

- The soundness or clarity of the program's framework
- Program fidelity (i.e., adherence to original program operation guidelines)
- The strength of the evaluation's design
- The empirical evidence demonstrating that the program prevents or reduces problem behaviors

Programs in the database fall into one of the following classifications:

Level 1 (Exemplary)—In general, when implemented with a high degree of fidelity, these programs demonstrate robust empirical findings using a reputable conceptual framework and an evaluation design of the highest quality (experimental). Programs in this category are designated exemplary or model programs and are considered very effective.

Level 2 (Effective)—In general, when implemented with sufficient fidelity these programs demonstrate adequate empirical findings using a sound conceptual framework and an evaluation design of high quality (quasi-experimental).

Level 3 (Promising)—In general, when implemented with minimal fidelity, these programs demonstrate promising (perhaps inconsistent) empirical findings using a reasonable conceptual framework and a very limited evaluation design (single group pre- and post-test) that requires causal confirmation using more appropriate experimental techniques. None of these programs is reviewed here. (Interested readers should consult the Strategic Planning Tool.)

Table 10.2 illustrates evidence-based programs that scored in the first two levels.[2]

Historically, gang intervention and prevention programs have been designed for boys, thereby neglecting special program needs for girls. As Moore and Hagedorn (2001) note, only one major federal program in U.S. history has specifically addressed female gang involvement. However, there are numerous research-based gender-neutral programs that have proved effective in delinquency reduction. These also are likely to be effective with girls, and several of these hold excellent potential for girls in the gang context. It is exceedingly good news from meta-analyses that general juvenile delinquency prevention and treatment programs are about equally effective with girls and boys (Lipsey, 2009). These evidence-based primary services include cognitive-behavioral therapy, individual counseling, family counseling, behavior management, and others. Zahn, Day, Mihalic, and Tichavsky's (2009) separate review of programs for girls identified two model programs that have produced comparable cross-gender outcomes and incorporate these services, Multidimensional Treatment Foster Care (MTFC) and Multisystemic Therapy (MST). But achieving optimal outcomes will

Table 10.2 Effective Gang and Gang-Related Programs

Program Name	Parent Program	Rating*	Repository**	Key References
Gang Resistance Education and Training	NA	L-2	OJJDP SPT OJJDP MPG	Esbensen, Peterson, et al., 2001
Montreal Preventive Treatment Program [z]	NA	L-1	OJJDP SPT OJJDP MPG	Tremblay et al., 1996
Aggression Replacement Training	NA	L-2	OJJDP SPT OJJDP MPG	Goldstein et al., 1998
Aggressive Behavioral Control Program [Y]	NA	L-2	OJJDP SPT OJJDP MPG	Di Placido et al., 2006
Building Resources for the Intervention and Deterrence of Gang Engagement	OJJDP Comp Gang Model-Riverside, CA	L-2	OJJDP SPT OJJDP MPG	Spergel et al., 2006
Operation New Hope[z]	NA	L-2	OJJDP SPT OJJDP MPG	Josi & Sechrest, 1999
Hardcore Gang Investigations Unit	NA	L-2	OJJDP SPT OJJDP MPG	Dahmann, 1983, 1995
Tri-Agency Resource Gang Enforcement Team	NA	L-2	OJJDP SPT OJJDP MPG	Kent et al., 2000
Chicago Alternative Policing Strategy[z]	NA	L-2	OJJDP SPT OJJDP MPG	Skogan & Hartnett, 1997
	OJJDP Comprehensive Gang Prevention, Intervention, and Suppression Model	L-2	OJJDP SPT OJJDP MPG	Spergel et al., 2006
Gang Violence Reduction Program	OJJDP Comp Gang Model-Chicago	L-2	OJJDP SPT OJJDP MPG	Spergel, 2007
Gang Reduction Program [Y]	OJJDP Comp Gang Model-Los Angeles	L-2		Hayeslip & Cahill, 2009
Gang Intervention Program	OJJDP Comp Gang Model-Mesa, AZ	L-2	X	Spergel et al., 2006
Gang Reduction and Intervention Program [Y]	OJJDP Comp Gang Model-Richmond, VA	L-2		Hayeslip & Cahill, 2009

(Continued)

(Continued)

Program Name	Parent Program	Rating*	Repository**	Key References
The Boston Midcity Project[Y]	NA	L-2	[X]	W. Miller, 1962
CeaseFire-Chicago	NA	L-2	OJJDP SPT OJJDP MPG	Skogan et al., 2008
Operation Ceasefire-Los Angeles[Y]	Boston Operation Ceasefire	L-2	OJJDP SPT	Tita, Riley, et al., 2005

* Legend: L-1=model or exemplary program; L-2=effective program. Criteria for all program ratings are provided in the OJJDP Model Program Guide (http://www2.dsgonline.com/mpg/ratings.aspx) and in the OJJDP Strategic Planning Tool (http://www.nationalgangcenter.gov/SPT/Planning-implementation/Review.

**OJJDP SPT= OJJDP Strategic Planning Tool (National Gang Center, http://www.nationalgangcenter.gov/SPT/Program-Matrix); OJJDP MPG= OJJDP Model Programs Guide (Development Services Group, http://www2.dsgonline.com/mpg/)

[X] Neither of these two programs is in the OJJDP Model Program Guide or the OJJDP Strategic Planning Tool database because they no longer exist.

[Y] Ratings of these five programs are preliminary. Among these, only the Aggressive Behavioral Control Program and Operation Ceasefire-Los Angeles have been added to the Strategic Planning Tool database at this point.

[Z] These three programs did not intentionally target gangs, gang members, or youth at risk of gang involvement; nevertheless, each of them demonstrated evidence of effectiveness in at least one of these intervention levels.

depend on implementing and maintaining them with high fidelity, and success in implementing them in a gender-sensitive manner. Bloom, Owen, and Covington (2006) relay that this requirement involves "creating an environment through site selection, staff selection, program development, content, and material that reflects an understanding of the realities of women's lives" (p. 2). Among the well-evaluated gang programs, the Comprehensive Gang Model proved to be particularly effective for girls.

Evidence-Based Gang Prevention Programs

Only two gang prevention programs are evidence based at this time, Gang Resistance Education and Training, (L-2) and the Montreal Preventive Treatment Program (L-1). Both of these programs were reviewed in Chapter 9. In addition, several delinquency and youth violence prevention programs were noted in Chapter 9 that have potential for preventing gang involvement. Because poor school climates may contribute to gang formation and emergence, programs that improve school climate are needed for a complete continuum. Two excellent candidates were reviewed in Chapter 9, PeaceBuilders (for elementary and middle schools, K–8), and Responding in Peaceful and Positive Ways (for students ages 10 to 14 in middle and junior high schools). Both of these programs are evidence based.

Evidence-Based Gang Intervention Programs

Aggression Replacement Training (ART) (L-2) is a 10-week, 30-hour cognitive-behavioral program administered to groups of 8 to 12 juvenile offenders three times per week. ART has three main curriculum components—structured learning training, which teaches social skills; anger control training, which teaches youth a variety of ways to manage their anger; and moral education, which helps youth develop a higher level of moral reasoning. In the juvenile court setting, ART can be implemented by court probation staff or private contractors, after they receive formal ART training. A juvenile offender is eligible for ART if it is determined—from the results of a formal assessment tool administered by juvenile court staff—that the youth has a moderate to high risk for reoffense and has a problem with aggression or lacks skills in prosocial functioning. Using repetitive learning techniques, offenders develop skills to control anger and use more appropriate behaviors. In addition, guided group discussion is used to correct antisocial thinking that leads to problem situations.

ART has produced impressive results working with gangs in Brooklyn, New York, communities. More rigorous evaluations have assessed the effectiveness of ART as an intervention for incarcerated juvenile delinquents. In these studies, ART enhanced prosocial skill competency and overt prosocial behavior, reduced the level of rated impulsiveness, decreased the frequency and intensity of acting-out behaviors, and enhanced the participants' levels of moral reasoning. In a Washington State Institute for Public Policy study, Barnoski (2004) found that when ART is delivered competently, the program reduced felony recidivism and was found to be cost-effective. For the 21 courts in which ART service providers were rated as either competent or highly competent, the 18-month felony recidivism rate was just 19%. This is a 24% reduction in felony recidivism compared with the control group, which is statistically significant.

One treatment program stands out for having reduced recidivism among imprisoned gang members. The Aggressive Behavioral Control Program (L-2) has demonstrated evidence of effectiveness in rehabilitation of young inmates in Canadian prisons (Di Placido, Simon, Witte, Gu, & Wong, 2006). Housed in the maximum-security Regional Psychiatric Centre, the program provides high-intensity, cognitive-behavioral therapy designed specifically for high-risk, high-need offenders, those who demonstrate low responsiveness to rehabilitative services. Prison gang inmates (average age 25) were included in the eligible target group and the evaluation demonstrated program effectiveness with them as well. The program also produced favorable cost benefits.

One of the OJJDP Comprehensive Gang Model sites evolved into mainly a gang intervention program. Thus OJJDP officials changed the original name of the Riverside (California) Comprehensive Gang Model to Building Resources for the Intervention and Deterrence of Gang Engagement (BRIDGE) in 1999 and focused its 5-year (1995–2000) operation on two areas of the city with high rates of gang crime (National Gang Center, 2010a). Formation of an intervention team was a key factor in the success of the program. It consisted of several core members, including the project coordinator, police officers, probation and parole officers, the outreach worker, the social service provider, and others. Case management involved the intervention team's development and implementation of a treatment plan. Spergel and colleagues (2006) report the project effectively reduced arrests for both serious and nonserious violent crimes after program participation and in comparison with a control group. Program youth also had fewer repeat drug arrests. The largest reduction in total violent crime arrests occurred when probation officers, police officers, outreach workers, and job and school personnel integrated their services for youth.

One correctional aftercare program has produced positive short-term effects for gang members: the Operation New Hope program (formerly called Lifeskills '95) (L-2), which was implemented in California's San Bernardino and Riverside counties. This program was designed for high-risk, chronic juvenile offenders released from the California Youth Authority. In addition to reintegrating these youths into communities, the New Hope program aimed to reduce their need for gang participation and affiliation as a support mechanism. An evaluation of the program's results found that participating youths were far less likely to have frequent associations with former gang associates than were members of the control group. In addition, youths assigned to the control group were about twice as likely as program participants to have been arrested, to be unemployed, and to have abused drugs or alcohol frequently since their release. The outcomes for this program suggest that some recidivism-reduction impact may be realized by giving special attention to gang members in reentry programming.

In addition to individualized case management of comprehensive reentry plans, probation, parole, and other staff working with returning gang-affiliated inmates must exercise special skills in steering them from gang life (Howell, 2011). The first challenge is separating them from prison gangs in which they had been members. Parolee safety and security of others is a major consideration. The prison gang mantra "blood in, blood out" is not an empty promise in many cases. At the same time, inmates' dependence on prison gangs and street gangs in the urban areas to which they return must be broken. Scott (2004) suggests that this may be a formidable challenge because gang members have a continuing identification as such. To further complicate matters, family members with whom they are reunited may also be gang members. Identification and immediate engagement of returning gang-affiliated inmates in alternatives to gang life, job training or employment, needed services, and a substitute support network are paramount.

To increase the validity of gang member classifications, juvenile correctional facilities, jails, and state prison systems should develop and establish a method for assessing and classifying individuals who are gang affiliated or gang involved, or who are at high risk of becoming gang involved, at the time of their intake and throughout their detention or incarceration. Training should be provided for institutional intake and classification staffs on how to identify, assess, and classify inmates who are at high risk of becoming gang affiliated or involved. It also is important to create and maintain a complete electronic record or paper trail about gang-involved individuals and ensure

In Focus 10.2
Criteria for Identifying a Security Threat Group Member

1. Self-admission

2. Gang tattoo or branding

3. Possession of Security Threat Group (STG) paraphernalia (e.g., STG lessons, rules, oaths, etc.)

4. Information from an outside law enforcement agency

5. Information obtained during a Special Investigations Division investigation

6. Correspondence from other known STG members

7. Group gang photo

8. Other (e.g., active participation in STG-related activity)

Source: New Jersey Department of Corrections (American Probation and Parole Association and Institute for Intergovernmental Research, 2011)

These are excellent criteria for the purpose of documenting active street gang members, and these are used in more advanced state prison systems. However, classification protocols typically require that only two of these criteria must be met to identify an inmate as a Security Threat Group (STG) member, the more general classification that includes gang members.

that any significant new information about them is communicated to other key personnel and agencies throughout their incarceration, release to the community, community supervision, and discharge from supervision processes.

Two promising juvenile delinquency programs are worthy of mention here. A correctional program developed for violent juvenile offenders also proved equally effective for rehabilitating gang members. The Wisconsin Mendota Juvenile Treatment Center (MJTC) provided cognitive–behavioral services to the most disturbed boys in the state's secure correctional facilities, both Aggression Replacement Training (Goldstein, Glick, & Gibbs, 1998) and the "decompression" treatment model (Caldwell & Van Rybroek, 2001). Caldwell, Vitacco, and Van Rybroek (2006) found the MJTC youths had less than half the recidivism rates of the comparison group and the program produced benefits of $7.18 for every dollar of costs. Given the effectiveness of Aggression Replacement Training, its use in the treatment regimen suggests that MJTC is likely to be effective with gang members.

A court-based nongang program also holds promise for the rehabilitation of gang members, the Multidisciplinary Team (MDT) Home Run program in San Bernardino County, California. This is an excellent example of a nongang program that successfully served gang members with an intervention team. The program's MDT resembles an intervention team recommended in the Comprehensive Gang Model. Four professionals make up each MDT: (1) a probation officer, (2) a licensed therapist, (3) a social worker, and (4) a public health nurse. Probation officers refer high-risk youthful offenders (including gang members) and their families. Each team provides intensive and comprehensive wraparound services for 6 months. Schram and Gaines's evaluation (2005) compared outcomes for randomly selected gang and nongang offenders from among all youths served in the MDT program. Both gang and nongang members reported improvements in school performance, family functioning, and decreases in alcohol and substance abuse, as well as delinquency involvement. Arciaga (2007) asserts that programs likely would be more effective with gang members if an outreach worker were added to the MDT to assist gang members in negotiating a path away from gangs. Outreach workers and school resource officers need to coordinate their efforts.

A North Carolina community-based residential program for high-risk juvenile offenders developed by Methodist Home for Children (MHC) has demonstrated effectiveness, according to Strom, Colwell, Dawes, and Hawkins (2010), in rehabilitating juvenile delinquents, including gang members, who were candidates for long-term confinement. Nearly one in five (19%) MHC youth were reported to be gang associates. The MHC therapeutic Model of Care incorporates both individual and family counseling in its multipurpose homes that provide residential services for youth referred through the state's Department of Juvenile Justice and Delinquency Prevention. This is a very promising program model for gang members that blends them into the treatment group but provides attentive services during reentry to avoid reengagement with hometown gangs.[3]

School Resource Officers. The role of school resource officers (SROs) in promoting school safety is not always clear. In a recent North Carolina study, Langberg, Fedders, and Kukorowski (2011) concluded that the presence of armed law enforcement officers in middle and high schools "is a misguided approach that is financially unsound and educationally imprudent. Research shows that alternative policies and programs are available to create more positive school environments and ensure student and teacher safety" (p. 13). However, SROs who are well trained to deliver the GREAT curriculum are effective in reducing student victimization (Esbensen, Osgood, Taylor, Peterson, & Freng, 2001), and possibly gang joining as well (see the discussion of GREAT in Chapter 9).

Because juvenile courts in North Carolina are overwhelmed with school referrals for minor offenses, Weisel and Howell (2007) suggest the potential benefits of SROs are sometimes stifled. Furthermore, Lassiter (2009) contends that myths about student violence lead to excessive school referrals to juvenile courts. In 2010, Howell, Lassiter, and Anderson (2011) found school referrals represented more than 40% of all court referrals. Langberg and colleagues' (2011) comprehensive SRO study made the following recommendations:

- Teachers and administrators should have readily available, high-quality alternatives to suspensions, arrests, and court referrals, such as mediation, community service, restitution, and mental health programs.
- All SROs, security investigators, and other security personnel should be required to undergo mandatory, intensive, ongoing trainings, including instruction on:
  - Legal standards for searches and seizures of students in schools
  - Positive behavior interventions and supports (PBIS)
  - Adolescent development science
  - Working with students who have disabilities and other special needs
  - Cultural competency
  - De-escalating student misbehavior without using physical force
  - Using safe restraint techniques
  - Long-term consequences of court involvement and arrests
- School administrators and school police should adhere to clear limitations reflecting the status of SROs as fully authorized and armed law enforcement personnel.
- SROs should be prohibited from carrying guns and Tasers on school campuses.
- Students who commit minor offenses in schools should not be routinely arrested and referred to court—only as a last resort.
- Clear, standardized, well-publicized complaint procedures should be established for students, parents, teachers, and administrators to use when SROs behave inappropriately.
- The public should have access to more complete, easy-to-understand data about SROs.

Langberg and colleagues (2011) relay two program examples to illustrate effective ways of implementing several of these recommendations. First, in Jefferson County (Birmingham), Alabama, a family court judge, school superintendent, police chief, and community members developed a set of graduated consequences for certain offenses so that youth would not be arrested and referred to court as a first option. Under the protocol, a warning would be issued for a first offense, the student and a parent may be required to attend a school workshop for a second offense, and a third offense may be referred to court. The advocacy efforts soon led to a 50% reduction in family court referrals. Ultimately, the collaborative agreement is expected to reduce court referrals from Birmingham schools by more than 80%.

Second, in Clayton County (metropolitan Atlanta), Georgia, a cooperative agreement among the school district, prosecutor's office, juvenile court, and local police departments ensures that misdemeanor delinquent acts such as fighting and disorderly conduct will not result in delinquent or criminal charges unless the student commits a third similar offense during the school year, and provided that the school principal reviews the student's behavior plan. Warnings are issued to students after a first offense, and for a second offense, a referral is made to mediation or school conflict resolution programs. Langberg and colleagues (2011) relay the following results: "As a result of these reforms, juvenile court referrals have been reduced by 47% in three years, Black students have been referred significantly less frequently, the relationships between students and police officers have improved, and graduation rates have increased by 20%" (p. 10).

Gang Suppression

As seen in Table 10.2, three targeted gang suppression strategies have proved effective: the Hardcore Gang Investigations Unit, Tri-Agency Resource Gang Enforcement Team, and Operation Ceasefire-Los Angeles. Other rated-effective programs referenced in this table had suppression components, including Building Resources for the Intervention and Deterrence of Gang Engagement (Riverside), Gang Violence Reduction Program (Chicago), the Gang Reduction Program (Los Angeles), and CeaseFire Chicago.

The Hardcore Gang Investigations Unit (L-2) (initially called Operation Hardcore, is a prosecutorial gang suppression program that was created by the Los Angeles District Attorney's Office in 1979 and is still operating today. Its distinctive features include reduced caseloads, additional investigative support, resources for assisting victims, and vertical prosecution—in which the prosecutor who files a case remains responsible for it throughout the prosecution process. In an independent evaluation of Operation Hardcore, Dahmann (1983, 1995) showed that fewer dismissals; more convictions or adjudications, including more convictions or adjudications on the most serious charge; and a higher rate of state prison commitments or secure confinement dispositions were achieved for cases subject to the program than for cases undergoing the normal prosecutorial process. Dahmann concluded, "Operation Hardcore program has obtained demonstrable improvements in the criminal justice handling of gang defendants and their cases" (p. 303). Operation Hardcore remains a highly regarded program.

The Tri-Agency Resource Gang Enforcement Team (TARGET) (L-2) program in Orange County, California, represents a multiagency approach to targeting current gang members with suppression measures while also targeting entire gangs with police suppression. Each team in the TARGET program consists of gang investigators, a probation officer, a deputy district attorney, and a district attorney investigator. This program uses a three-pronged strategy: (1) selective incarceration of the most violent and repeat older gang offenders in the most violent gangs; (2) enforcement of probation controls (graduated sanctions and intensive supervision) on younger, less violent gang offenders; and (3) arrests of gang leaders in hot spots of gang activity. A major aim of the TARGET program is to reduce gang crime by selectively incarcerating the most violent and repeat gang offenders (based on their criminal records) in the most violent gangs in Orange County. Once these offenders are identified, they are monitored closely for new offenses and undergo intensive supervision when on probation for violation of probation terms and conditions. The TARGET program produced a sharp increase in the incarceration of gang members and a cumulative 47% decrease in gang crime over a 7-year period (Kent, Donaldson, Wyrick, & Smith, 2000) and has been given credit for reducing the overall level of gang crime in a targeted hot spot to near zero (Wiebe, 1998).

Operation Ceasefire in Los Angeles (L-2) is modeled after Boston's Operation Ceasefire.[4] Beginning in 1999, it was implemented in the Hollenbeck area, home to

29 criminally active street gangs among a "dense network" of violent rivalries (Tita, Riley, & Greenwood, 2003, 2005). Like Boston, Los Angeles set a manageable and measurable objective, focusing on the specific problem of gun violence. Again, as in Boston, Los Angeles adopted a menu of "sticks" and "carrots." Sticks were a range of sanctions or "levers" used to encourage gang members to desist from violence by "retailing" the message to gang members that (1) all of them would be held accountable for violence committed by any one of them and that (2) violent crime would have consequences (long prison sentences). Services were offered as an alternative incentive to turn away from crime. These "carrots" included job training and development, substance abuse treatment, and tattoo removal. These were offered to targeted offenders through or by police and probation officers and various other city agencies; however, other services were provided in the target area by a number of community-based organizations, including Homeboy Industries/Jobs for a Future, a local job referral and training center.

The research team sought to determine whether the intervention reduced three categories of offenses—violent crime, gang crime, and gun crime—and whether it did so both during the time the intervention was taking place—the suppression phase—and during the months following—the deterrence phase. Although the findings were mixed, importantly, in Boyle Heights, overall violent crime fell significantly during the suppression phase, and the decline was even stronger in the following months—the deterrence phase (37% overall, compared with 24% in the remainder of Boyle Heights). Gang crime also fell in Boyle Heights during the suppression phase, although it began to rise during the deterrence phase. Gun crime, however, did not decline. Tita's (2003) research team attributed the near-term reduction in gang violence observed in the Boyle Heights area to the gang-based "street intervention."

Community policing, a softer form of police suppression, implemented in the Chicago Alternative Policing Strategy (CAPS) (L-2), has demonstrated evidence of effectiveness in reducing adult crimes. CAPS was field-tested in five experimental police districts in Chicago. In each of the districts, patrol officers were divided on a rotating basis into beat teams and rapid response teams. Beat teams were able to spend most of their time working their beats and with community organizations, while rapid response teams concentrated their efforts on excess 911 calls. Beat meetings were held on a monthly basis, and each district formed a civilian advisory committee to commanders and help mobilize problem-solving resources. Districts were responsible for identifying local priorities, planning strategies to address them, and then executing their plans. The evaluators gave an "excellent" rating to 4 of the 15 beats. Another 5 were found to be reasonably successful programs, 2 were of questionable effectiveness, and 4 received failing marks. Skogan and Hartnett (1997) observed gang problems were reduced in two of the three experimental police districts, but once CAPS was implemented citywide, Skogan and Steiner (2004) found successes with gang problems were largely limited to perceptions of African American residents.

The use of singular suppression tactics in combating gangs and gun crime has earned only "a mixed report card" (Decker, 2003, p. 290). Both Klein (1995) and

Papachristos (2001) believe the positive effects of suppression tend to be short-lived. For example, Braga and colleagues (2001) and Braga and Pierce (2005) reported serious gang problems returned and homicides increased after the successes attributed to the Boston Ceasefire project, dubbed the "Boston Miracle" (Kennedy, 2007). But suppression strategies were successfully implemented in Operation Ceasefire Los Angeles, and Tita, Riley, and Greenwood (2003) reported that "tailoring the intervention against an activity, such as gun violence, rather than an affiliation, such as gang membership, helped make it possible for the community to support the intervention" (p. 47).

In the most detailed study to date of a singular suppression initiative, Papachristos (2001) found that successful federal prosecution of members of Chicago's Gangster Disciples had the short-term effect of reducing this gang to a loosely arranged delinquent group in the neighborhood as members aligned themselves with other criminal gangs. However, the destabilization of the Gangster Disciples created a power void that was filled by competing gangs—mainly the Vice Lords and Latin Kings. Papachristos calls this unintended effect "gang succession" (p. 5).

A formidable obstacle to reducing gang violence is the huge cache of guns that is readily available to gangs. Research shows that beginning in the 1970s, gangs' access to more lethal weapons grew (Block & Block, 1993; W. Miller, 1982/1992). The most reliable estimate of privately owned guns in the United States is 258 million, of which up to 90 million are handguns (Wellford, Pepper, & Petrie, 2005). Two-thirds of samples of incarcerated adolescent offenders in four states said they would encounter "no trouble at all" in securing a gun upon release, either by borrowing one or otherwise getting one off the street (Sheley & Wright, 1993, 1995). Studies also consistently show gun availability fuels lethal gang violence (Block & Block, 1993; Cook & Ludwig, 2006). It has been estimated, according to Klein and colleagues (2006), that more than 90% of U.S. gang homicides involve firearms. Therefore, Ludwig (2005), along with Wellford and fellow researchers (2005), contend that "supernormal" reductions in gang firearm homicide are not on the horizon for the United States in the absence of effective access limitations, but Ludwig (2005) and McGarrell and associates (2009) suggest that Project Safe Neighborhoods (PSN) could produce worthwhile results in small geographical locations by coupling targeted patrols, gang units, and aggressive prosecution in areas where gangs account for a large share of gun violence; that is, in violent hotspots or gang set spaces. However, the most cost-effective strategy may well be intervention with "violence interrupters" in potential gang-related shootings with the CeaseFire Chicago model (described below), because Cohen, Piquero, and Jennings (2010) demonstrate enormous cost savings from successful interventions before criminal justice system, victim, and incarceration costs are incurred.

The Utility of Gang Intelligence Databases. Gang intelligence databases hold a great deal of potential as a tool for targeting chronic and violent gang offenders. In an Arizona study, Katz, Webb, and Schaefer (2000) compared the criminal histories of gang members and associates documented by the Mesa Police Department's Gang

Unit with nondocumented delinquent youth in the files of the Maricopa County Juvenile Probation Department. This comparison revealed that documented gang members and associates of gang members were significantly more criminally active than their undocumented delinquent counterparts. This research suggests that crime control benefits can accrue from maintaining accurate and useable intelligence information on chronic and violent gang offenders. However, Barrows and Huff (2009) caution such databases can be misused, if unaffiliated persons are inadvertently entered into the intelligence system. Therefore, it is very important to ensure that gang members are accurately classified.[5] State and local agencies that receive federal funds to operate criminal intelligence systems must comply with specific regulations, including those in 28 CFR Part 23. Technical assistance reviews of criminal intelligence systems and their compliance with the regulation have been provided at no charge through funding from the U.S. Department of Justice, Bureau of Justice Assistance. In addition, such a review could address special procedures for handling criminal intelligence information on juveniles. More frequent purging of the information on juveniles should also be considered, as suggested by research, presented in Chapter 3, which shows most adolescents who join a gang remain in it for less than 1 year. One process for protecting non gang juveniles is to "firewall" them in the criminal intelligence systems, restricting the dissemination of potentially incorrect information on their gang involvement. Weisel and Howell (2007) insist that inclusion of juveniles becomes particularly important where gang intelligence databases are not regularly purged. Wright (2005) suggests that a properly managed database might attend to due process requirements but that a predocumentation or postdocumentation hearing is the best remedy.

Intelligence-Led Policing. For more than 30 years, the Bureau of Justice Assistance has supported intelligence-led policing throughout a national network called the Regional Information Sharing Systems® (RISS). It is the sensitive but unclassified secure communications backbone for implementation of a nationwide criminal intelligence sharing capability. The Global Justice Information Sharing Initiative recently was created to support the broad-scale exchange of pertinent justice and public safety information. Global is a "group of groups," representing more than 30 independent organizations, spanning the spectrum of law enforcement, judicial, correctional, and related bodies. In addition to managing RISS and Global, the Institute for Intergovernmental Research provides the following training opportunities for the law enforcement community on combating gangs:

- *Basic training for street gang investigators*—Participants are provided useful information about the different types of gangs throughout the United States and receive specific, in-depth information about gangs in their region. They learn about collecting gang intelligence, investigative techniques, suppression strategies, case-building strategies, and legal considerations in prosecuting gang crimes.
- *Gangs in Indian Country*—The previous basic training curriculum is expressly tailored for tribal law enforcement officers and state and local officers who deal with Native American gangs.

- *Advanced gang investigations*—Participants learn more sophisticated investigative, intelligence, and suppression tools to investigate gang crimes and suppress gang activity.
- *Gang unit supervision*—Through the review and evaluation of best-practice strategies, participants are better equipped to develop the most appropriate organizational and management strategies for their department's gang unit.
- *Anti-gang seminar for law enforcement chief executives*—This 1-day seminar engages police chiefs, sheriffs, and other law enforcement chief executives in discussions of a variety of gang-related topics relevant to law enforcement executives.
- Routinely sharing expertise and information across law-enforcement agencies can make a significant difference in the use of advanced technology and strategies.

Evidence-Based Comprehensive Gang Models

Use of the term *comprehensive* means the use of multiple program strategies. Results of three such program models that have been proven effective are presented here. The first one, introduced in Chapter 9 and commonly called the OJJDP Comprehensive Gang Model, has been adapted to several settings. Ceasefire Chicago, the second comprehensive anti-gang model, has addressed community violence generally, while targeting gang homicide. A third such model was implemented many years ago in Boston, called A "Total Community" Delinquency Control Project.

The OJJDP Comprehensive Gang Model. In 1987, the OJJDP launched a Juvenile Gang Suppression and Intervention Research and Development Program. In the first phase, Spergel (1990) and also Spergel and Curry (1993) conducted a comprehensive national assessment of organized agency and community group responses to gang problems in the United States. It remains the only national assessment of efforts to combat gangs.

In the second phase, the Comprehensive, Community-Wide Gang Program Model was developed (Spergel 1995; Spergel, Chance et al. 1992, 1994; Spergel & Curry, 1993) as the final product of the gang research and development program. In several tests and demonstrations of this model over the past 15 years, this program model has been called the Little Village Gang Violence Reduction Program; the Spergel Model; the Comprehensive Gang Model; the Comprehensive Gang Prevention, Intervention, and Suppression Model; and the Gang Reduction Program.

In 1993, the initial version of the Comprehensive Gang Model was implemented in the Little Village neighborhood of Chicago, a low-income and working-class community that is approximately 90% Mexican American (Spergel, 2007). The Little Village Gang Violence Reduction Program targeted mainly older members (ages 17 to 24) of two of the area's most violent Mexican American gangs, the Latin Kings and the Two Six. The program targeted and provided services to individual gang members (rather than to the gangs as groups). Specifically, the Little Village program targeted more than 200 of the "shooters" (i.e., the influential members or leaders of the two gangs). As a whole, these two gangs accounted for about 75% percent of felony gang violence in the Little Village community. The process evaluation (Spergel et al., 2006) revealed that it was implemented with high fidelity on most "Program Implementation Characteristics" (Table 10.3), as indicated by the scored "Levels of Implementation" for the Chicago site.

Table 10.3 Program Implementation Characteristics: Degree of Importance and Levels of Implementation

Program Implementation Characteristics		Degree of Importance to Program Success[†]	Levels of Implementation by Project Site[‡]					
			Chicago	Mesa	Riverside	Bloomington-Normal	San Antonio	Tucson
Program Elements (Structure)	City/county leadership	***	2	4	4	1	1	1
	Steering committee	**	1	4	3	1	1	0
	Interagency street team/coordination	***	4	4	3	0	0	0
	Grassroots involvement	*	3	1	1	0	1	0
	Social services: youth work, individual counseling, family treatment, and recreation	**	3	3	3	2	3	3
	Criminal justice participation	***	4	4	4	1	1	0
	School participation	**	1	3	3	3	2	0
	Employment and training	**	3	1	4	3	1	0
	Lead agency/ management/ commitment	***	4	4	4	0	0	0
Strategies	Social intervention: outreach and crisis intervention	**	4	3	3	1	1	0
	Community mobilization: interagency and grassroots	**	1	3	2	1	0	0
	Provision of social opportunities: education, job, and culture	**	3	2	2	2	1	0
	Suppression	***	4	4	3	0	0	0
	Organizational change and development	***	2	4	4	0	0	0

(Continued)

(Continued)

Program Implementation Characteristics		Degree of Importance to Program Success[†]	Levels of Implementation by Project Site[‡]					
			Chicago	Mesa	Riverside	Bloomington-Normal	San Antonio	Tucson
Operating Principle	Targeting gang members/at-risk gang youth	***	4	2	3	1	3	3
	Balance of service	***	4	3	3	0	0	0
	Intensity of service	*	4	3	3	1	0	0
	Continuity of service	**	2	1	2	2	0	2

Source: Spergel, Wa, and Sosa, 2006, pp. 216–217; National Gang Center, 2010a, Table A1, p.43

[†]Importance of characteristic to success: *** = extremely, ** = moderately, * = somewhat

[‡]Levels of implementation: 4 = excellent, 3 = good, 2 = fair, 1 = poor, 0 = none

The priority goal of the project was to reduce the extremely high level of gang violence among youth who were already involved in the two gangs; drug-related activity was not specifically targeted. The main goal was to be accomplished by a combination of outreach work, an intervention team, case management, youth services, and suppression. Outreach youth workers would attempt to prevent and control gang conflicts in specific situations and to persuade gang youth to leave the gang as soon as possible. Virtually all of the outreach youth workers were former members of the two target gangs. Their program-related activities included crisis intervention, family and individual counseling and referrals for services, and surveillance and suppression activities. Altogether, Spergel (2007) observed a good balance of services and suppression.

Each of the intervention team members applied their own skills in providing services themselves to project youth and supplying contacts for others. In due course, the team (mainly outreach youth workers, police, and probation officers) convened biweekly and exchanged information on violence that was occurring (or about to occur) in the community. Suppression contacts, made mainly by project police, reduced the youth's interest in and attachment to the gang. Services such as job placement reduced target youth's time spent with other gang members.

Spergel (2007) examined the effects of the Little Village project on the approximately 200 hardcore gang youth targeted for services during the period in which the program served them. The following are some key findings:

- Self-reports of criminal involvement showed that the program reduced serious violent and property crimes, and the frequency of various types of offenses including robbery, gang intimidations, and drive-by shootings.
- The program was more effective with older, more violent gang offenders than with younger, less violent offenders.

- Active gang involvement was reduced among project youth, mostly for older members, and this change was associated with less criminal activity.
- Most youth in both targeted gangs improved their educational and employment status during the program period.
- Employment was associated with a general reduction in youth's criminal activity, especially in regard to reductions in drug dealing.

Spergel (2007) next compared arrests among project youth versus two control groups; one which received minimal services and the other which received no services from project workers. This comparison revealed the following:

- Program youth had significantly fewer total violent crime and drug arrests.
- The project had no significant effect on total arrests, property arrests, or other minor crime arrests.

Spergel (2007) also compared communitywide effects of the project on arrests in Little Village versus other nearby communities with high rates of gang crime. His analysis compared arrests in the periods before and during which the program was implemented and revealed the following:

- The project was less effective in its overall impact on the behavior of the target gangs as a whole—that is, changing the entrenched pattern of gangbanging and gang crime among the target gangs—than in reducing crime among targeted members. Gang violence was on the upswing during the project period (1992 to 1997) in this general area of Chicago, one of the deadliest gang-violence areas of the city, but the increase in homicides and other serious violent gang crimes was lower among the Latin Kings and Two Six compared with the other Latino and African American gangs in the area.
- Similarly, the increase in serious violent gang crimes was lower in Little Village than in all other comparable communities. Residents and representatives of various organizations perceived a significant reduction in overall gang crime and violence in Little Village during the program period.

In summary, although mixed, the outcomes for the Little Village project are consistent for violent crimes across analyses at all three impact levels: (1) individual, (2) group (gang), and (3) community (especially in the views of residents). The evaluation suggested that a youth outreach (or social intervention) strategy may be more effective in reducing the violent behavior of younger, less violent, gang youth. A combined youth outreach and police suppression strategy might be more effective with older, more criminally active and violent gang youth, particularly with respect to drug-related crimes.

Interactive and collaborative project outreach worker efforts, combining suppression, social support, and provision of social services, were shown to be most effective in changing criminal involvement of gang members. Larger program dosages (multiple providers and greater frequency and duration of services) proved to be important and were associated with reduced levels of arrests for violent crimes. Four types of services or sanctions predicted successful outcomes among program youth: suppression (particularly by police), job referrals by youth outreach workers, school referrals (mainly by outreach youth workers), and program dosage (contacts by all workers together).

The Six-Site Comprehensive Gang Model Evaluation

OJJDP implemented and tested the effectiveness of the Comprehensive Gang Model in other settings, each of which helped in gauging the utility of the model, and none of these sites was chosen by the program developer. Notwithstanding the name change, OJJDP made no substantive changes to the model. However, OJJDP embedded the model in a strategic planning and implementation process, which recast the model in several respects. First, the importance of a comprehensive assessment of local gang problems was strongly recommended prior to implementing any of the components of the model. Second, the "targeting" process called for in the original model was expanded to consist of a strategic planning process, based on the assessment results. Third, just five of the numerous strategies that were recommended in the model were emphasized (discussed in the preceding section).

Outcome data are currently available on two of the subsequent OJJDP initiatives, and those are described here. In the first of two main applications of the model, OJJDP chose five competitively selected sites to implement the Comprehensive Gang Model: Mesa, AZ; Riverside, CA; Bloomington-Normal, IL; San Antonio, TX; and Tucson, AZ. Each of the OJJDP demonstration projects was funded in 1995, and they shared the main goals of the Comprehensive Gang Model: (1) to reduce youth gang crime, especially violence, in targeted communities; and (2) to improve the capacity of the community—including its institutions and organizations—to prevent, intervene with, and suppress the youth gang problem through the targeted application of interrelated strategies of community mobilization, opportunities provision, social intervention, suppression, and organizational change and development.

In order to develop a composite picture of the implementation process and impact outcomes of the initial Comprehensive Gang Model implementations, Spergel, Wa, and Sosa (2006) combined the Chicago Little Village evaluation with the studies of the five urban sites. Three of the six communities either made fatal planning mistakes (e.g., selecting a lead agency that failed to perform) or encountered fundamental implementation difficulties because key agencies in the community simply were unwilling to work together. But when it was well implemented in three of the sites (in Chicago, Mesa, and Riverside), Spergel, Wa, and Sosa (2006) concluded that the Comprehensive Gang Model effectively guided these communities in developing services and strategies that contributed to reductions in both gang violence (in three sites) and drug-related offenses (in two sites). Key factors in implementation success are reflected in the following discussion. (For a site-by-site summary of the degree of fidelity to the Comprehensive Gang Model, see Table 10.3).

Although none of the sites met all of the program implementation guidelines, the three sites (Chicago, Mesa, and Riverside) that showed the largest reductions in violence and drug-related crimes complied with more of them. At the successful sites, a key factor was length of time that clients were engaged in the program. When youth were in the program for 2 or more years, there were fewer arrests for all types of offenses. In general, arrest reductions were greater among older youth and females than among younger youth and males. General deterrence effects (across the project area) were not as strong as the program effects for individual youth. Nevertheless, these three sites were somewhat successful in integrating police suppression with service-oriented strategies. In summary, the Spergel research team's (2006) evaluation indicates that, when properly implemented, a combination of prevention, intervention, and suppression strategies was successful in reducing the gang problem.

Spergel's (2010) later preliminary analysis across all six sites using the propensity-score matching statistical procedure to construct more exact treatment and control groups (nearly 900 youth) showed even more positive in support of the comprehensive, communitywide approach. In the multivariate analyses, program youth, compared to equivalent nonserved comparison youth, reduced their felony violence arrests by 60% in the inclusive six-site sample, and by 220% among youth who had violence arrests either in the preprogram or program period. Drug arrests were even more highly significantly reduced by program youth compared to nonserved equivalent youth—3.4 times in the inclusive six-site sample.

Older youth, 19 and older, did better than younger youth in reducing violence arrests, but not drug arrests. Females generally did better than males in reducing arrests, except for drug arrests. There was little difference in changes in arrest patterns for youth of different racial/ethnic backgrounds. A very significant finding was that youth who were in the program longer did better than those who were in for shorter periods. The comprehensive gang-program approach significantly hastened aging out of the gang structure and process.

The analysis of program strategy revealed that collaboration among social-development workers was more effective in reducing arrests, particularly for younger youth engaged, or about to be engaged, in violent behavior. Social-development (opportunities provision) contacts alone were especially effective in reducing drug-use and drug-selling arrests. However, the interaction of social-development and suppression strategies was more effective with chronic offenders and older youth than were social-development contacts alone, particularly for those youth with preprogram arrests for property crime. Overall, larger program dosages (multiple providers and greater frequency and duration of services) proved to be important and were associated with reduced levels of arrests for violent crimes, suggesting that an appropriate mix, interaction, and dosage of multiple strategies is required for different categories of program youth in a particular community, at different times and in different circumstances.

Launched in 2000, the Gang-Free Schools and Communities Initiative funded a school component within the Comprehensive Gang Model framework. The distinctive features of this program are the planning and implementation of special or enhanced programs within the school setting and the linking of the school component to community-based gang prevention, intervention, and suppression activities. Four sites—Houston, TX; Pittsburgh, PA; Miami-Dade County, FL; and East Cleveland, OH—participated in this program. The Pittsburgh and Houston Gang-Free Schools projects were most effective of the four sites in preliminary findings of an independent evaluation that was not completed. Nevertheless, these are two very promising school-based programs that continue to operate successfully.

The next funded version of the OJJDP Comprehensive Gang Model, the 2003 Gang Reduction Program, was implemented in Los Angeles, CA; Richmond, VA; Milwaukee, WI; and North Miami Beach, FL. Each test site faced a different gang problem. Independent evaluators Hayeslip and Cahill (2009) found that two of these sites proved effective during the time frame of the independent evaluation, Los Angeles and Richmond, and North Miami Beach (better known as the PanZOu Project) was assessed as "largely a success" (p. 268). Follow-up data collected by the PanZOu Project showed a 22% reduction in violent, property, and drug-related crimes after PanZOu began targeting services in this community. Outcomes for the Milwaukee program did not demonstrate evidence of effectiveness. Unfortunately, no outcome data were collected on clients served in the four programs.

The Los Angeles Gang Reduction Program (LA GRP) (L-2) site is located in the Boyle Heights area, 3 miles east of downtown Los Angeles. This area is home to a large immigrant population made up mostly of residents from Mexico and Central America. Five major gangs inhabit the target area. The LA GRP program model included prevention, intervention, and suppression components. Overall, Hayeslip and Cahill (2009) found the LA GRP to achieve full and effective implementation. Findings revealed that shots-fired calls declined significantly post-implementation, as did violent gang crimes. The trends for all gang crimes were similar in the target area and the comparison area, although gang crimes decreased more in the target area after the implementation.

The Richmond Gang Reduction and Intervention Program (GRIP) (L-2) target area consists of two police reporting sectors in south Richmond, Virginia. The target area is a suburban-type community of single-family homes and apartments. The area is transitioning from a middle-class to a working-class population, with an increase in Hispanic residents. Traditional "homegrown" African American gangs are the most prevalent gang presence in terms of membership and criminal activity. The Richmond GRIP has four program components including prevention, intervention, suppression, and reentry. Hayeslip and Cahill (2009) measured crime outcomes using police data on four different measures: serious violence, gang-related incidents, gang-related violent incidents, and drug incidents. Serious violent incidents, gang incidents, and serious violent gang crime all declined sharply following implementation of GRIP, while trends in the comparison area were relatively flat during the same period. In contrast, drug-related incidents increased in the target area and declined in the comparison area, contrary to the evaluation hypotheses.

A distinctly different comprehensive community approach to preventing and reducing gang violence was formed in Chicago in 1999, named CeaseFire Chicago (L-2). The unique feature of this model is in its singular focus on preventing street-level shootings. Undergirded by the public health model, the program approaches violence as an infectious disease. In the words of the founder, Gary Slutkin, "For violence, we're trying to interrupt the next event, the next transmission, the next violent activity" (as cited in Kotlowitz, 2008, p. 1). Hence the broader program goal is to change community norms about violence, both in the wider community and among its clients in the Campaign to Stop the Shooting. On the first front, the program mobilizes two key groups into action in the community, especially the clergy and residents who could be stirred to direct action, to promote the new norm that violence is unacceptable. On the second front, CeaseFire Chicago focuses on risky activities by a small number of carefully selected members of the community (mostly gang members), those with a high chance of either "being shot or being a shooter" in the immediate future. Change agents are mobilized to address key immediate causes of violence including norms regarding violence, on-the-spot decision making by individuals at risk of triggering violence, and the perceived risks and costs of involvement in violence among the targeted population. At the focal point of the intervention, outreach workers called "violence interrupters" (most of whom were former gang members) work alone or in pairs on the street mediating conflicts between gangs on the streets and in hospital emergency rooms, intervening to stem the cycle of retaliatory violence. Outreach workers also counsel young clients and connect them to a range of services.

Skogan and associates' (2008) independent evaluation found that the project significantly reduced homicides and shootings in six of the seven CeaseFire sites, in some of the most violent communities in Chicago. A project report (Ransford, Kane, Metzger, Quintana, & Slutkin, 2010) noted that "CeaseFire worked to decrease violence with 40 of the most violent gangs in Chicago, both with outreach clients and conflict mediations" (p. 94).

W. B. Miller's (1962) evaluation of a comprehensive Boston program is perhaps the most rigorous gang program evaluation ever conducted. (This program is not in the OJJDP Strategic Planning Tool only because the program no longer exists.) For 3 years, project staff in the Midcity Project (established in Roxbury, Boston, in 1954) provided intensive services to 400 members of seven gangs. This "total community" project consisted of three major program components, community organization, family service, and gang work. The project aimed to open channels of access to legitimate opportunities, especially in the education and employment areas. The project plan was comprehensive and unusually well implemented. Nevertheless, Miller's (1962) evaluation reported only a "negligible impact" (p. 187). However, the actual impact was larger than Miller's statistical analysis suggested, based on Hodgkinson and colleagues' (2009) recent systematic review of comprehensive gang programs.

A Comprehensive Framework for Integration of Delinquency and Gang Programs and Strategies

The Comprehensive Strategy for Serious, Violent, and Chronic Juvenile Offenders was introduced in Chapter 9, along with a brief indication of how its prevention component can provide coverage of the service needs of youth at risk for delinquency and gang involvement. Similarly, the graduated sanctions component of the Comprehensive Strategy can guide placement of court-referred youth in the array of community, juvenile court, and corrections programs suitable for both of these system clients. The key management tools for this purpose are validated risk and needs assessment instruments, a disposition matrix that guides placements in a manner that protects the public, and protocols for developing comprehensive treatment plans that facilitate matching effective services with offender treatment needs and evaluating those services on an ongoing basis (Lipsey et al., 2010). Use of this forward-looking administrative framework and advanced management tools should increase the capacity of state and local juvenile justice systems to more effectively prevent delinquency and gang participation.

At intake, all court-referred youth should be assessed for gang involvement in the course of administering a validated delinquency risk assessment instrument. If gang membership or regular association with gang members is suggested, an in-depth assessment for validation of gang membership should be performed to make a security risk classification before placement. It is important to determine a youth's level of gang involvement because this is associated with risk of recidivism. To ensure the safety of everyone, continuing validation of gang participation (and desistance progress) should be noted at each transition point in the juvenile justice system, adjudication, detention, and placement in a secure juvenile correctional facility. This designation must be removed when periodic reassessment indicates that the youth is no longer an active gang member.

Determining gang membership among youth referred to the juvenile justice system is often problematic. Juvenile court intake workers and court counselors rarely receive training that prepares them to assess referred youth's gang involvement. Formal risk assessment instruments are sometimes used by courts, but these typically include only one pertinent question, whether or not the referred youth is a gang member or associates with gang members. Much more information about the level of gang membership or risk of involvement would be very useful in the course of developing individualized, comprehensive treatment plans. Court, community program, and correctional staff should be provided training on assessing levels of gang involvement shown in Table 10.4.

Table 10.4 Individuals' Levels of Gang Involvement

Level I. Fantasy

- Knows about gangs primarily from the media (music, movies, and literature)
- May or may not know about "real" gangs
- May or may not like, respect, or admire a gang, gang member, or the gang lifestyle

Level II. At-Risk

- Knows about gangs and gang members firsthand
- Occasionally casually associates with gang members

- Lives in or near gang areas (turf)
- May like or admire gangs or gang members as individuals
- May like or admire the gang lifestyle, but does not actively participate

Level III. Associate

- Knows and likes gang members firsthand
- Regularly associates with gang members
- Considers gangs and related activity as normal, acceptable, and admirable
- Finds many things in common with gang members
- Is seriously thinking about joining a gang

Level IV. Gang Member

- Is officially a gang member
- Associates almost exclusively with gang members to the exclusion of family and former friends
- Participates in gang crimes and most other related activities
- Is not considered hard-core by fellow gang members or others
- Has substantially rejected the authority or value system of family and society

Level V. Hard-Core Gang Member

- Totally committed to the gang and gang lifestyle
- Totally rejects anyone or any value system other than the gang
- Is considered hard-core by self, other gang members, and authorities
- Will commit any act with the approval of a demand from the gang
- Does not accept any authority other than the gang

Source: Johnson, E. H. (1987). *Handbook on crime and delinquency prevention.* New York: Greenwood Press, pp. 1–12.

Court-referred active gang members should be supervised more closely in more restrictive placements, and these youth as well as those at high risk of gang involvement should be considered for case management by intervention teams and outreach workers. A management information system (MIS) recommended in the OJJDP Strategic Planning Tool should be used to execute individualized service plans. High-risk youth in both settings should be given priority for placement, and recommended intervention and suppression services and strategies should be made available for these clients on a priority basis. Reducing their recidivism and disengaging youth from gangs is of paramount importance because of their typically higher offending rate and more violent offenses.

Concluding Observations

There is good news herein about the availability of effective gang programs. The evidence base for gang programming has grown enormously in the past decade. In

1998, there were no known effective gang programs. There are now 12 gang programs with acceptable scientific support of effectiveness, and several other nongang programs have demonstrated evidence of effectiveness in preventing or reducing gang activity. However, there are three important caveats. First, none of them has showed dramatic effects. Some criminologists expect large impacts from gang programs including Klein and Maxson (2006), but it is unreasonable to expect uniformly large recidivism reductions with gang members, especially not among stable gang members, because of what Vigil (2006) calls their "multiple marginality"; typically a large number of risk factors often appear in multiple developmental domains (discussed in Chapter 5), and the extremely criminogenic environments in which most of them reside throughout program participation and afterwards (discussed in Chapter 1). Even marginally reduced gang crime and individual involvement would be of great practical value in most communities. A meta-analysis is needed to determine the average effect sizes of the effective programs reviewed here, though we can rest assured that this statistic will not be large. But on a brighter note, four of the evidence-based gang programs were demonstrated to be effective in Chicago and Los Angeles (two in each city), indisputably the reigning capitals of gang violence. This is a major source of encouragement, particularly given that anti-gang program development is yet in its infancy. Only one gang research and development program has been undertaken in recent history; the Comprehensive Gang Model. However, it has produced considerable evidence of effectiveness in five controlled studies, none of which was chosen by its developer.

The second qualification on the good news reported here is that no program has conclusively prevented gang joining and none has prevented the formation of gangs. But has any program demonstrated effectiveness in preventing the formation of delinquent groups? The first GREAT model failed to prevent gang joining, although preliminary results of the second iteration of it suggesting that it does are based only on a 1-year follow-up. Thus, Finn Esbensen cautions that "it's too soon to know" and it will be several years before his GREAT evaluation will be completed. There also is some basis for optimism that early prevention and intervention programs can prevent gang joining, such as the Montreal Preventive Treatment Program. The body of developmental research contains many examples of early interventions that target precursor behaviors and risk factors prevent later deviance from developing (OJJDP Model Programs Guide, Box 9.2).

The third important caveat is that only one of the evidence-based gang programs, the Comprehensive Gang Model, has demonstrated conclusive evidence of effectiveness with girls and older females. Across multiple program sites, services were actually more effective with females than with males. However, the research does not provide guidance with respect to specific services that are effective with them. But meta-analyses show that juvenile delinquency prevention and treatment programs are about equally effective with girls and boys. For example, two model delinquency reduction programs, Multidimensional Treatment Foster Care (MFTC) and Multisystemic Therapy, have produced comparable cross-gender outcomes. In Chapter 9, attention was drawn to a model gender-sensitive program that has demonstrated effectiveness with very aggressive girls under age 12 (and in building protective mother-daughter relations): the Toronto-based Earlscourt Girls Connection. In addition, neighborhood and community residential centers hold very promising potential for meeting girls' needs in ganglands, for whom moving out to opportunity is not an immediate option.

To optimize the chances for success, communities and neighborhoods that have gangs need to undertake a comprehensive assessment that begins with risk factors. The National Gang Center has developed an assessment protocol that any community can use to assess its gang problem, which guides its development of a gang prevention, intervention, and suppression continuum of programs and strategies. Resource materials that assist communities in developing an action plan to implement the Comprehensive Gang Model are also available. Prevention programs are needed to target youths at risk of gang involvement, to reduce the number of youths who join gangs; intervention programs and strategies are needed to provide sanctions and services for younger youths who are actively involved to separate them from gangs; and law enforcement suppression strategies are needed to target the most violent gangs and older, criminally active gang members. A balanced and integrated approach is most likely to be effective.

Another theme in the evidence-based gang programs reviewed here is the positive results from a number of multidisciplinary or comprehensive programs. The integration of strategies (e.g., provision of support in separating from gangs with suppression that discourages further involvement in gang crime) and services (multiple service providers and high dosage levels) has emerged as a predominant feature of effective gang programs. In addition, Hodgkinson and colleagues' (2009) preliminary systematic review of comprehensive gang programs suggests that the key features of effective programs are (1) case management (an intervention team), (2) community involvement in the planning and delivery of interventions, and (3) expertise sharing among involved agencies. The National Gang Center (2010 a) also finds that outreach workers play a critically important role.

In the systematic review of comprehensive gang programs internationally, Hodgkinson's team (2009) found the higher quality studies that produced positive effects included one or more of the following mechanisms of change:

- A case management strategy for service provision that was personalized to individual offenders
- Community involvement in the planning and delivery of interventions
- Expertise sharing between agencies
- Provision of incentives to gang members to change their offending behavior, including educational opportunities, tattoo removal, and financial assistance (pp. 2–3)

The next frontier is the integration of program strategies, services, outreach, and sanctions deployed by multiple agencies in a systematic way for gang members. Although criminal justice systems lag behind, many juvenile courts and correctional systems are using structured decision-making tools to assess risk and treatment needs, make program placements accordingly, and develop comprehensive treatment plans that link offenders to evidence-based programs. Special attention needs to be given to gang involvement in the risk calculus, but only in objective decision making based on actual risk. Many of the existing evidence-based delinquency rehabilitation programs will work with gang members because they share common risk factors and treatment needs.

DISCUSSION TOPICS

1. How would you go about helping an active gang member negotiate his or her way out of a serious gang?

2. Why is law enforcement suppression sometimes effective when integrated with services?

3. If you were a state or federal program manager in charge of designing an anti-gang program, describe the key components you would require.

4. Revisit the Chapter 9 class exercise you carried out for prevention programs using the OJJDP Strategic Planning Tool (www.nationalgangcenter.gov/SPT) to build an ideal program continuum. Now, add intervention and suppression programs and strategies.

5. How would you go about tailoring your continuum of intervention and suppression programs and strategies to adolescent girls and young women?

RECOMMENDATIONS FOR FURTHER READING

Gang Program Evaluations

Howell, J. C. (1998). Promising programs for youth gang violence prevention and intervention. In R. Loeber and D. P. Farrington (Eds.), *Serious and violent juvenile offenders: Risk factors and successful interventions* (pp. 284–312). Thousand Oaks, CA: Sage.

Howell, J. C. (2000). *Youth gang programs and strategies.* Washington, DC: U.S. Department of Justice, Office of Juvenile Justice and Delinquency Prevention.

Howell, J. C. (2010a). Gang prevention: An overview of current research and programs. *Juvenile Justice Bulletin.* Washington, DC: U.S. Department of Justice, Office of Juvenile Justice and Delinquency Prevention.

Howell, J. C. (2010b). Lessons learned from gang program evaluations: Prevention, intervention, suppression, and comprehensive community approaches. In R. J. Chaskin (Ed.), *Youth gangs and community intervention: Research, practice, and evidence* (pp. 51–75). New York: Columbia University Press.

Gun Availability and Gang Violence

Cook, P. J., & Ludwig, J. (2006). The social costs of gun ownership. *Journal of Public Economics, 90,* 379–391.

Lizotte, A. J., Tesoriero, J. M., Thornberry, T. P., & Krohn, M. D. (1994). Patterns of adolescent firearms ownership and use. *Justice Quarterly, 11,* 51–73.

Sheley, J. F., & Wright, J. D. (1993). Gun acquisition and possession in selected juvenile samples. *Research in Brief.* Washington, DC: National Institute of Justice and Office of Juvenile Justice and Delinquency Prevention.

Sheley, J. F., & Wright, J. D. (1995). *In the line of fire: Youth, guns and violence in urban America.* Hawthorne, NY: Aldine De Gruyter.

Watkins, A., Huebner, B., & Decker, S. (2008). Patterns of gun acquisition, carrying and use among juvenile and adult arrestees: Evidence from a high-crime city. *Justice Quarterly, 25,* 674–700.

Program Structures Versus Program Services

Howell, J. C. (2009). *Preventing and reducing juvenile delinquency: A comprehensive framework* (2nd ed.). Thousand Oaks, CA: Sage.

Lipsey, M. W., Howell, J. C., Kelly, M. R., Chapman, G., & Carver, D. (2010). *Improving the effectiveness of juvenile justice programs: A new perspective on evidence-based practice.* Washington, DC: Georgetown University, Center for Juvenile Justice Reform.

Desistance From Gang Activity

Decker, S. H., & Lauritsen, J. L. (1996). Breaking the bonds of membership: Leaving the gang. In C. R. Huff (Ed.), *Gangs in America* (pp. 103–122). Thousand Oaks, CA: Sage.

Kissner, J., & Pyrooz, D. C. (2009). Self-control, differential association, and gang membership: A theoretical and empirical extension of the literature. *Journal of Criminal Justice, 37,* 478–487.

Peterson, D., Taylor, T. J., & Esbensen, F. (2004). Gang membership and violent victimization. *Justice Quarterly, 21,* 793–815.

Pyrooz, D. C., Decker, S. H., & Webb, V. J. (Forthcoming). The ties that bind: Desistance from gangs. *Crime and Delinquency.*

Vigil, J. D. (1988). *Barrio gangs: Street life and identity in Southern California.* Austin: University of Texas Press.

First Steps/Assessment

National Gang Center. (2010a). *Best practices to address community gang problems: OJJDP's Comprehensive Gang Model.* Washington, DC: Author.

Evidence-Based Gang Programs

Level 3 (Promising Programs)

Howell, J. C. (2000). *Youth gang programs and strategies.* Washington, DC: U.S. Department of Justice, Office of Juvenile Justice and Delinquency Prevention.

Howell, J. C. (2009). *Preventing and reducing juvenile delinquency: A comprehensive framework* (2nd ed.). Thousand Oaks, CA: Sage.

Singular Suppression

Braga, A. A., Pierce, G. L., McDevitt, J., Bond, B. J., & Cronin, S. (2008). The strategic prevention of gun violence among gang-involved offenders. *Justice Quarterly, 25,* 132–162.

Bureau of Justice Assistance. (1997). Urban street gang enforcement. Washington, DC: U.S. Department of Justice, Bureau of Justice Assistance.

Bynum, T. S., & Varano, S. P. (2003). The anti-gang initiative in Detroit: An aggressive enforcement approach to gangs. In S. H. Decker (Ed.), *Policing gangs and youth violence* (pp. 214–238). Belmont, CA: Wadsworth/ Thompson Learning.

Ludwig, J. (2005). Better gun enforcement, less crime. *Criminology and Public Policy, 4,* 677–716.

Petrosino, A., Turpin-Petrosino, C., & Buehler, J. (2003). Scared Straight and other juvenile awareness programs for preventing juvenile delinquency: A systematic review of the randomized experimental evidence. *The Annals of the American Academy of Political and Social Science, 589,* 41–62.

Gang Intelligence Databases

Barrows, J., & Huff, C. R. (2009). Gangs and public policy: Constructing and deconstructing gang databases. *Criminology and Public Policy, 8,* 675–703.

Bjerregaard, B. (2003). Anti-gang legislation and its potential impact: The promises and the pitfalls. *Criminal Justice Policy Review, 14,* 171–192.

Implementation Problems

Decker, S. H., & Curry, G. D. (2002b). "I'm down for my organization": The rationality of responses to delinquency, youth crime and gangs. In A. R. Piquero & S. G. Tibbits (Eds.), *Rational choice and criminal behavior* (pp. 197–218). New York: Routledge.

Effective Integration Strategies

Office of Juvenile Justice and Delinquency Prevention. (2009b). *OJJDP Comprehensive Gang Model: Planning for implementation.* Washington, DC: U.S. Department of Justice, Office of Juvenile Justice and Delinquency Prevention.

Project Safe Neighborhoods

Healy, G. (2002). *There goes the neighborhood: The Bush-Ashcroft plan to "help" localities fight gun crime.* Policy Analysis No. 440. Washington, DC: Heritage Foundation.
Katz, C. M., & Webb, V. J. (2006). *Policing gangs in America.* New York: Cambridge University Press.
Ludwig, J. (2005). Better gun enforcement, less crime. *Criminology and Public Policy, 4,* 677–716.

Notes

1. One such initiative is underway at the University of Maryland. This program of research involves testing the adaptability of "blueprint" delinquency programs to gang members, which bears close watch.

2. The term *evidence based,* refers to programs rated either as "exemplary" (L–1) or "effective" (L-2). Tennessee legislation makes a useful distinction in characterizing L-1 programs as evidence based, and L-2 programs as research based (Tennessee Code Annotated, Title 37, Chapter 5, Part 1, Section 1, 2007).

3. Personal communication, Ben Sanders, Methodist Home for Children, March 22, 2011.

4. This widely acclaimed effective program only rated "promising" in the Model Program Guide owing mainly to the absence of a control group in the evaluation design.

5. Wright (2005) argues that placing a person in gang or criminal intelligence databases increases the likelihood of a criminal conviction and a longer sentence, and this may violate the due process requirements of the Fifth and Fourteenth Amendments. Minorities are more likely than Whites and other groups to have official records for the same crimes. Others, including Barrows and Huff (2009), Bjerregaard (2003), and S. Moore (2007), assert that constitutional (due process) issues come into play when a nebulous status such as *gang member* is used as evidence for enhanced criminal sentences.

Glossary

Age blocks: Middle childhood, ages 7 to 9; late childhood, ages 10 to 12; early adolescence, ages 13 to 15; late adolescence, ages 16 to 19; early adulthood, ages 20 to 25 (Loeber et al., 2008).

Anglo: Early Americans largely of English heritage and English territories (including the Scotch and Irish), and also Dutch, German, Swedish, and Scandinavian peoples (Pincus & Ehrlich, 1999, pp. 223–228). Also, in contemporary usage, *Anglo* is used to distinguish Americans (non-Hispanics) from Mexican Americans, Hispanics, or Latinos (Telles & Ortiz, 2008).

Backup: "Having a network of friends to protect a person from attack, with force if necessary" (Sullivan, 2006, p. 25).

Barrio: A neighborhood in a Mexican community (Vigil, 1988); also Puerto Rican neighborhoods or communities, such as East Harlem, known as El Barrio (Bourgois, 2003).

Block gangs: When adolescents in a bona fide gang strongly identify with their street block which they represent (see also the definition for *represent*).

Chicano: "An identity that symbolized cultural and political autonomy for Mexican Americans rather than assimilation and acceptance as white," particularly in association with the Chicano movement (Telles & Ortiz, 2008, pp. 92–93). This term also is used "to reflect the multiple heritage experience of Mexicans in the U.S., comprised of Indian, Spanish, Mexican, and Anglo backgrounds" (Vigil, 1998, pp. 1, 251–255).

Cholo: Marginal; living in a period between dominant races or cultures; a subcultural style of dress, walk, talk, values, and norms that reflect the mixed nature of Mexican, American, and street experiences (Vigil, 1993).

Claim: To declare one's gang affiliation or membership; to represent a gang.

Dissed: Disrespected.

Drug-related dispute: An argument associated with drug transactions, use, or a combination (Valdez et al., 2009).

Gang retaliation or revenge: Retribution associated with specific gang incident (Valdez, et al., 2009).

Gang rivalry: An ongoing feud between two gangs (Valdez et al., 2009).

Gang solidarity: Expression of shared goals, norms, and aims among gang members (camaraderie) (Valdez et al., 2009).

Gang sweeps: Police teams are deployed to a specific area with the main objective of arresting gang members (Klein, 1995, 2004).

Gangsta: Gang member.

Gangsta rap: Music popularized in the early 1990s that extols the gang lifestyle (Ro, 1996).

Go down: To attack or declare war on another gang (Perkins, 1987).

Green light: A Mexican Mafia mechanism for enforcing "taxing" (extorting) a percentage of barrio gangs' street-level drug sales. If a barrio did not pay up, barrio members in prison would be threatened or killed—green lighted (Vigil, 2007, p. 69).

Heart: Courage (Perkins, 1987).

Heater: Gun (Perkins, 1987).

Hispanic: "This term is used particularly by federal and state bureaucracies to refer to persons who reside in the United States who were born in, or trace their ancestry back to, one of 23 Spanish-speaking nations" (Moore & Pinderhughes, 1993, p. xi). Many of these individuals prefer to use the term *Chicano, Latino,* or *Mexican American.*

Homeboy: A fellow gang member; also a friend in the neighborhood.

Hood: The neighborhood.

Hybrid gang culture: Gang culture characterized by a mixture of graffiti and symbols (that are copied from other gangs); less concern over turf or territory; members of mixed race or ethnicity; members may belong to more than one gang; members may switch from one gang to another (Starbuck et al., 2001).

Jap: To ambush or attack someone (Perkins, 1987).

Klika: Spanish word for clique, or cohort within the gang (Vigil, 1988).

Latino: See *Hispanic.*

Locality: The major types of named place units found in the United States including cities, suburban areas, and counties (W. Miller, 2001).

Loco: Generally unrestrained conduct (J. Moore, 1991).

Locos: Crazies; usually in the gang context; shorthand for *vatos locos* (Vigil, 1988).

Locura: "A mind-set complex that values crazy or unpredictable thinking and acting" (Vigil, 1988, p. 178).

Muy loco: Getting drunk, loaded with drugs, and fighting (J. Moore, 1991).

Nor, Norteño: Northern California gangs (Al Valdez, 2007).

Odds ratio: A measure of the strength of the relationship between two variables (i.e., the number of times more likely).

O.G: Original Gangster; long-time gang member.

Pachucos: A transitional form of gang that evolved out of the *palomilla* cohorting tradition (age-graded groupings), the youngest of which might be considered "boy gangs" (Bogardus, 1926); more generally, the *pachuco* (subcultural) lifestyle of early Mexican Americans (J. Moore, 1978; Vigil, 1988, 1990).

Packing: Carrying a firearm (Perkins, 1987).

Palomilla: A Mexican term for an age or sex cohort (Vigil, 1988).

Persistent gang problem: Consistent reports of gang problems across all survey years (Egley et al., 2004). See *variable gang problem.*

Piece: A weapon, usually a gun (Perkins, 1987).

To play the dozen: To speak disparagingly of another person's family (Perkins, 1987).

Punk: "A boy who won't fight; a boy with no gang affiliation" (Perkins, 1987).

Racketeering: "Patterns of craft governance found in trades like construction, laundry, and kosher foods, where unions and associations enforced standard prices and wages through strikes, boycotts, and violence" (Cohen, 2003, p. 576).

Rep: "Reputation, usually a fighting reputation" (Perkins, 1987, p. 77).

Represent: "To assert a social claim to a social group or a very specific local area, such as a city block or a housing project, with the clear implication that the members of the group or the youthful residents of the area would provide backup" (Sullivan, 2006, p. 26).

Rolled on: "When several girls ambush one girl or a large group ambushes a smaller group" (Ness, 2010, p. 71).

Rolling out: A gang exit rite that entails a physical beating by several gang members (Valdez et al., 2009).

Rumble: Gang fight.

Set tripping: An insult or a challenge, often the catalyst for gang-related violence (Arciaga, Sakamoto, & Jones, 2010).

Shorty: Younger gang member (Papachristos, 2001).

Square: "One who doesn't like gangs" (Perkins, 1987, p. 78).

Sur, Sureños: Southern California gangs (Al Valdez, 2007).

Throw down: Challenge another gang or gang member.

Trajectories (in criminology): The classification of individuals according to their pattern of offending over time (Lacourse et al., 2003; Piquero, 2008); also displays of communities' distinct crime and gang histories across time and space (Griffiths & Chavez, 2004; Tita et al., 2005), and dimensions of cities' gang problem histories (Howell et al., 2011).

Turf: "A gang's own territory; also street or sidewalk" (Perkins, 1987).

V.L. (Vata Loca): Crazy dudette (female gangster) (Al Valdez, 2007).

V.L. (Vato Loco): Crazy dude (male gangster) (Al Valdez, 2007).

Variable gang problem: Reported gang activity in one or more years but not in every year (Egley et al., 2004). See *persistent gang problem.*

Vata loco: "Literally, crazy guy; refers to wild and violent behavior" (Vigil, 1988, p. 180).

Veterano: "Veteran; an older gang member who has been through it all and is now a role model for younger members" (Vigil, 1988, p. 180).

Walked into the gang: An informal initiation ritual for Mexican youth who have lived in the barrio most of their lives and grown up with others in street socialization. In these instances, the ritual "is a pro forma affair—a couple of punches by close friends and they are in" (Vigil, 2007, p. 58).

Warlord: "Prime minister of gang or one who plans strategy" (Perkins, 1987, p. 79).

Zip gun: "A homemade gun capable of firing a .22 caliber bullet, made with a car [radio] aerial and a wooden stock or cap pistol frame" (Perkins, 1987, p. 79).

References

Abbott, A. (1997). Of time and space: The contemporary relevance of the Chicago school. *Social Forces, 75,* 1149–1182.

Acoca, L. (1998). Outside/inside: The violation of American girls at home, on the streets, and in the juvenile justice system. *Crime & Delinquency, 44,* 561–589.

Acoca, L. (1999). Investing in girls: A 21st century strategy. *Juvenile Justice, 6,* 3–13.

Acuna, R. (1981). *Occupied America.* New York: Harper & Row.

Adamshick, P. Z. (2010). The lived experience of girl-to-girl aggression in marginalized girls. *Qualitative Health Research, 20,* 541–555.

Adamson, C. (1998). Tribute, turf, honor and the American street gang: Patterns of continuity and change since 1820. *Theoretical Criminology, 2,* 57–84.

Adamson, C. (2000). Defensive localism in white and black: A comparative history of European-American and African-American youth gangs. *Ethnic and Racial Studies, 23,* 272–298.

Adler, F. (1975a). The rise of the female crook. *Psychology Today, 9,* 42–46, 112–114.

Adler, F. (1975b). *Sisters in crime: The rise of the new female criminal.* New York: McGraw-Hill.

Adler, P., Ovando, C., & Hocevar, D. (1984). Family correlates of gang membership: An exploratory study of Mexican-American youth. *Hispanic Journal of Behavioral Sciences, 6,* 65–76.

The Advancement Project. (2007). *Citywide gang activity reduction strategy: Phase III report.* Los Angeles: Author.

Agnew. R. (2005). *Why do criminals offend? A general theory of crime and delinquency.* Los Angeles: Roxbury Publishing Company.

Alba, R., & Ne, V. (2003). *Remaking the American mainstream: Assimilation and contemporary immigration.* Cambridge, MA: Harvard University Press.

Aldridge, J., Shute, J., Ralphs, R., & Medina, J. (2009). Blame the parents? Challenges for parent-focused programmes for families of gang-involved young people. *Children & Society,* 1–11.

Alonso, A. A. (2004). Racialized identities and the formation of black gangs in Los Angeles. *Urban Geography, 25,* 658–674.

Alvarez, A., & Bachman, R. (1997). Predicting the fear of assault at school and while going to and from school in an adolescent population. *Violence and Victims, 12,* 69–85.

American Bar Association. (2001). *Zero tolerance report.* Chicago: American Bar Association, Criminal Justice Section.

American Probation and Parole Association and Institute for Intergovernmental Research. (2011). *Guidelines to gang reentry.* Lexington, KY: Author. Available on compact disk from the APPA: appa@csg.org.

Anbinder, T. (2001). *Five points.* New York: Free Press.

Anderson, E. (1998). The social ecology of youth violence. *Crime and Justice, 24,* 65–104.

Anderson, E. (1999). *Code of the street: Decency, violence, and the moral life of the inner city.* New York: W. W. Norton.

Anderson, N. (1926). *The hobo.* Chicago: University of Chicago Press.

Arbreton, A. J. A., & McClanahan, W. (2002). *Targeted outreach: Boys & Girls Clubs of America's approach to gang prevention and intervention.* Philadelphia: Public/Private Ventures.

Arciaga, M. (2001). *Evolution of prominent youth subcultures in America.* Tallahassee, FL: Institute for Intergovernmental Research, National Gang Center.

Arciaga, M. (2007). Multidisciplinary gang intervention teams. *NYGC Bulletin No. 3.* Tallahassee, FL: National Youth Gang Center.

Arciaga, M., Sakamoto, W., & Jones, E. F. (2010). Responding to gangs in the school setting. *NYGC Bulletin No. 5.* Tallahassee, FL: Institute for Intergovernmental Research, National Gang Center.

Arredondo, G. F. (2004). Navigating ethno-racial currents: Mexicans in Chicago, 1919–1939. *Journal of Urban History, 30,* 399–427.

Asbury, H. (1927). *Gangs of New York: An informal history of the underworld*. New York: Knopf.

Augimeri, L. K., Koegl, C. J., Levene, K., & Slater, N. (2006). *A comprehensive strategy: Children under 12 in conflict with the law, "The Forgotten Group."* Toronto, CA: Center for Children Committing Offenses, Child Development Institute.

Auletta, K. (1982). *The underclass*. New York: Random House.

Baldry, A. C., & Farrington, D. P. (2007). Effectiveness of programs to prevent school bullying. *Victims and Offenders, 2,* 183–204.

Ball, R. A., & Curry, G. D. (1995). The logic of definition in criminology: Purposes and methods for defining "gangs." *Criminology, 33,* 225–245.

Bandes, S. (2000). To reform the LAPD, more civilian pressure is necessary. *Los Angeles Times* (November 19): M6.

Bankston, C. L. (1998). Youth gangs and the new second generation: A review essay. *Aggression and Violent Behavior, 3,* 33–45.

Barnoski, R. (2004). *Outcome evaluation of Washington State's research-based programs for juvenile offenders.* Document No. 04–01–1201. Olympia: Washington State Institute for Public Policy. Retrieved from http://www.wsipp.wa.gov/

Barrows, J., & Huff, C. R. (2009). Gangs and public policy: Constructing and deconstructing gang databases. *Criminology and Public Policy, 8,* 675–703.

Batchelor, S. (2009). Girls, gangs and violence: Assessing the evidence. *Probation Journal, 56,* 399–414.

Batchelor, S., Joe-Laidler, K., & Myrtinen, H. (2010). *The other half: Girls in gangs. Small Arms Survey 2010.* Geneva, Switzerland: Small Arms Survey.

Battin, S. R., Hill, K. G., Abbott, R. D., Catalano, R. F., & Hawkins, J. D. (1998). The contribution of gang membership to delinquency beyond delinquent friends. *Criminology, 36,* 93–115.

Battin-Pearson S. R., Thornberry, T. P., Hawkins, J. D., & Krohn, M. D. (1998). Gang membership, delinquent peers, and delinquent behavior. *Juvenile Justice Bulletin. Youth Gang Series.* Washington, DC: Office of Juvenile Justice and Delinquency Prevention.

Beck, A., Gilliard, D., Greenfeld, L., Harlow, C., & Hester, T. (1993). *Survey of state prison inmates, 1991.* Washington, DC: U.S. Department of Justice, Bureau of Justice Statistics.

Beck, A. J., Harrison, P. M., & Guerino, P. (2010). Sexual victimization in juvenile facilities reported by youth: 2008–2009. *Special Report.* Washington, DC: U.S. Department of Justice, Bureau of Justice Statistics.

Becker, H. S. (1999). The Chicago school, so-called. *Qualitative Sociology, 22,* 3–12.

Beckett, M., Hawken, A., and Jacknowitz, A. (2001). *Accountability for after-school care: Devising standards and measuring adherence to them.* Santa Monica, CA: RAND.

Bell, J., & Lim, N. (2005). Young once, Indian forever: Youth gangs in Indian Country. *The American Indian Quarterly, 29,* 626–650.

Bellair, P. E., & McNulty, T. L. (2009). Gang membership, drug selling, and violence in neighborhood context. *Justice Quarterly, 26,* 644–669.

Benda, B. B., Corwyn, R. F., & Toombs, N. J. (2001). Recidivism among adolescent serious offenders: Prediction of entry into the correctional system for adults. *Criminal Justice and Behavior, 28,* 588–613.

Bendixen, M., Endresen, I. M., & Olweus, D. (2006). Joining and leaving gangs: Selection and facilitation effects on self-reported antisocial behaviour in early adolescence. *European Journal of Criminology, 3,* 85–114.

Bennett, W., DiIulio, J., & Waters, J. (1996). *Body count.* New York: Simon & Schuster.

Bennett, T., & Holloway, K. (2004). Gang membership, drugs and crime in the United Kingdom. *British Journal of Criminology, 44,* 305–323.

Bernard, W. (1949). *Jailbait.* New York: Greenberg.

Bernard, T. (1992). *The cycle of juvenile justice.* New York: Oxford University Press.

Bernburg, J. G., Krohn, M. D., & Rivera, C. J. (2006). Official labeling, criminal embeddedness, and subsequent delinquency: A longitudinal test of labeling theory. *Journal of Research in Crime and Delinquency, 43,* 67–88.

Best, J., & Hutchinson, M. M. (1996). The gang initiation rite as a motif in contemporary crime discourse. *Justice Quarterly, 13,* 383–404.

Bingenheimer, J. B., Brennan, R. T., and Earls, F. J. (2005). Firearm violence exposure and serious violent behavior. *Science, 308,* 1323–1326.

Bjerregaard, B. (2002a). Operationalizing gang membership: The impact measurement on gender differences in gang self-identification and delinquent involvement. *Women and Criminal Justice, 13,* 79–100.

Bjerregaard, B. (2002b). Self-definitions of gang membership and involvement in delinquent activities. *Youth and Society, 34,* 31–54.

Bjerregaard, B. (2003). Anti-gang legislation and its potential impact: The promises and the pitfalls. *Criminal Justice Policy Review, 14,* 171–192.

Bjerregaard, B. (2010). Gang membership and drug involvement: Untangling the complex relationship. *Crime and Delinquency, 56,* 3–34.

Bjerregaard, B., & Lizotte, A. J. (1995). Gun ownership and gang membership. *Journal of Criminal Law and Criminology, 86,* 37–58.

Bjerregaard, B., & Smith, C. (1993). Gender differences in gang participation, delinquency, and substance use. *Journal of Quantitative Criminology, 9,* 329–355.

Bloch, H. A. (1963). The juvenile gang: A cultural reflex. *The Annals of the American Academy of Political and Social Sciences, 347,* 20–29.

Bloch, H. A., & Niederhoffer, A. (1958). *The gang: A study in adolescent behavior.* New York: Philosophical Library.

Block, C. R., & Block, R. (1993). Street gang crime in Chicago. *Research in Brief.* Washington, DC: U.S. Department of Justice, National Institute of Justice.

Block, C. R., Christakos, A., Jacob, A., & Przybylski, R. (1996). *Street gangs and crime: Patterns and trends in Chicago.* Chicago: Illinois Criminal Justice Information Authority.

Block, R. (1977). *Violent crime.* Lexington, MA: Lexington Books.

Block, R. (2000). Gang activity and overall levels of crime: A new mapping tool for defining areas of gang activity using police records. *Journal of Quantitative Criminology, 16,* 369–383.

Bloom, B., Owen, B., & Covington, S. (2006). *A summary of research, practice and guiding principles for women offenders.* Washington, DC: National Institute of Corrections.

Blumstein, A. (1995a). Violence by young people: Why the deadly nexus? *National Institute of Justice Journal* (August), 1–9.

Blumstein, A. (1995b). Youth violence, guns, and the illicit-drug industry. *Journal of Criminal Law and Criminology, 86,* 10–36.

Blumstein, A. (1996). Youth violence, guns, and illicit drug markets. *Research Preview.* Washington, DC: U.S. Department of Justice, National Institute of Justice.

Bogardus, E. S. (1926). *The city boy and his problems: A survey of boy life in Los Angeles.* Los Angeles: House of Ralston.

Bogardus, E. S. (1943). Gangs and Mexican-American youth. *Sociology and Social Research* (September), 55–56.

Bolithi, W. (1930). The psychosis of the gang. *Survey,* 501–506.

Bonta, J. (1996). Risk–needs assessment and treatment. In A. T. Harland (Ed.), *Choosing correctional options that work* (pp. 18–32). Thousand Oaks, CA: Sage.

Bookin-Weiner, H., & Horowitz, R. (1983). The end of the gang: Fact or fiction? *Criminology, 21,* 585–602.

Bordua, D. J. (1961). Delinquent Subcultures: Sociological interpretations of gang delinquency. *The Annals of the American Academy of Political and Social Sciences, 338,* 119–136.

Bourgois, P. (2001). *In search of respect: Selling crack in El Barrio* (2nd ed.). New York: Cambridge University Press.

Bowker, L. H. (1978a). Gangs and prostitutes: Two cases of female crime. In. L. H. Bowker (Ed.), *Women, crime, and the criminal justice system.* Lexington, MA: Lexington Books.

Bowker, L. H. (Ed.). (1978b). *Women, crime and the criminal justice system.* Lexington, MA: Lexington Books.

Bowker, L. H. (1998). *Masculinities and violence.* Thousand Oaks, CA: Sage.

Bowker, L. H., & Klein, M. W. (1983). The etiology of female juvenile delinquency and gang membership: A test of psychological and social structural explanations. *Adolescence, 18,* 739–751.

Boyle, J. (1977). *A sense of freedom.* London: Pan Books.

Bradshaw, P. (2005). Terrors and young teams: Youth gangs and delinquency in Edinburgh. In S.H. Decker & F. M. Weerman (Eds.), *European street gangs and troublesome youth groups* (pp. 193–218). Lanham, MD: AltaMira Press.

Braga, A. A. (2004). Gun violence among serious young offenders. *Problem-Oriented Guides for Police. Problem-Specific Guides Series* (Guide No. 23). Washington, DC: Office of Community Oriented Policing Services.

Braga, A. A., Kennedy, D. M., & Tita, G. E. (2002). New approaches to the strategic prevention of gang and group-involved violence. In C. R. Huff (Ed.), *Gangs in America III* (pp. 271–285). Thousand Oaks, CA: Sage.

Braga, A. A., Kennedy, D. M., Waring, E. J., & Piehl, A. M. (2001). Problem-oriented policing, deterrence, and youth violence: An evaluation of Boston's Operation Ceasefire. *Journal of Research in Crime and Delinquency, 38,* 195–225.

Braga, A. A., Papachristos, A. V., & Hureau, D. M. (2010). The concentration and stability of gun violence at micro places in Boston, 1980–2008. *Journal of Quantitative Criminology, 26,* 33–53.

Braga, A. A., & Pierce, G. L. (2005). Disrupting illegal firearms markets in Boston: The effects of Operation

Ceasefire on the supply of new handguns to criminals. *Criminology and Public Policy, 4,* 717–748.

Brame, B., Nagin, D. S., & Tremblay, R. E. (2001). Developmental trajectories of physical aggression from school entry to late adolescence. *Journal of Child Psychology and Psychiatry and Allied Disciplines, 42,* 503–512.

Brezina, T., Agnew, R., Cullen, F. T., & Wright, J. P. (2004). The code of the street: A quantitative assessment of Elijah Anderson's subculture of violence thesis and its contribution to youth violence research. *Youth Violence and Juvenile Justice, 2,* 303–328.

Briggs, X., Popkin, S. J., & Goerin, J. (2010). *Moving to opportunity: The story of an American experiment to fight ghetto poverty.* Oxford, UK: Oxford University Press.

Broidy, L. M., Tremblay, R. E., Brame, B., Fergusson, D., Horwood, J. L., Laird, R., et al. (2003). Developmental trajectories of childhood disruptive behaviors and adolescent delinqeuncy: A six-site, cross-national study. *Developmental Psychology, 39,* 222–245.

Bronfenbrenner, U. (1979). *The ecology of human development: Experiments by nature and design.* Cambridge, MA: Harvard University Press.

Brotherton, D. C. (1996). The contradictions of suppression: Notes from a study of approaches to gangs in three public high schools. *Urban Review, 28,* 95–117.

Brown, W. K. (1977). Black female gangs in Philadelphia. *International Journal of Offender Therapy and Comparative Criminology, 21,* 221–228.

Brown, W. K. (1999). Black female gangs in Philadelphia. In M. Chesney-Lind & J. Hagedorn (Eds.), *Female gangs in America: Essays on girls, gangs, and gender* (pp. 57–63). Chicago: Lake View Press.

Brownstein, H. (1996). *The rise and fall of a violent crime wave: Crack cocaine and the social construction of a crime problem.* Guilderland, NY: Harrow and Heston.

Bruhns, K., & Wittman, S. (2002). '*Ich mein, mit der Gewalt kannst du dir Respect verschaffen': Mädchen und junge Frauen in gewaltbereiten Jugendgruppen.* Opladen, Germany: Leske & Budrich.

Bryant, D. (1989). Communitywide responses crucial for dealing with youth gangs. *Juvenile Justice Bulletin.* Washington, DC: U.S. Department of Justice, Office of Juvenile Justice and Delinquency Prevention.

Bucerius, S. M. (2010). Fostering academic opportunities to counteract social exclusion. In N. A. Frost, J. D. Freilich, & T. R. Clear (Eds.), *Contemporary issues in criminal justice policy: Policy proposals from the American Society of Criminology* (pp. 235–245). Belmont, CA: Wadsworth.

Bullock, K., & Tilley, N. (2002). *Shooting, gangs and violent incidents in Manchester: Developing a crime reduction strategy.* London: Home Office.

Burgess, E. W. (1925). The growth of the city: An introduction to a research project. In R. E. Park & E. W. Burgess (Eds.), *The city* (pp. 47–62). Chicago: University of Chicago Press.

Burke, J. D., Loeber, R., & Birmaher, B. (2002). Oppositional defiant disorder and conduct disorder: A review of the past 10 years, Part II. *Journal of the American Academy of Child and Adolescent Psychiatry, 41,* 1275–1293.

Bursik, R. J., Jr., & Grasmick, H. G. (1993). *Neighborhoods and crime: The dimensions of effective community control.* New York: Lexington.

Burton, F., & West, B. (2009). *When the Mexican drug trade hits the border.* Stratfor Global Intelligence (April). Retrieved from http://www.stratfor.com/weekly/20090415_when_mexican_drug_trade_hits_border

Bynum, T. S., & Varano, S. P. (2003). The anti-gang initiative in Detroit: An aggressive enforcement approach to gangs. In S. H. Decker (Ed.), *Policing gangs and youth violence* (pp. 214–238). Belmont, CA: Wadsworth/Thompson Learning.

Cairns, R. B., & Cairns, B. D. (1994). *Lifelines and risks: Pathways of youth in our time.* New York: Cambridge University Press.

Caldwell, M. F., & Van Rybroek, G. (2001). Efficacy of a decompression treatment model in the clinical management of violent juvenile offenders. *International Journal of Offender Therapy and Comparative Criminology, 45,* 469–477.

Caldwell, M. F., Vitacco, M., & Van Rybroek, G. J. (2006). Are violent delinquents worth treating? A cost-benefit analysis. *Journal of Research in Crime and Delinquency, 43,* 148–168.

Camp, G. M, & C. G. Camp (Eds.). (1985). *Prison gangs: Their extent, nature and impact on prisons.* Washington, DC: U.S. Department of Justice.

Camp, C. G., & G. M. Camp. (1988). *Management strategies for combating prison gang violence.* South Salem, NY: Criminal Justice Institute.

Campbell, A. (1984/1991). *The girls in the gang: A report from New York City.* New York: Basil Blackwell.

Campbell, A. (1990). Female participation in gangs. In C. R. Huff (Ed.), *Gangs in America* (pp. 163–182). Newbury Park, CA: Sage.

Campbell, A. (1999). Self-definition by rejection: The case of gang girls. In M. Chesney-Lind & J. Hagedorn (Eds.), *Female gangs in America: Essays on girls, gangs, and gender* (pp. 100–117). Chicago: Lake View Press.

Carbone-Lopez, K., Esbensen, F., & Brick, B. T. (2010). Correlates and consequences of peer victimization: Gender differences in direct and indirect forms of bullying. *Youth Violence and Juvenile Justice, 8,* 332–350.

Cartwright, D. S., Tomson, B., & Schwartz, H. (1975). *Gang delinquency.* Monterey, CA: Brooks/Cole.

Caspi, A., Lahey, B. B., & Moffitt, T. E. (2003). *The causes of conduct disorder and serious juvenile delinquency.* London: Guilford Press.

Catalano, R. F., & Hawkins, J. D. (1996). The social development model: A theory of antisocial behavior. In J. D. Hawkins (Ed.), *Delinquency and crime: Current theories* (pp. 149–197). New York: Cambridge University Press.

Cavan, R. (1927). *Suicide.* Chicago: University of Chicago Press.

Cepeda, A., & Valdez, A. (2003). Risk behaviors among young Mexican American gang associated females: Sexual relations, partying, substance use, and crime. *Journal of Adolescent Research, 18,* 90–106.

Chesney-Lind, M. (1989). Girls' crime and women's place: Toward a feminist model of female delinquency. *Crime & Delinquency, 35,* 5–29.

Chesney-Lind, M. (1993). Girls, gangs, and violence: Reinventing the liberated female crook. *Humanity and Society, 17,* 321–344.

Chesney-Lind, M. (1997). *The female offender: Girls, women, and crime.* Thousand Oaks, CA: Sage.

Chesney-Lind, M. (1999). Girls, gangs, and violence: Reinventing the liberated female crook. In M. Chesney-Lind & J. Hagedorn (Eds.), *Female gangs in America: Essays on girls, gangs, and gender* (pp. 295–310). Chicago: Lake View Press.

Chesney-Lind, M., & Hagedorn, J. (Eds.). (1999). *Female gangs in America.* Chicago: Lake View Press.

Chettleburgh, M. C. (2003). *Results of the 2002 Canadian Police Survey on Youth Gangs.* Toronto, Canada: Astwood Strategy Corporation.

Chettleburgh, M. C. (2007). *Young thugs: Inside the dangerous world of Canadian street gangs.* New York: HarperCollins.

Chicago Crime Commission. (2006). *The Chicago Crime Commission gang book.* Chicago: Author.

Chin, K. (1996). Gang violence in Chinatown. In C. R. Huff (Ed.), *Gangs in America* (pp. 157–184). Newbury Park, CA: Sage.

Chin, K. (2000). *Chinatown gangs: Extortion, enterprise and ethnicity.* New York: Oxford University Press.

Clampet-Lundquist, S., Edin, K., Kling, J. R., & Duncan, G. J. (2006). *Moving at-risk teenagers out of high-risk neighborhoods: Why girls fare better than boys.* Working paper, Industrial Relations Series. Princeton, NJ: Princeton University.

Clarke, R. V. (1995). Situational crime prevention. In M. Tonry & D. Farrington (Eds.), *Building a safer society: Strategic approaches to crime prevention* (pp. 91–150). Chicago: University of Chicago Press.

Cloward, R. A., & Ohlin, L. E. (1960). *Delinquency and opportunity: A theory of delinquent gangs.* New York: Free Press.

Cockburn, A., & St. Clair, J. (1998). *Whiteout: The CIA, drugs and the press.* London: Verso.

Cohen, A. K. (1955). *Delinquent boys: The culture of the gang.* Glencoe, IL: The Free Press.

Cohen, A. W. (2003). The racketeer's progress: Commerce, crime, and the law in Chicago, 1900–1940. *Journal of Urban History, 29,* 575–596.

Cohen, B. (1969). The delinquency of gangs and spontaneous groups. In T. Sellin & M. E. Wolfgang (Eds.), *Delinquency: Selected studies* (pp. 61–111). New York: John Wiley & Sons.

Cohen, J., Cork, D., Engberg, J., & Tita, G. E. (1998). The role of drug markets and gangs in local homicide rates. *Homicide Studies, 2,* 241–262.

Cohen, J., & Tita, G. E. (1999). Spatial diffusion in homicide: Exploring a general method of detecting spatial diffusion processes. *Journal of Quantitative Criminology, 15,* 451–493.

Cohen, L. E., & Felson, M. (1979). Social change and crime rate trends: A routine activity approach. *American Sociological Review, 44,* 588–608.

Cohen, M. A., Piquero, A. R., & Jennings, W. G. (2010). Estimating the costs of bad outcomes for at-risk youth and the benefits of early childhood interventions to reduce them. *Criminal Justice Policy Review, 21,* 391–434.

Cohen, S. (1980). *Folk devils and moral panics: The creation of the Mods and Rockers.* New York: Basil Blackwell.

Coie, J. D., & Dodge, K. A. (1998). The development of aggression and antisocial behavior. In N. Eisenberg (Ed.), *Handbook of child psychology: Vol. 3. Social, emotional, and personality development* (5th ed., pp. 779–861). New York: Wiley.

Coleman, J. S. (1988). Social capital in the creation of human capital. *American Journal of Sociology, 94,* 95–120.

Coleman, J. S. (1990). *Foundations of social theory.* Cambridge, MA: Harvard University Press.

Cook, P. J., & Laub, J. H. (1998). The unprecedented epidemic of youth violence. In M. Tonry & M. H. Moore (Eds.), *Youth violence* (pp. 27–64). Chicago: University of Chicago Press.

Cook, P. J., & Ludwig, J. (2006). The social costs of gun ownership. *Journal of Public Economics, 90,* 379–391.

Cooley, C. H. (1909). *Social organization.* New York: Scribners.

Cork, D. (1999). Examining space-time interaction in city-level homicide data: Crack markets and the diffusion of guns among youth. *Journal of Quantitative Criminology, 15,* 379–406.

Cornell, D., & Sheras, P. (2006). *Guidelines for responding to student threats of violence.* Longmont, CO: Sopris West.

Coughlin, B. C., & Venkatesh, S. A. (2003). The urban street gang after 1970. *Annual Review of Sociology, 29,* 41–64.

Covey, H. C. (2010). *Street gangs throughout the world.* Springfield, IL: Charles C Thomas.

Craig, W. M., Vitaro, F., Gagnon, C., & Tremblay, R. E. (2002). The road to gang membership: Characteristics of male gang and non-gang members from ages 10 to 14. *Social Development, 11,* 53–68.

Cressey, P. (1932). *The taxi dance hall.* Chicago: University of Chicago Press.

Criminal Intelligence Service Canada. (2006). *Annual report on organized crime.* Ottawa, Ontario: Criminal Intelligence Service Canada.

Cruz, J. M. (2010). Central American maras: From youth street gangs to transnational protection rackets. *Global Crime, 11,* 379–398.

Cummings, L. L. (1994). Fighting by the rules: Women street fighters in Chihuahua, Mexico, *Sex Roles, 30,* 189–198.

Cummings, S. (1993). Anatomy of a wilding gang. In S. Cummings & D. J. Monti (Eds.), *Gangs* (pp. 49–74). Albany: State University of New York Press.

Cummings, S., & Monti, D. J. (Eds.). (1993). *Gangs: The origins and impact of contemporary youth gangs in the United States.* Albany: State University of New York Press.

Cureton, S. R. (2009). Something wicked this way comes: A historical account of Black gangsterism offers wisdom and warning for African American leadership. *Journal of Black Studies, 40,* 347–361.

Currie, E. (1998). *Crime and punishment in America.* New York: Henry Holt.

Curry, G. D. (1998). Female gang involvement. *Journal of Research on Crime and Delinquency, 35,* 100–118.

Curry, G. D. (1999). Responding to female gang involvement. In M. Chesney-Lind & J. Hagedorn (Eds.), *Female gangs in America: Essays on girls, gangs, and gender* (pp. 133–153). Chicago: Lake View Press.

Curry, G. D. (2000). Self-reported gang involvement and officially recorded delinquency. *Criminology, 38,* 1253–1274.

Curry, G. D., & Decker, S. H. (2003). *Confronting gangs: Crime and community* (2nd ed.). Los Angeles: Roxbury.

Curry, G. D., Decker, S. H., & Egley, A. Jr. (2002). Gang involvement and delinquency in a middle school population. *Justice Quarterly, 19,* 275–292.

Curry, G. D., Egley, A., & Howell, J. C. (2004). *Youth gang homicide trends in the National Youth Gang Survey: 1999–2003.* Paper presented at the Annual Meeting of the American Society of Criminology, November. Nashville, TN.

Curry, G. D., & Spergel, I. A. (1988). Gang homicide, delinquency and community. *Criminology, 26,* 381–405.

Curry, G. D., & Spergel, I. A. (1992). Gang involvement and delinquency among Hispanic and African-American adolescent males. *Journal of Research in Crime and Delinquency, 29,* 273–291.

Dahmann, J. (1983). *Prosecutorial response to violent gang criminality—An evaluation of operation hardcore.* Washington, DC: National Institute of Justice.

Dahmann, J. (1995). Operation Hardcore: A prosecutorial response to violent gang criminality. In M. A. Klein, C. L. Maxson, & J. Miller (Eds.), *The modern gang reader* (pp. 301–303). Los Angeles: Roxbury Publishing Company.

Danelo, D. J (2009). Disorder on the border. *Proceedings,* (October), 45–47.

Daniels, S. (1987). Prison gangs: Confronting the threat. *Corrections Today, 66,* 126, 162.

Davis, A. F., & Haller, M. H. (1973). *The people of Philadelphia: A history of ethnic groups and lower-class life, 1790–1940.* Philadelphia: Temple University Press.

Davis, M. (2006). *City of quartz: Excavating the future in Los Angeles* (2nd ed.). New York: Verso.

Davis, M. S., & Flannery, D. J. (2001). The institutional treatment of gang members. *Corrections Management Quarterly, 5*(1), 38–47.

Dawley, D. (1992). *A nation of lords: The autobiography of the vice lords* (2nd ed.). Prospect Heights, IL: Waveland Press.

Deane, L., Bracken, D. C., & Morrissette, L. (2007). Desistance within an urban Aboriginal gang. *The Journal of Community and Criminal Justice, 54,* 125–141.

Debarbieux, E., & Baya, C. (2008). An interactive construction of gangs and ethnicity: The role of school segregation in France. In F. van Gemert, D. Peterson, & I.-L. Lien (Eds.), *Street gangs, migration and ethnicity* (pp. 211–226). Portland, OR: Willan Publishing.

Decker, S. H. (1996). Deviant homicide: A new look at the role of motives and victim-offender relationships. *Journal of Research in Crime and Delinquency, 33,* 427–449.

Decker, S. H. (2003). *Policing gangs and youth violence.* Belmont, CA: Wadsworth/Thompson Learning.

Decker, S. H. (2007). Youth gangs and violent behavior. In D. J. Flannery, A. T. Vazsonyi, & I. D. Waldman (Eds.), *The Cambridge handbook of violent behavior and aggression* (pp. 388–402). Cambridge, MA: Cambridge University Press.

Decker, S. H., Bynum, T., & Weisel, D. L. (1998). Gangs as organized crime groups: A tale of two cities. *Justice Quarterly, 15,* 395–423.

Decker, S. H., & Curry, G. D. (2000). Addressing key features of gang membership: Measuring the involvement of young members. *Journal of Criminal Justice, 28,* 473–482.

Decker, S. H., Katz, C. M., & Webb, V. J. (2008). Understanding the black box of gang organization: Implications for involvement in violent crime, drug sales, and violent victimization. *Crime and Delinquency, 54,* 153–172.

Decker, S. H., & Kempf-Leonard, K. (1991). Constructing gangs: The social definition of youth activities. *Criminal Justice Policy Review, 5,* 271–291.

Decker, S. H., & Lauritsen, J. L. (1996). Breaking the bonds of membership: Leaving the gang. In C. R. Huff (Ed.), *Gangs in America* (pp. 103–122). Thousand Oaks, CA: Sage.

Decker, S. H., & Pyrooz, D. C. (2010a). Gang violence worldwide: Context, culture, and country. *Small Arms Survey 2010.* Geneva, Switzerland: Small Arms Survey.

Decker, S. H., & Pyrooz, D. C. (2010b). On the validity and reliability of gang homicide: A comparison of disparate sources. *Homicide Studies, 14,* 359–376.

Decker, S. H., & Van Winkle, B. (1996). *Life in the gang: Family, friends, and violence.* New York: Cambridge University Press.

Decker, S. H., & Weerman, F. M. (Eds.). (2005). *European street gangs and troublesome youth groups.* Lanham, MD: AltaMira Press.

Dennehy, G., & Newbold, G. (2001). *The girls in the gang.* Auckland, New Zealand: Reed Publishing.

Deschenes, E. P., & Esbensen, F. (1999). Violence among girls: Does gang membership make a difference? In M. Chesney-Lind & J. Hagedorn (Eds.), *Female gangs in America: Essays on girls, gangs, and gender* (pp. 277–294). Chicago: Lake View Press.

Deuchar, R. (2009). *Gangs, marginalised youth and social capital.* Sterling, VA: Trentham Books.

Deutsch, L. (2000). Los Angeles police officer gets five years. *USA Today,* p. 6A.

Di Placido, C., Simon, T. L., Witte, T. D., Gu, D., & Wong, S. C. P. (2006). Treatment of gang members can reduce recidivism and institutional misconduct. *Law and Human Behavior, 30,* 93–114.

Diamond, A. J. (2001). Rethinking culture on the streets: Agency, masculinity, and style in the American city. *Journal of Urban History, 27,* 669–685.

Diamond, A. J. (2005). Gangs. In the *Encyclopedia of Chicago.* Chicago, IL: Chicago Historical Society. Retrieved from http://www.encyclopedia.chicago history.org/pages/497.html

DiIulio, J. J., Jr. (1995a). Arresting ideas. *Policy Review, 74,* 12–16.

DiIulio, J. J., Jr. (1995b, November 27). The coming of the super-predators. *Weekly Standard,* pp. 23–28.

DiIulio, J. J. (1997). Jail alone won't stop juvenile super-predators. *The Wall Street Journal,* (June 11), A23.

Dimitriadis, G. (2006). The situation complex: Revisiting Frederic Thrasher's *The Gang: A Study of 1,313 Gangs in Chicago. Cultural Studies and Critical Methodologies, 6,* 335–353.

Dodge, K. A. (2001). The science of youth violence prevention: Progressing from developmental epidemiology to efficacy to effectiveness to public policy. *American Journal of Preventive Medicine, 20*(1S), 63–70.

Dupere, V., Lacourse, E., Wilms, J. D., Vitaro, F., & Tremblay, R. E. (2007). Affiliation to youth gangs during adolescence: The interaction between childhood psychopathic tendencies and neighborhood disadvantage. *Journal of Abnormal Child Psychology, 35,* 1035–1045.

Dwyer, K., & Osher, D. (2000). *Safeguarding our children: An action guide.* Washington, DC: U.S. Departments of Education and Justice, American Institutes for Research.

Early, P. (1991). *The hot house: Life inside Leavenworth Prison.* New York: Bantam Books.

The Economist. (2010). Organized crime in Mexico: Under the volcano. *The Economist* (October 16), 29–31.

The Economist. (2011). The rot spreads: Organized crime in Central America. *The Economist* (January 22), 45–46.

Eckhart, D. (2001). Civil cases related to prison gangs: A survey of federal cases. *Corrections Management Quarterly, 5*(1), 60–65.

Eddy, P., Sabogal, H., & Walden, S. (1988). *The cocaine wars.* New York: W. W. Norton.

Egley, A., Jr. (2005). *Highlights of the 2002–2003 National Youth Gang Surveys* (OJJDP Fact Sheet, June 2005–01). Washington, DC: U.S. Department of Justice, Office of Juvenile Justice and Delinquency Prevention.

Egley, A., Jr., & Howell, J. C. (2010). *Gang activity, subgroups, and crime.* Paper presented at the annual meeting of the American Society of Criminology, San Francisco, November.

Egley, A., Jr., & Howell, J. C. (2011). *Highlights of the 2009 National Youth Gang Survey.* Washington, DC: Office of Juvenile Justice and Delinquency Prevention.

Egley, A., Jr., Howell, J. C., Curry, G. D., & O'Donnell, C. E. (2007). *Are the newer gangs different?* Paper presented at the annual meeting of the American Society of Criminology, Atlanta, November.

Egley, A., Jr., Howell, J. C., & Major, A. K. (2004). Recent patterns of gang problems in the United States: Results from the 1996–2002 National Youth Gang Survey. In F-A. Esbensen, S. G. Tibbetts, & L. Gaines (Eds.), *American youth gangs at the millennium* (pp. 90–108). Long Grove, IL: Waveland Press.

Egley, A., Jr. Howell, J. C., & Major, A. K. (2006). *National Youth Gang Survey: 1999–2001.* Washington, DC: U.S. Department of Justice, Office of Juvenile Justice and Delinquency Prevention.

Egley, A. E., O'Donnell, C. E., & Howell, J. C. (2009). *Over a decade of National Youth Gang Survey research: What have we learned?* Paper presented at the annual meeting of the American Society of Criminology, November. Philadelphia, PA.

Egley, A., Jr., & Ritz, C. E. (2006). *Highlights of the 2004 National Youth Gang Survey* (OJJDP Fact Sheet). Washington, DC: U.S. Department of Justice, Office of Juvenile Justice and Delinquency Prevention.

Eitle, D., Gunkel, S., & Gundy, K. V. (2004). Cumulative exposure to stressful life events and male gang membership. *Journal of Criminal Justice, 32,* 95–111.

Elder, G. H., Jr. (Ed.). (1985). *Life course dynamics: Trajectories and transitions, 1968–1980.* Ithaca, NY: Cornell University Press.

Elder, G. H. Jr., (1997). The life course and human development. In R. M. Lerner (Ed.), *Handbook of child psychology, Volume 1: Theoretical models of human development* (pp. 939–991). New York: Wiley.

Erickson, P. G., Butters, J. E., Cousineau, M., Harrison, L., & Korf, D. (2006). Girls and weapons: An international study of the perpetration of violence. *Urban Health, 83,* 788–801.

Esbensen, F. (2000). Preventing adolescent gang involvement: Risk factors and prevention strategies. *Juvenile Justice Bulletin. Youth Gang Series.* Washington, DC: U.S. Department of Justice, Office of Justice Programs, Office of Juvenile Justice and Delinquency Prevention.

Esbensen, F., Brick, B. T., Melde, C., Tusinski, K., & Taylor, T. J. (2008). The role of race and ethnicity in gang membership. In F. V. Genert, D. Peterson, & I. Lien, *Street gangs, migration and ethnicity* (pp. 117–139). Portland, OR: Willan.

Esbensen, F., & Deschenes, E. P. (1998). A multi-site examination of gang membership: Does gender matter? *Criminology, 36,* 799–828.

Esbensen, F., Deschenes, E. P., & Winfree, L. T. (1999). Differences between gang girls and gang boys: Results from a multi-site survey. *Youth and Society, 31,* 27–53.

Esbensen, F., & Huizinga, D. (1993). Gangs, drugs, and delinquency in a survey of urban youth. *Criminology, 31,* 565–589.

Esbensen, F., Huizinga, D., & Weiher, A. W. (1993). Gang and non-gang youth: Differences in explanatory variables. *Journal of Contemporary Criminal Justice, 9,* 94–116.

Esbensen, F., & Lynskey, D. P. (2001). Youth gang members in a school survey. In M. W. Klein, H. Kerner, C. L. Maxson, & E. Weitekampf (Eds.), *The Eurogang paradox: Street gangs and youth groups in the U.S. and Europe* (pp. 93–113). Amsterdam: Kluwer Academic Publishers.

Esbensen, F., Osgood, D. W., Taylor, T. J., Peterson, D., & Freng, A. (2001). How great is G.R.E.A.T.? Results from a longitudinal quasi-experimental design. *Criminology and Public Policy, 1,* 87–118.

Esbensen, F., Peterson, D., Freng, A., & Taylor, T. J. (2002). Initiation of drug use, drug sales, and violent offending among a sample of gang and nongang youth. In C. R. Huff (Ed.), *Gangs in America III* (pp. 37–50). Thousand Oaks, CA: Sage.

Esbensen, F., Peterson, D., Taylor, T. J., & Freng, A. (2010). *Youth violence: Sex and race differences in offending, victimization, and gang membership.* Philadelphia: Temple University Press.

Esbensen, F., Peterson, D., Taylor, T. J., Freng, A., Osgood, D. W., Carson, D. C., & Matsuda, K. N. (2011). Evaluation and evolution of the Gang Resistance Education and Training (G.R.E.A.T.) program. *Journal of School Violence, 10,* 53–70.

Esbensen, F., Peterson, D., Taylor, T. J., & Osgood, D. W. (2010). *Results from a multi-site evaluation of the G.R.E.A.T. program.* St. Louis, MO: University of Missouri—St. Louis, Department of Criminology and Criminal Justice.

Esbensen, F., & Tusinski, K. (2007). Youth gangs in the print media. *Journal of Criminal Justice and Popular Culture, 14,* 21–38.

Esbensen, F., & Winfree, L. T. (1998). Race and gender differences between gang and non-gang youths: Results from a multi-site survey. *Justice Quarterly, 15,* 505–526.

Esbensen, F., Winfree, L. T., He, N., & Taylor, T. J. (2001). Youth gangs and definitional issues: When is a gang a gang, and why does it matter? *Crime and Delinquency, 47,* 105–130.

Fagan, A. A., Hanson, K., Hawkins, J. D., & Arthur, M. W. (2008). Implementing effective community-based prevention programs in the community youth development study. *Youth Violence and Juvenile Justice, 6,* 256–278.

Fagan, J., & Chin, K. L. (1989). Initiation to crack: A tale of two epidemics. *Contemporary Drug Problems, 16,* 579–618.

Farrington, D. P. (1986). Stepping stones to adult criminal careers. In D. Olweus, J. Block, & M. R. Yarrow (Eds.), *Development of antisocial and prosocial behavior* (pp. 359–384). New York: Academic Press.

Farrington, D. P. (2003). Developmental and life-course criminology: Key theoretical and empirical issues—The 2002 Sutherland Award Address. *Criminology, 41,* 221–255.

Farrington, D. P. (2005). *Integrated developmental and life-course theories of offending.* New Brunswick, NJ: Transaction Publishing.

Farrington, D. P. (2006). Building developmental and life-course theories of offending. In F. T. Cullen, J. P. Wright, & K. R. Blevins (Eds.), *Taking stock: The status of criminological theory* (pp. 335–364). New Brunswick, NJ: Transaction Publishing.

Federal Bureau of Investigation. (2009). *National Gang Threat Assessment: 2009.* Washington, DC: U.S. Department of Justice, Federal Bureau of Investigation.

Felson, M. (2006). The street gang strategy. In M. Felson (Ed.), *Crime and Nature* (pp. 305–324). Thousand Oaks, CA: Sage.

Felson, M., & Cohen, L. E. (1980). Human ecology and crime: A routine activity approach. *Human Ecology, 8,* 389–406.

Ferguson, C. J., Miguel, C. S., Kilburn, J. C., & Sanchez, P. (2007). The effectiveness of school-based anti-bullying programs: A meta-analytic review. *Criminal Justice Review, 32,* 4001–4414.

Fernandez, M. E. (1998). An urban myth sees the light again. *The Washington Post,* p. B2.

Finestone, H. (1976). The delinquent and society: The Shaw and McKay tradition. In J. F. Short (Ed.), *Delinquency, crime, and society* (pp. 23–49). Chicago: University of Chicago Press.

Finkelhor, D., Turner, H. A., Ormrod, R., & Hamby, S. L. (2009). Violence, abuse and crime exposure in a national sample of children and youth. *Pediatrics, 124,* 1411–1123.

Fishman, L. T. (1995). The Vice Queens: An ethnographic study of Black female gang behavior. In M. W. Klein, C. L. Maxson, & J. Miller (Eds.), *Modern gang reader* (pp. 83–92). Los Angeles: Roxbury.

Fishman, L. T. (1999). Black female gang behavior: An historical and ethnographic perspective. In M. Chesney-Lind & J. Hagedorn (Eds.), *Female gangs in America: Essays on girls, gangs, and gender* (pp. 64–84). Chicago: Lake View Press.

Fishman, M. (1978). Crime waves as ideology. *Social Problems, 25,* 531–543.

Flannery, D. J., Alexander T., Vazsonyi, A. K. Liau, S. G., Powell, K. E., Atha, H., Vesterdal, W., & Embry, D. D. (2003). Initial behavior outcomes for the PeaceBuilders universal school-based violence prevention program. *Developmental Psychology, 39,* 292–308.

Flannery, D. J., Singer, M. I., Van Dulmen, M., Kretschmar, J. M., & Belliston, L. M. (2007). Exposure to violence, mental health, and violent behavior. In D. Flannery, A. Vazonsyi, & I. Waldman (Eds.), *Cambridge handbook of violent behavior* (pp. 306–321). Cambridge, MA: Cambridge University Press.

Flannery, D. J., Vazsonyi, A. T., & Waldman, I. D. (2007). *Cambridge handbook of violent behavior.* Cambridge, MA: Cambridge University Press.

Fleisher, M. S. (1989). *Warehousing violence*. Newbury Park, CA: Sage.

Fleisher, M. S. (1995). *Beggars and thieves: Lives of urban street criminals*. Madison: University of Wisconsin Press.

Fleisher, M. S. (1998). *Dead end kids: Gang girls and the boys they know*. Madison: University of Wisconsin Press.

Fleisher, M. S. (2002). Doing field research on diverse gangs: Interpreting youth gangs as social networks. In C. R. Huff (Ed.), *Gangs in America* (3rd ed., pp. 199–217). Thousand Oaks, CA: Sage.

Fleisher, M. (2006a). *Societal and correctional context of prison gangs*. Cleveland, OH: Case Western Reserve University, Mandel School of Applied Social Sciences.

Fleisher, M. S. (2006b). Youth gang social dynamics and social network analysis: Applying degree centrality measures to assess the nature of gang boundaries. In J. F. Short & L. A. Hughes (Eds.), *Studying youth gangs* (pp. 86–99). Lanham, MD: AltaMira Press.

Fleisher, M. S., & Decker, S. (2001). An overview of the challenge of prison gangs. *Corrections Management Quarterly, 5*, 1–9.

Fong, R. S., & Fogel, R. E. (1994–95, Winter). A comparative analysis of prison gang members, security threat group inmates and general population prisoners in the Texas Department of Corrections. *Journal of Gang Research, 2*, 1–12.

Franco, C. (2008a). *The MS-13 and 18th Street gangs: Emerging transnational gang threats?* (CRS Report RL34233, updated January 30, 2008). Washington, DC: Congressional Research Service, Library of Congress.

Franco, C. (2008b). *Youth gangs: Background, legislation, and issues* (CRS Report RL33400, updated January 25, 2008). Washington, DC: Congressional Research Service, Library of Congress.

Fremon, C. (1995). *Father Greg and the homeboys*. New York: Hyperion.

Fremon, C. (2004). *G-Dog and the homeboys: Father Greg Boyle and the gangs of East Los Angeles*. Albuquerque: University of New Mexico Press.

Furfey, P. H. (1928). *The gang age: A study of the preadolescent boy and his recreational needs*. New York: Macmillian.

Gaes, G., Wallace, S., Gilman, E., Klein-Saffran, J., & Suppa, S. (2002). The influence of prison gang affiliation on violence and other prison misconduct. *The Prison Journal, 82*, 359–385.

Gannon, T. M. (1967). Dimensions of current gang delinquency. *Journal of Research in Crime and Delinquency, 4*, 119–131.

Garot, R. (2007). "Where you from!": gang identity as Performance. *Journal of Contemporary Ethnography, 36*, 50–84.

Garot, R. (2010). *Who you claim: Performing gang identity in school and on the streets*. New York: New York University Press.

Garot, R., & Katz, J. (2003). Provocative looks: Gang appearance and dress codes in an inner-city alternative school. *Ethnography, 4*, 421–454.

Gatti, U., Tremblay, R. E., Vitaro, F., & McDuff, P. (2005). Youth gangs, delinquency and drug use: A test of selection, facilitation, and enhancement hypotheses. *Journal of Child Psychology and Psychiatry, 46*, 1178–1190.

Geis, G. (1965). *Juvenile gangs: Report to the President's Committee on Juvenile Delinquency and Youth Crime*. Washington, DC: U.S. Government Printing Office.

Gilfoyle, T. J. (2003). Scorsese's *Gangs of New York*: Why myth matters. *Journal of Urban History, 29*, 620–630.

Giordano, P. C. (1978). Girls, guys, and gangs: The changing social context of female delinquency. *Journal of Criminal Law and Criminology, 69*, 126–132.

Glesmann, C., Krisberg, B., & Marchionna, S. (2009). *Youth in gangs: Who is at risk? Focus*. Oakland, CA: National Council on Crime and Delinquency.

Goldstein, A. P., & Glick, B. (1994). *The prosocial gang: Implementing Aggression Replacement Training*. Thousand Oaks, CA: Sage.

Goldstein, A. P., Glick, B., & Gibbs, J. C. (1998). *Aggression Replacement Training: A comprehensive intervention for aggressive youth* (Rev. ed.). Champaign, IL: Research Press.

Golub, A., & Johnson, B. D. (1997). Crack's decline: Some surprises among U.S. cities. *Research in Brief*. Washington, DC: National Institute of Justice.

Gordon, R. A. (1967). Social level, disability, and gang interaction. *American Journal of Sociology, 73*, 42–62.

Gordon, R. A. (2010). *Drugs, gangs, and guns: What explains Pittsburgh youths' rising and falling participation in these activities during the 1990s*. Presentation at the Methodology Center, Penn State University, April 21.

Gordon, R. A., Lahey, B. B., Kawai, E., Loeber, R., Stouthamer-Loeber, M., & Farrington, D. P. (2004). Antisocial behavior and youth gang membership: Selection and socialization. *Criminology, 42*, 55–88.

Gordon, R. M. (1994). *Incarcerating gang members in British Columbia: A preliminary study.* Victoria, BC: Ministry of the Attorney General.

Gorman-Smith, D., & Loeber, R. (2005). Are developmental pathways in disruptive behaviors the same for girls and boys? *Journal of Child and Family Studies, 14,* 15–27.

Gottfredson, D. C., Cross, A., & Soule, D. A. (2007). Distinguishing characteristics of effective and ineffective after-school programs to prevent delinquency and victimization. *Criminology and Public Policy, 6,* 289–318.

Gottfredson, G. D., & Gottfredson, D. C. (2001). *Gang problems and gang programs in a national sample of schools.* Ellicott City, MD: Gottfredson Associates.

Gottfredson, G. D., Gottfredson, D. C., Payne, A. A., & Gottfredson, N. C. (2005). School climate predictors of disorder: Results from a national study of delinquency prevention in schools. *Journal of Research in Crime and Delinquency, 42,* 412–444.

Graves, K. N., Ireland, A., Benson, J., DiLuca, K., Chiu, K., Johnston, K., Dunn, L., McCoy, S., & Sechrist, S. (2010). *Guilford County gang assessment: OJJDP Comprehensive Gang Assessment Model.* Greensboro, NC: Center for Youth, Family, and Community Partnerships, University of North Carolina at Greensboro.

Grayson, G. W. (2009). *Mexico's struggle with drugs and thugs.* New York: Foreign Policy Association.

Greene, J., & Pranis, K. (2007). *Gang wars: The failure of enforcement tactics and the need for effective public safety strategies.* Washington, DC: Justice Policy Institute.

Griffin, M. L., & Hepburn, J. R. (2006). The effect of gang affiliation on violent misconduct among inmates during the early years of confinement. *Criminal Justice and Behavior, 33,* 419–448.

Griffiths, E., & Chavez, J. M. (2004). Communities, street guns and homicide trajectories in Chicago, 1980–1995: Merging methods for examining homicide trends across space and time. *Criminology, 42,* 941–975.

Grogger, J., & Willis, M. (1998). *The introduction of crack cocaine and the rise in urban crime rates.* National Bureau of Economic Research Working Paper No. W6353. Cambridge, MA: National Bureau of Economic Research.

Gugliotta, G., & Leen, J. (1989). *Kings of cocaine.* New York: Simon & Schuster.

Hagan, J. (1995). Rethinking crime and theory and policy: The new sociology of crime and disrepute. In H. D. Barlow (Ed.), *Crime and public policy* (pp. 317–339). Boulder, CO: Westview Press.

Hagan, J., & Foster, H. (2000). Making corporate and criminal America less violent: Public norms and structural reforms. *Contemporary Sociology, 29,* 44–53.

Hagedorn, J. M. (1988). *People and folks: Gangs, crime and the underclass in a Rustbelt city.* Chicago: Lake View Press.

Hagedorn, J. M. (1994). Homeboys, dope fiends, legits, and new jacks. *Criminology, 32,* 197–217.

Hagedorn, J. M. (1998). Gang violence in the postindustrial era. In M. Tonry & M. H. Moore (Eds.), *Youth violence* (pp. 365–420). Chicago: University of Chicago.

Hagedorn, J. M. (2008). *A world of gangs: Armed young men and gangsta culture.* Minneapolis: University of Minnesota Press.

Hagedorn, J. M., & Rauch, B. (2007). Housing, gangs, and homicide: What we can learn from Chicago. *Urban Affairs Review, 42,* 435–456.

Hall, G. S. (1904). *Adolescence.* New York: Appleton.

Hallfors, D., & Godette, D. (2002). Will the "Principles of Effectiveness" improve prevention practice? Early findings from a diffusion study. *Health Education Research, 17,* 461–470.

Hallsworth, S., & Young, T. (2008). Gang talk and gang talkers: A critique. *Crime Media Culture, 42,* 175–195.

Hanson, K. (1964). *Rebels in the streets: The story of New York's girl gangs.* Englewood Cliffs, NJ: Prentice Hall.

Hardman, D. G. (1967). Historical perspectives of gang research. *Journal of Research in Crime and Delinquency, 4,* 5–26.

Hardman, D. G. (1969). Small town gangs. *Journal of Criminal Law, Criminology and Police Science, 60,* 173–181.

Hare, R. D., Hart, S. D., & Harpur, T. J. (1991). Psychopathy and the *DSM-IV* criteria for antisocial personality disorder. *Journal of Abnormal Psychology, 100,* 391–398.

Harris, M. C. (1988). *Cholas: Latino girls and gangs.* New York: AMS Press.

Harrison, L. E. (1999). How cultural values shape economic success. In F. L. Pincus and H. J. Ehrlich (Eds.) *Race and Ethnic Conflict* (pp. 97–109). Boulder, CO: Westview Press.

Hartman, D. A., & Golub, A. (1999). The social construction of the crack epidemic in the print media. *Journal of Psychoactive Drugs, 31,* 423–433.

Haskins, J. (1974). *Street gangs: Yesterday and today.* Wayne, PA: Hastings Books.

Haviland, A. M., & Nagin, D. S. (2005). Causal inferences with group-based trajectory models. *Psychometrika, 70,* 1–22.

Haviland, A. M., Nagin, D. S., Rosenbaum, P. R., & Tremblay, R. E. (2008). Combining group-based trajectory modeling and propensity score matching for causal inferences in nonexperimental longitudinal data. *Developmental Psychology, 44,* 422–436.

Hawkins, J. D. (Ed.). (1996). *Delinquency and crime: Current theories.* New York: Cambridge University Press.

Hawkins, J. D., Oesterle, S., Brown, E. C., Arthur, M. W., Abbott, R. D., Fagan, A. A., & Catalano, R. F. (2009). Results of a type 2 translational research trial to prevent adolescent drug use and delinquency: A test of Communities That Care. *Archives of Pediatrics and Adolescent Medicine, 163*(9), 789–798.

Hayden, T. (2005). *Street wars: Gangs and the future of violence.* New York: New Press.

Hayeslip, D., & Cahill, M. (2009). *Community collaboratives addressing youth gangs: Final evaluation findings from the Gang Reduction Program.* Washington, DC: Urban Institute.

Haymoz, S., & Gatti, U. (2010). Girl members of deviant youth groups, offending behavior and victimisation: Results from the ISRD2 in Italy and Switzerland. *European Journal on Criminal Policy and Research, 16,* 167–182.

Haynie, D. L., Steffensmeier, D., & Bell, K. E. (2007). Gender and serious violence: Untangling the role of friendship sex composition and peer violence. *Youth Violence and Juvenile Justice, 5,* 235–253.

Healy, G. (2002). *There goes the neighborhood: The Bush-Ashcroft plan to "help" localities fight gun crime. Policy Analysis No. 440.* Washington, DC: Heritage Foundation.

Hemphill, S. A., Toumborou, J. W., Herrenkohl, T. L., McMorris, B. J., & Catalano, R. F. (2006). The effect of school suspensions and arrests on subsequent adolescent behavior in Australia and the United States. *Journal of Adolescent Health, 39,* 736–744.

Hill, K. G., Chung, I. J., Guo, J., & Hawkins, J. D. (2002). *The impact of gang membership on adolescent violence trajectories.* Paper presented at the International Society for Research on Aggression, XV World Meeting, Montreal, Canada, July.

Hill, K. G., Hawkins, J. D., Catalano, R. F., Kosterman, R., Abbott, R., & Edwards, T. (1996). *The longitudinal dynamics of gang membership and problem behavior: A replication and extension of the Denver and Rochester gang studies in Seattle.* Paper presented at the annual meeting of the American Criminological Society, Chicago, November.

Hill, K. G., Howell, J. C., Hawkins, J. D., & Battin-Pearson, S. R. (1999). Childhood risk factors for adolescent gang membership: Results from the Seattle Social Development Project. *Journal of Research in Crime and Delinquency, 36,* 300–322.

Hill, K. G., Lui, C., & Hawkins, J. D. (2001). Early precursors of gang membership: A study of Seattle youth. *Juvenile Justice Bulletin. Youth Gang Series.* Washington, DC: U.S. Department of Justice, Office of Juvenile Justice and Delinquency Prevention.

Hipp, J. R., Tita, G. E., & Boggess, L. N. (2009). Intergroup and intragroup violence: Is violent crime an expression of group conflict or social disorganization? *Criminology, 47,* 521–564.

Hipwell, A. E., Keenan, K., Bean, T., Loeber, R., & Stouthamer-Loeber, M. (2008). Reciprocal influences between girls' behavioral and emotional problems and caregiver mood and parenting style: A six-year prospective analysis. *Journal of Abnormal Child Psychology, 36,* 663–677.

Hipwell, A. E., Loeber, R., Stouthamer-Loeber, M., Keenan, K., White, H. R., & Kroneman, L. (2002). Characteristics of girls with early onset of disruptive and antisocial behavior. *Criminal Behaviour and Mental Health, 12,* 99–118.

Hipwell, A. E., Pardini, D. A., Loeber, R., Sembower, M. A., Keenan, K., & Stouthamer-Loeber, M. (2007). Callous-unemotional behaviors in young girls: Shared and unique effects relative to conduct problems. *Journal of Clinical Child and Adolescent Psychology, 36,* 293–304.

Hipwell, A. E., White, H. R., Loeber, R., Stouthamer-Loeber, M., Chung, T., & Sembower, M. (2005). Young girls' expectancies about the effects of alcohol, future intentions and patterns of use. *Journal of Studies on Alcohol, 66,* 630–639.

Hodgkinson J., Marshall, S., Berry, G., Reynolds, P., Newman, M., Burton, E., Dickson, K., & Anderson, J. (2009). Reducing gang-related crime: A systematic review of 'comprehensive' interventions. *Summary report.* London: EPPI-Centre, Social Science Research Unit, Institute of Education, University of London.

Horowitz, R. (1983). *Honor and the American dream: Culture and identity in a Chicano community.* New Brunswick, NJ: Rutgers University Press.

Horowitz, R. (1990). Sociological perspectives on gangs: Conflicting definitions and concepts. In

R. Huff (Ed.), *Gangs in America* (pp. 37–54). Newbury Park, CA: Sage.

Horowitz, R., & Schwartz, G. (1974). Honor, normative ambiguity, and gang violence. *American Sociological Review, 39,* 238–251.

Houston Intelligence Support Center. (2010). *Houston high-intensity drug trafficking area gang threat assessment.* Houston, TX: Author.

Houston Multi-Agency Gang Task Force (n.d.). *Report gang crime tips.* Retrieved from http://www.stophoustongangs.org/

Howell, J. C. (1998). Youth gangs: An overview. *Juvenile Justice Bulletin. Youth Gang Series.* Washington, DC: U.S. Department of Justice, Office of Juvenile Justice and Delinquency Prevention.

Howell, J. C. (1999). Youth gang homicides: A literature review. *Crime and Delinquency, 45,* 208–241.

Howell, J. C. (2000). *Youth gang programs and strategies.* Washington, DC: U.S. Department of Justice, Office of Juvenile Justice and Delinquency Prevention.

Howell, J. C. (2003a). Diffusing research into practice using the Comprehensive Strategy for Serious, Violent, and Chronic Juvenile Offenders. *Youth Violence and Juvenile Justice: An Interdisciplinary Journal, 1,* 219–245.

Howell, J. C. (2003b). *Preventing and reducing juvenile delinquency: A comprehensive framework.* Thousand Oaks, CA: Sage.

Howell, J. C. (2004). Youth gangs: Prevention and intervention. In P. Allen-Meares & M. W. Fraser (Eds.), *Intervention with children and adolescents: An interdisciplinary perspective* (pp. 493–514). Boston: Allyn & Bacon.

Howell, J. C. (2006). The impact of gangs on communities. *NYGC Bulletin No. 2.* Tallahassee, FL: National Youth Gang Center.

Howell, J. C. (2007). Menacing or mimicking? Realities of youth gangs. *The Juvenile and Family Court Journal, 58,* 9–20.

Howell, J. C. (2009). *Preventing and reducing juvenile delinquency: A comprehensive framework* (2nd ed.). Thousand Oaks, CA: Sage.

Howell, J. C. (2010a). Gang prevention: An overview of current research and programs. *Juvenile Justice Bulletin.* Washington, DC: U.S. Department of Justice, Office of Juvenile Justice and Delinquency Prevention.

Howell, J. C. (2010b). Lessons learned from gang program evaluations: Prevention, intervention, suppression, and comprehensive community approaches. In R. J. Chaskin (Ed.), *Youth gangs and community intervention: Research, practice, and evidence* (pp. 51–75). New York: Columbia University Press.

Howell, J. C., & Curry, G. D. (2009). Mobilizing communities to address gang problems. *NYGC Bulletin No. 4.* Tallahassee, FL: National Youth Gang Center.

Howell, J. C., & Decker, S. H. (1999). The youth gangs, drugs, and violence connection. *Juvenile Justice Bulletin. Youth Gang Series.* Washington, DC: Office of Juvenile Justice and Delinquency Prevention.

Howell, J. C., & Egley, A., Jr. (2005a). Gangs in small towns and rural counties. *NYGC Bulletin No. 1.* Tallahassee, FL: National Youth Gang Center.

Howell, J. C., & Egley, A., Jr. (2005b). Moving risk factors into developmental theories of gang membership. *Youth Violence and Juvenile Justice, 3*(4), 334–354.

Howell, J. C., Egley, A., Jr. & Gleason, D. K. (2002). Modern day youth gangs. *Juvenile Justice Bulletin. Youth Gang Series.* Washington, DC: U.S. Department of Justice, Office of Juvenile Justice and Delinquency Prevention.

Howell, J. C., Egley, A., Jr., Tita, G., & Griffiths, E. (2011). *U.S. gang problem trends and seriousness.* Tallahassee, FL: Institute for Intergovernmental Research, National Gang Center.

Howell, J. C., & Gleason, D. K. (1999). Youth gang drug trafficking. *Juvenile Justice Bulletin. Youth Gang Series.* Washington, DC: U.S. Department of Justice, Office of Juvenile Justice and Delinquency Prevention.

Howell, J. C., & Lynch, J. (2000). Youth gangs in schools. *Juvenile Justice Bulletin. Youth Gang Series.* Washington, DC: U.S. Department of Justice, Office of Juvenile Justice and Delinquency Prevention.

Howell, J. C., & Moore, J. P. (2010). History of street gangs in the United States. *National Gang Center Bulletin No. 4.* Tallahassee, FL: Institute for Intergovernmental Research, National Gang Center.

Howell, J. C., Moore, J. P., & Egley, A., Jr. (2002). The changing boundaries of youth gangs. In C. R. Huff (Ed.), *Gangs in America* (3rd ed., pp. 3–18). Thousand Oaks, CA: Sage.

Howell, M. Q., Lassiter, W., & Anderson, C. (2011). *North Carolina Department of Juvenile Justice and Delinquency Prevention annual report, 2010.* Raleigh: North Carolina Department of Juvenile Justice and Delinquency Prevention.

Hoyt, F. C. (1920). Gang in embryo. *Scribners Magazine,* 68, 146–154.

Huebner, B. M., Varano, S. P., & Bynum, T. S. (2007). Gangs, guns, and drugs: Recidivism among serious, young offenders. *Criminology & Public Policy,* 6, 187–221.

Huff, C. R. (1989). Youth gangs and public policy. *Crime and Delinquency, 35,* 524–537.

Huff, C. R. (1990). Denial, overreaction, and misidentification: A postscript on public policy. C. R. Huff (Ed.), *Gangs in America* (pp. 310–317). Newbury Park, CA: Sage.

Huff, C. R. (1993). Gangs in the United States. In A. Goldstein & C. R. Huff (Eds.), *The gang intervention handbook* (pp. 3–20). Champaign, IL: Research Press.

Huff, C. R. (1996). The criminal behavior of gang members and non-gang at-risk youth. In C. R. Huff (Ed.), *Gangs in America* (2nd ed., pp. 75–102). Thousand Oaks, CA: Sage.

Huff, C. R. (1998). *Comparing the criminal behavior of youth gangs and at-risk youth. Research in Brief.* Washington, DC: U.S. Department of Justice, Office of Justice Programs, National Institute of Justice.

Hughes, L. A. (2005). Studying youth gangs: Alternative methods and conclusions. *Journal of Contemporary Criminal Justice, 21,* 98–119.

Hughes, L. A. (2006). Studying youth gangs: The importance of context. In J. F. Short & L. A. Hughes (Eds.), *Studying Youth Gangs* (pp. 37–46). Lanham, MD: AltaMira Press.

Hughes, L. A. (2007). Youth street gangs. In M. P. McShane & F. P. Williams, III (Eds.), *Youth violence and juvenile delinquency. Vol. I: Juvenile offenders and victims* (pp. 41–60). Westport, CT: Praeger.

Hughes, L. A., & Short, J. F. (2005). Disputes involving gang members: Micro-social contexts. *Criminology, 43,* 43–76.

Huizinga, D. (1997). *Gangs and the volume of crime.* Paper presented at the annual meeting of the Western Society of Criminology.

Huizinga, D., & Lovegrove, P. (2009). *Summary of important risk factors for gang membership.* Boulder, CO: Institute for Behavioral Research.

Huizinga, D., & Schumann, K. (2001). Gang membership in Bremen and Denver: Comparative longitudinal data. In M. W. Klein, H. Kerner, C. L. Maxson, & E. Weitekampf (Eds.), *The Eurogang paradox: Street gangs and youth groups in the U.S. and Europe* (pp. 231–246). Amsterdam: Kluwer Academic Publishers.

Huizinga, D., Weiher, A. W., Espiritu, R., & Esbensen, F. (2003). Delinquency and crime: Some highlights from the Denver Youth Survey. In T. P. Thornberry & M. D. Krohn (Eds.), *Taking stock of delinquency: An overview of findings from contemporary longitudinal studies* (pp. 47–91). New York: Kluwer Academic/Plenum Publishers.

Hunt, G., & Joe-Laidler, K. (2001). Situations of violence in the lives of girl gang members. *Health Care for Women International, 22,* 363–384.

Hutchison, R. (1993). Blazon nouveau: Gang graffiti in the barrios of Los Angeles and Chicago. In S. Cummings & D. J. Monti (Eds.), *Gangs* (pp. 137–171). Albany: State University of New York Press.

Hutchison, R., & Kyle, C. (1993). Hispanic street gangs in the Chicago Public Schools. In S. Cummings & D. J. Monti (Eds.), *Gangs* (pp. 113–136). Albany: State University of New York Press.

Hutson, H. R., Anglin, D., Kyriacou, D. N., Hart, J., & Spears, K. (1995). The epidemic of gang-related homicides in Los Angeles County from 1979 through 1994. *Journal of the American Medical Association, 274,* 1031–1036.

Institute of Medicine. (2008). *Preventing mental, emotional, and behavioral disorders among young people.* Washington, DC: National Academy Press.

Ireland, T. O., Smith, C. A., & Thornberry, T. P. (2002). Developmental issues in the impact of child maltreatment on later delinquency and drug use. *Criminology, 40,* 359–400.

Irwin, J. (1980). *Prisons in turmoil.* Boston: Little, Brown.

Jackson, P. I. (1991). Crime, youth gangs, and urban transition: The social dislocations of postindustrial economic development. *Justice Quarterly, 8,* 379–397.

Jackson, P. G., & Rudman, C. (1993). Moral panic and the response to gangs in California. In S. Cummings & D. Monti (Eds.), *Gangs* (pp. 257–275). Albany: State University of New York Press.

Jacobs, J. B. (1974). Street gangs behind bars. *Social Problems, 21,* 395–409.

Jacobs, J. B. (1977). *Stateville: The penitentiary in mass society.* Chicago: University of Chicago Press.

Jacobs, J. B. (2001). Focusing on prison gangs. *Corrections Management Quarterly, 5*(1), vi–vii.

Jankowski, M. S. (1991). *Islands in the street: Gangs in American urban society.* Berkeley: University of California Press.

Joe, K. A., & Chesney-Lind, M. (1995). Just every mother's angel: An analysis of gender and ethnic variations in youth gang membership. *Gender & Society, 9,* 408–430.

Johansson, P., & Kempf-Leonard, K. (2009). A gender-specific pathway to serious, violent, and chronic offending? Exploring Howell's risk factors for serious delinquency. *Crime & Delinquency, 55,* 216–240.

Johnson, E. H. (1987). *Handbook on crime and delinquency prevention.* New York: Greenwood Press.

Johnson, K. (2006). Police tie jump in crime to juveniles: Gangs, guns, add up to increased violence. *USA Today* (July 13), pp. A1, A3.

Johnson, R. (1996). *Hard time: Understanding and reforming the prison.* Belmont, CA: Wadsworth.

Johnstone, J. W. (1981). Youth gangs and black suburbs. *Pacific Sociological Review, 24,* 355–375.

Jones, N. (2004). It's not where you love, it's how you love: Young women negotiate conflict and violence in the inner city. *The Annals of the American Academy of Political and Social Science, 595,* 49–62.

Jones, N. (2008). Working the 'Code': On girls, gender, and inner-city violence. *Australian and New Zealand Journal of Criminology, 41,* 63–83.

Josi, D., & Sechrest, D. K. (1999). A pragmatic approach to parole aftercare: Evaluation of a community reintegration program for high-risk youthful offenders. *Justice Quarterly, 16*(1), 51–80.

Kalb, L. M., & Loeber, R. (2003). Child disobedience and noncompliance: A review. *Pediatrics, 111,* 641–652.

Kaplan, H. B., & Damphouse, K. R. (1997). Negative social sanctions, self-derogation, and deviant behavior: Main and interactive effects in longitudinal perspective. *Deviant Behavior, 18,* 1–26.

Karoly, L. A., Greenwood, P. W., Everingham, S. S., Houbé, J., Kilburn, M. R., & Rydell, C. P. (1998). *Investing in our children.* Santa Monica, CA: RAND.

Katz, C. M., & Fox, A. (2011). *The reliability of NYGS data.* Paper presented at the annual meeting of the American Society of Criminology, Washington, DC, November.

Katz, C. M., & Schnebly, S. M. (2011). Neighborhood variation in gang member concentration. *Crime and Delinquency, 57,* 377–407.

Katz, C. M., & Webb, V. J. (2006). *Policing gangs in America.* New York: Cambridge University Press.

Katz, C. M., Webb, V. J., & Schaefer, D. (2000). The validity of police gang intelligence lists: Examining differences in delinquency between documented gang members and non-documented delinquent youth. *Police Quarterly, 3,* 413–437.

Katz, J., & Jackson-Jacobs, C. (2004). The criminologists' gang. In C. Sumner (Ed.), *The Blackwell companion to criminology* (pp. 91–124). Malden, MA: Blackwell.

Keenan, K. (2001). Uncovering preschool precursor problem behaviors. In R. Loeber & D. P. Farrington (Eds.), *Child delinquents: Development, intervention, and service needs* (pp. 117–134). Thousand Oaks, CA: Sage.

Keenan, K., Hipwell, A. E., Chung, T., Stepp, S., Stouthamer-Loeber, M., McTigue, K., & Loeber, R. (2010). The Pittsburgh girls studies: Overview and initial findings. *Journal of Clinical Child and Adolescent Psychology, 39,* 506–521.

Keiser, R. L. (1969). *The Vice Lords: Warriors of the street.* New York: Holt, Rinehart & Winston.

Kelley, B. T., Loeber, R., Keenan, K., & DeLamatre, M. (1997). Developmental pathways in boys' disruptive and delinquent behavior. *Juvenile Justice Bulletin.* Washington, DC: Office of Juvenile Justice and Delinquency Prevention.

Kennedy, D. M. (2007). Going to scale: *A national structure for building on proved approaches to preventing gang violence. A discussion document.* New York: Center for Crime Prevention and Control, John Jay College of Criminal Justice.

Kennedy, D. M., & Braga, A. A. (1998). Homicide in Minneapolis. *Homicide Studies, 2*(3), 263–290.

Kennedy, D. M., Piehl, A. M., & Braga, A. A. (1996). Youth violence in Boston: Gun markets, serious youth offenders, and a use-reduction strategy. *Law and Contemporary Problems, 59* (Special Issue), 147–196.

Kent, D. R., Donaldson, S. I., Wyrick, P. A., & Smith, P. J. (2000). Evaluating criminal justice programs designed to reduce crime by targeting repeat gang offenders. *Evaluation and Program Planning, 23,* 115–124.

Kim, Y. S., Leventhal., B. L., Koh, Y.-J., Hubbard, A., & Boyce, W. T. (2006). School bullying and youth violence. *Archives of General Psychiatry, 63,* 1035–1041.

Kingston, B., Huizinga, D., & Elliott, D. S. (2009). A test of social disorganization theory in high-risk urban neighborhoods. *Youth & Society, 41,* 53–79.

Kissner, J., & Pyrooz, D. C. (2009). Self-control, differential association, and gang membership: A theoretical and empirical extension of the literature. *Journal of Criminal Justice, 37,* 478–487.

Klein, M. W. (1969). Violence in American juvenile gangs. In D. J. Mulvihill & M. M. Tumin (Eds.), *Crimes of violence* (pp. 1427–1460). Washington,

DC: National Commission on the Causes and Prevention of Violence.

Klein, M. W. (1971). *Street gangs and street workers.* Englewood Cliffs, NJ: Prentice Hall.

Klein, M. W. (1995). *The American street gang.* New York: Oxford University Press.

Klein, M. W. (2002). Street gangs: A cross-national perspective. In C. R. Huff (Ed.), *Gangs in America III* (pp. 237–254). Thousand Oaks, CA: Sage.

Klein, M. W. (2004). *Gang cop: The words and ways of Officer Paco Domingo.* Walnut Creek, CA: AltaMira Press.

Klein, M. W. (2008). Foreword. In F. van Gemert, D. Peterson, & I.-L. Lien (Eds.), *Street gangs, migration and ethnicity* (pp. xi–xv). Portland, OR: Willan Publishing.

Klein, M. W., & Crawford, L. Y. (1967). Groups, gangs and cohesiveness. *Journal of Research in Crime and Delinquency, 4,* 63–75.

Klein, M. W., Kerner, H., Maxson, C. L., & Weitekampf, E. (Eds.). (2001). *The Eurogang paradox: Street gangs and youth groups in the U.S. and Europe.* Amsterdam: Kluwer Academic Publishers.

Klein, M. W., & Maxson, C. L. (1994). Gangs and cocaine trafficking. In D. MacKenzie & C. Uchida (Eds.), *Drugs and crime: Evaluating public policy initiatives* (pp. 42–58). Thousand Oaks, CA: Sage.

Klein, M. W., & Maxson, C. L. (2006). *Street gang patterns and policies.* New York: Oxford University Press.

Klein, M. W., Maxson, C. L., & Cunningham, L. C. (1991). Crack, street gangs, and violence. *Criminology, 29,* 623–650.

Klein, M. W., Weerman, F. M., & Thornberry, T. P. (2006). Street gang violence in Europe. *European Journal of Criminology, 3,* 413–437.

Knox, G. W. (1998). *An introduction to gangs* (4th ed.). Peotone, IL: New Chicago Press.

Kontos, L., Brotherton, D., & Barrios, L. (2003). *Gangs and society: Alternative perspectives.* New York: Columbia University Press.

Kosterman, R., Hawkins, J. D., Hill, K. G., Abbott, R. D., Catalano, R. F., & Guo, J. (1996, November). *The developmental dynamics of gang initiation: When and why young people join gangs.* Paper presented at the annual meeting of the American Society of Criminology, Chicago.

Kotlowitz, A. (1992). *There are no children here: The story of two boys growing up in the other America.* New York: Anchor Books.

Kotlowitz, A. (2008). Blocking the transmission of violence. *New York Times Magazine,* (May 4), 1–9.

Krohn, M. D., & Thornberry, T. P. (2008). Longitudinal perspectives on adolescent street gangs. In A. Liberman (Ed.), *The long view of crime: A synthesis of longitudinal research* (pp. 128–160). New York: Springer.

Krohn, M. D., Thornberry, T. P., Rivera, C., & Le Blanc, M. (2001). Later careers of very young offenders. In R. Loeber & D. P. Farrington (Eds.), *Child delinquents: Development, intervention, and service needs* (pp. 67–94). Thousand Oaks, CA: Sage.

Kroneman, L., Loeber, R., & Hipwell, A. E. (2004). Is neighborhood context differently related to externalizing problems and delinquency for girls compared with boys? *Clinical Child and Family Psychology Review, 7,* 109–122.

Kumpfer, K. L., & Alvarado, R. (1998). Effective family strengthening interventions. *Juvenile Justice Bulletin.* Washington, DC: U.S. Department of Justice, Office of Juvenile Justice and Delinquency Prevention.

Kupersmidt, J. B., Coie, J. D., & Howell, J. C. (2003). Building resilience in children exposed to negative peer influences. In K. I. Maton, C. J. Schellenbach, B. J. Leadbeater, C. J. Schellenbach, & A. L. Solarz (Eds.), *Investing in children, youth, families and communities: Strengths-based research and policy* (pp. 251–268). Washington, DC: American Psychological Association.

Kutash, K., Duchnowski, A. J., & Lynn, N. (2006). *School-based mental health: An empirical guide for decision-makers.* Tampa: University of South Florida, The Louis de la Parte Florida Mental Health Institute, Department of Child & Family Studies, Research and Training Center for Children's Mental Health.

Lacourse, E., Nagin, D. S., Tremblay, R. E., Vitaro, F., & Claes, M. (2003). Developmental trajectories of boys' delinquent group membership and facilitation of violent behaviors during adolescence. *Development and Psychopathology, 15,* 183–197.

Lacourse, E., Nagin, D. S., Vitaro, F., Cote, S., Arseneault, L., & Tremblay, R. E. (2006). Prediction of early-onset deviant peer group affiliation. *Archives of General Psychiatry, 63,* 562–568.

Lahey, B. B., Gordon, R. A., Loeber, R., Stouthamer-Loeber, M., & Farrington, D. P. (1999). Boys who join gangs: A prospective study of predictors of first gang entry. *Journal of Abnormal Child Psychology, 27,* 261–276.

Landesco, J. (1932). Crime and the failure of institutions in Chicago's immigrant areas. *Journal of Criminal Law and Criminology,* July, 238–248.

Lane, J., & Meeker, J. W. (2000). Subcultural diversity and the fear of crime and gangs. *Crime and Delinquency, 46,* 497–521.

Lane, J., & Meeker, J. W. (2003). Women's and men's fear of gang crimes: Sexual and nonsexual assault as perceptually contemporaneous offenses. *Justice Quarterly, 20,* 337–371.

Langberg, J., Fedders, B., & Kukorowski, D. (2011). *Law enforcement officers in Wake County schools: The human, educational, and financial costs.* Durham, NC: Advocates for Children's Services.

Lassiter, W. L. (2009). Separating myths from reality. In W. L. Lassiter & D. C. Perry (Eds.), *Preventing violence and crime in America's schools: From put-downs to lock-downs* (pp. 13–24). Santa Barbara, CA: Praeger.

Lassiter, W. L., & Perry, D. C. (2009). *Preventing violence and crime in America's schools: From put-downs to lock-downs.* Santa Barbara, CA: Praeger.

Le Blanc, M. (1993). Prevention of adolescent delinquency, an integrative multilayered control theory based perspective. D.P. Farrington, R.J. Sampson, & P.O.H. Widstrom (Eds.), *Integrating individual and ecological aspects of crime* (pp. 279–322). Stockholm, SWE: National Council for Crime Prevention.

Le Blanc, M., & Lanctot, N. (1998). Social and psychological characteristics of gang members according to the gang structure and its subcultural and ethnic makeup. *Journal of Gang Research, 5,* 15–28.

Le Blanc, M., & Loeber, R. (1998). Developmental criminology updated. In M. Tonry (Ed.), *Crime and justice: An annual review of research* (Vol. 23, pp. 115–198). Chicago: University of Chicago Press.

Leinwald, D. (2007). DEA busts ring accused of sending tons of drugs to US. *USA Today,* (March 1), A11.

Leinwald, D. (2000). LAPD, neighborhood shaken. *USA Today* (*February 25*), 3A.

Lerman, P., & Pottick, K. J. (1995). *The parents' perspective: Delinquency, aggression, and mental health.* Chur, Switzerland: Harwood.

Levitt, S. D., & Venkatesh, S. A. (2001). Growing up in the projects: The economic lives of a cohort of men who came of age in Chicago public housing. *American Economic Review, 91,* 79–84.

Lewis, O. (1966). *La Vida: A Puerto Rican family in the culture of poverty-San Juan and New York.* New York: Random House.

Ley, D. (1975). The street gang in its milieu. In G. Gappert & H. M. Rose (Eds.), *The social economy of cities* (pp. 247–273). Beverly Hills, CA: Sage.

Li, L., & Joe-Laidler, K. (2009). *Girl gangs in Hong Kong.* Unpublished background paper. Geneva: Small Arms Survey.

Li, X., Stanton, B., Pack, R., Harris, C., Cottrell, L., & Burns, J. (2002). Risk and protective factors associated with gang involvement among urban African American adolescents. *Youth and Society, 34,* 172–194.

Lipsey, M. W. (2009). The primary factors that characterize effective interventions with juvenile offenders: A meta-analytic overview. *Victims and Offenders, 4,* 124–147.

Lipsey, M. W., & Chapman, G. (2011). *Standardized Program Evaluation Protocol (SPEP): A user's guide.* Nashville, TN: Peabody Research Institute, Vanderbilt University.

Lipsey, M. W., Howell, J. C., Kelly, M. R., Chapman, G., & Carver, D. (2010). *Improving the effectiveness of juvenile justice programs: A new perspective on evidence-based practice.* Washington, DC: Georgetown University, Center for Juvenile Justice Reform.

Lizotte, A. J., Howard, G. J., Krohn, M. D., & Thornberry, T. P. (1997). Patterns of illegal gun carrying among young urban males. *Valparaiso University Law Review, 31,* 375–393.

Lizotte, A. J., Krohn, M. D., Howell, J. C., Tobin, K., & Howard, G. J. (2000). Factors influencing gun carrying among young urban males over the adolescent-young adult life course. *Criminology, 38,* 811–834.

Lizotte, A. J., Tesoriero, J. M., Thornberry, T. P., & Krohn, M. D. (1994). Patterns of adolescent firearms ownership and use. *Justice Quarterly, 11,* 51–73.

Lobo, A. P., Flores, R. J. O., & Salvo, J. J. (2002). The impact of Hispanic growth on the racial/ethnic composition of New York City neighborhoods. *Urban Affairs Review, 37,* 703–727.

Loeber, R. (1990). Development and risk factors of juvenile antisocial behavior and delinquency. *Clinical Psychology Review, 10,* 1–41.

Loeber, R. (1996). Developmental continuity, change, and pathways in male juvenile problem behaviors and delinquency. In J. D. Hawkins (Ed.), *Delinquency and crime: Current theories* (pp. 1–27). New York: Cambridge University Press.

Loeber, R., Burke, J. D., & Pardini, D. A. (2009). Development and etiology of disruptive and delinquent behavior. *Annual Review of Clinical Psychology, 5,* 291–310.

Loeber, R., & Farrington, D. P. (Eds.). (1998). *Serious and violent juvenile offenders: Risk factors and successful interventions.* Thousand Oaks, CA: Sage.

Loeber, R., & Farrington, D. P. (Eds.). (2001). *Child delinquents: Development, interventions, and service needs.* Thousand Oaks, CA: Sage.

Loeber, R., & Farrington, D. P. (2011). *Who will kill and who will be killed? Development of young homicide offenders and victims.* New York: Springer.

Loeber, R., Farrington, D. P., & Petechuk, D. (2003). Child delinquency: Early intervention and prevention. *Juvenile Justice Bulletin.* Washington, DC: U.S. Department of Justice, Office of Juvenile Justice and Delinquency Prevention.

Loeber, R., Farrington, D. P., Stouthamer-Loeber, M., White, H. R., & Wei, E. (2008). *Violence and serious theft: Development and prediction from childhood to adulthood.* New York: Routledge.

Loeber, R., Farrington, D. P., Stouthamer-Loeber, M., Moffitt, T. E., Caspi, A., White, H. R., Wei, E. H., & Beyers, J. M. (2003). The development of male offending: Key findings from fourteen years of the Pittsburgh Youth Study. In T. P. Thornberry & M. D. Krohn (Eds.), *Taking stock of delinquency: An overview of findings from contemporary longitudinal studies* (pp. 93–136). New York: Kluwer Academic/Plenum Publishers.

Loeber, R., Keenan, K., & Zhang, Q. (1997). Boys' experimentation and persistence in developmental pathways toward serious delinquency. *Journal of Child and Family Studies, 6,* 321–357.

Loeber, R., Slot, W., & Stouthamer-Loeber, M. (2007). A cumulative, three-dimensional, development model of serious delinquency. In P-O Wikstrom & R. Sampson (Eds.), *The explanation of crime: Context, mechanisms and development series* (pp. 153–194). Cambridge, MA: Cambridge University Press.

Loeber, R., Wei, E., Stouthamer-Loeber, M., Huizinga, D., & Thornberry, T. P. (1999). Behavioral antecedents to serious and violent offending: Joint analyses from the Denver Youth Survey, Pittsburgh Youth Study and the Rochester Youth Development Study. *Studies on Crime and Crime Prevention, 8,* 245–263.

Loeber, R., Wung, P., Keenan, K., Giroux, B., Stouthamer-Loeber, M., Van Kammen, W. B., & Maughan, B. (1993). Developmental pathways in disruptive child behavior. *Development and Psychopathology, 5,* 103–133.

Lopez, D. A., & Brummett, P.O. (2003). Gang membership and acculturation: ARSMA-II and choloization. *Crime and Delinquency, 49,* 627–642.

Lopez, E. M., Wishard, A., Gallimore, R., & Rivera, W. (2006). Latino high school students' perception of gangs and crews. *Journal of Adolescent Research, 21,* 299–318.

Los Angeles Police Department. (2007). *2007 Gang Enforcement Initiative.* Los Angeles: Author.

Ludwig, J. (2005). Better gun enforcement, less crime. *Criminology and Public Policy, 4,* 677–716.

Lyman, M. D. (1989). *Gangland: Drug trafficking by organized criminals.* Springfield, IL: Charles C Thomas.

Lyons, W., & Drew, J. (2006). *Punishing schools: Fear and citizenship in American public education.* Ann Arbor: University of Michigan Press.

Major, A. K., Egley, A., Jr., Howell, J. C., Mendenhall, B., & Armstrong, T. (2004). Youth gangs in Indian Country. *Juvenile Justice Bulletin.* Washington, DC: U.S. Department of Justice, Office of Juvenile Justice and Delinquency Prevention.

Manwaring, M. G. (2005). *Street gangs: The new urban insurgency.* Carlisle, PA: Strategic Studies Institute, U.S Army College. Retrieved from www.StrategicStudiesInstitute.army.mil.

Manwaring, M. G. (2009a). *A "new" dynamic in the western hemisphere security environment: The Mexican Zetas and other private armies.* Carlisle, PA: Strategic Studies Institute, U.S Army College. Retrieved from www.StrategicStudiesInstitute.army.mil

Manwaring, M. G. (2009b). *State and nonstate associated gangs: Credible "midwives of new social orders."* Carlisle, PA: Strategic Studies Institute, U.S Army College. Retrieved from www.StrategicStudiesInstitute.army.mil

Marks, C. (1985). Black labor migration: 1910–1920. *Critical Sociology, 12,* 5–24.

Martinez, R., Rodriguez, J., & Rodriguez, L. (1998). *East Side stories: Gang life in East L.A.* New York: PowerHouse Books.

Martinez, R., Rosenfel, R., & Mares, D. (2008). Social disorganization, drug market activity, and neighborhood violent crime. *Urban Affairs Review, 43,* 846–874.

Maxson, C. L. (1995). Street gangs and drug sales in two suburban cities. *Research in Brief.* Washington, DC: U.S. Department of Justice, National Institute of Justice.

Maxson, C. L. (1998). Gang members on the move. *Juvenile Justice Bulletin. Youth Gang Series.* Washington, DC: Office of Juvenile Justice and Delinquency Prevention.

Maxson, C. L. (1999). Gang homicide: A review and extension of the literature. In D. Smith & M. Zahn (Eds.), *Homicide: A sourcebook of social research* (pp. 197–220). Thousand Oaks, CA: Sage.

Maxson, C. L., Gordon, M. A., & Klein, M. W. (1985). Differences between gang and nongang homicides. *Criminology, 23,* 209–222.

Maxson, C. L., & Klein, M. W. (1990). Street gang violence: Twice as great, or half as great? In C.R. Huff (Ed.), *Gangs in America* (pp. 71–100). Newbury Park, CA: Sage.

Maxson, C. L., Whitlock, M., & Klein, M. W. (1998). Vulnerability to street gang membership: Implications for prevention. *Social Service Review, 72*(1), 70–91.

McBride, W. (2005). Comment made on the National Gang Center listserv, November 22.

McCorkle, R. C., & Miethe, T. D. (2002). *Panic: The social construction of the street gang problem.* Upper Saddle River, NJ: Prentice Hall.

McGarrell, E. F., & Chermak, S. (2003). Problem solving to reduce gangs and drug-related violence in Indianapolis. In S. H. Decker (Ed.), *Policing gangs and youth violence* (pp. 77–101). Belmont, CA: Wadsworth/Thompson Learning.

McGarrell, E. F., Hipple, N. K., Corsaro, N., Bynum, T. S., Perez, H., Zimmermann, C. A., & Garmo, M. (2009). *Project Safe Neighborhoods—A national program to reduce gun crime: Final project report.* Lansing: Michigan State University.

McGloin, J. M. (2005). Policy and intervention considerations of a network analysis of street gangs. *Criminology and Public Policy, 4,* 607–636.

McGuire, C. (2007). *Central American youth gangs in the Washington D.C. area.* Working Paper. Washington, DC: Washington Office on Latin America.

McKinney, K. C. (1988). Juvenile gangs: Crime and drug trafficking. *Juvenile Justice Bulletin.* Washington, DC: U.S. Department of Justice, Office of Juvenile Justice and Delinquency Prevention.

Melde, C. (2009). Lifestyle, rational choice, and adolescent fear: A test of a risk-assessment framework. *Criminology, 47,* 781–812.

Melde, C., & Esbensen, F. (2011). Gang membership as a turning point in the life course. *Criminology, 49,* 513–552.

Melde, C., Taylor, T. J., & Esbensen, F. (2009). "I got your back": An examination of the protective function of gang membership in adolescence. *Criminology, 47,* 565–594.

Merton, R. K. (1957). *Social theory and social structure.* Glencoe, IL: Free Press.

Meyer, A. L., Farrell, A. D., Northup, W., Kung, E. M., & Plybon, L. (2000). *Promoting non-violence in middle schools: Responding in Peaceful and Positive Ways (RIPP).* New York: Plenum.

Miethe, T. D., & McCorkle, R. C. (1997). Gang membership and criminal processing: A test of the "master status" concept. *Justice Quarterly, 14,* 407–427.

Mihalic, S., Irwin, K., Elliott, D., Fagan, A., & Hansen, D. (2001). Blueprints for violence prevention. *Juvenile Justice Bulletin.* Washington, DC: U.S. Department of Justice, Office of Juvenile Justice and Delinquency Prevention.

Miller, B. J. (2008). The struggle over redevelopment at Cabrini-Green, 1989–2004. *Journal of Urban History, 34,* 944–960.

Miller, J. A. (2001). *One of the guys: Girls, gangs and gender.* New York: Oxford University Press.

Miller, J. A. (2008). *Getting played: African American girls, urban inequality, and gendered violence.* New York: New York University Press.

Miller, J. A., & Brunson, R. (2000). Gender dynamics in youth gangs: A comparison of males' and females' accounts. *Justice Quarterly, 17,* 419–448.

Miller, J. A., & Decker, S. H. (2001). Young women and gang violence: Gender, street offending, and violent victimization in gangs. *Justice Quarterly, 18,* 601–626.

Miller, W. B. (1958). Lower class culture as a generating milieu of gang delinquency. *Journal of Social Issues, 14,* 5–19.

Miller, W. B. (1962). The impact of a "total community" delinquency control project. *Social Problems, 10,* 168–191.

Miller, W. B. (1966). Violent crimes in city gangs. *The Annals of the American Academy of Political and Social Science, 364,* 96–112.

Miller, W. B. (1973). Race, sex, and gangs: The Molls. *Trans-Action, 11*(1), 32–35.

Miller, W. B. (1974a). American youth gangs: Fact and fantasy. In L. Rainwater (Ed.), *Deviance and liberty: A survey of modern perspectives on deviant behavior* (pp. 262–273). Chicago: Aldine.

Miller, W. B. (1974b). American youth gangs: Past and present. In A. Blumberg (Ed.), *Current perspectives on criminal behavior* (pp. 410–420). New York: Knopf.

Miller, W. B. (1975). *Violence by youth gangs and youth groups as a crime problem in major American cities.* Washington, DC: U.S. Department of Justice, Office of Juvenile Justice and Delinquency Prevention.

Miller, W. B. (1982/1992). *Crime by youth gangs and groups in the United States.* Washington, DC: U.S. Department of Justice, Office of Juvenile Justice and Delinquency Prevention.

Miller, W. B. (2001). *The growth of youth gang problems in the United States: 1970–1998.* Washington, DC: Office of Juvenile Justice and Delinquency Prevention.

Miller, W. B., Geertz, H., & Cutter, H. S. G. (1962). Aggression in a boys' street-corner group. *Psychiatry, 24,* 283–298.

Moffitt, T. E. (1993). Adolescence-limited and life-course-persistent antisocial behavior: A developmental taxonomy. *Psychological Review, 100,* 674–701.

Moffitt, T. E. (2007). A review of research on the taxonomy of life-course persistent versus adolescent-limited antisocial behavior. In D. Flannery, A. Vazonsyi, & I. Waldman (Eds.), *Cambridge handbook of violent behavior* (pp. 49–74). Cambridge, MA: Cambridge University Press.

Monti, D. J. (1991). The practice of gang research. *Sociological Practice Review, 1,* 29–39.

Monti, D. J. (1993). Gangs in more- and less-settled communities. In S. Cummings & D. J. Monti (Eds.), *Gangs: The origins and impact of contemporary youth gangs in the United States* (pp. 219–253). Albany: State University of New York Press.

Moore, J. W. (1978). *Homeboys: Gangs, drugs and prison in the barrios of Los Angeles.* Philadelphia: Temple University Press.

Moore, J. W. (1985). Isolation and stigmatization in the development of an underclass: The case of Chicano gangs in East Los Angeles. *Social Problems, 33,* 1–13.

Moore, J. W. (1988). Introduction: Gangs and the underclass: A comparative perspective. In J. M. Hagedorn, *People and Folks* (pp. 3–17). Chicago: Lake View Press.

Moore, J. W. (1991). *Going down to the barrio: Homeboys and homegirls in change.* Philadelphia: Temple University Press.

Moore, J. W. (1993). Gangs, drugs, and violence. In S. Cummings & D. J. Monti (Eds.), *Gangs* (pp. 27–46). Albany: State University of New York Press.

Moore, J. W. (1994). The *chola* life course: Chicana heroin users and the barrio gang. *International Journal of Addictions, 29,* 1115–1126.

Moore, J. W. (1998). Understanding youth street gangs: Economic restructuring and the urban underclass. In M. W. Watts (Ed.), *Cross-cultural perspectives on youth and violence* (pp. 65–78). Stamford, CT: JAI.

Moore, J. W. (2007a). Female gangs: Gender and globalization. In J. M. Hagedorn, *Gangs in the global city* (pp. 187–203). Chicago: University of Illinois Press.

Moore, J. W. (2007b). Foreword. In A. Valdez, *Mexican American girls and gang violence: Beyond risk* (pp. ix–xii). New York: Palgrave Macmillan.

Moore, J. W., & Hagedorn, J. M. (1996). What happens to girls in the gang? In C. R. Huff (Ed.), *Gangs in America,* (2nd ed., pp. 205–218). Thousand Oaks, CA: Sage.

Moore, J. W., & Hagedorn, J. M. (1999). What happens to girls in the gang? In M. Chesney-Lind & J. Hagedorn (Eds.), *Female gangs in America: Essays on girls, gangs, and gender* (pp. 176–186). Chicago: Lake View Press.

Moore, J. W., & Hagedorn, J. M. (2001). Female gangs. *Juvenile Justice Bulletin. Youth Gang Series.* Washington, DC: U.S. Department of Justice, Office of Juvenile Justice and Delinquency Prevention.

Moore, J. W., & Long, J. M. (1987). *Final report: Youth culture vs. individual factors in adult drug use.* Los Angeles: Community Systems Research.

Moore, J. W., & Pinderhughes, R. (1993). Introduction. In J. W. Moore & R. Pinderhughes (Eds.), *In the barrios: Latinos and the underclass debate* (pp. xi–xxxix). New York: Russell Sage Foundation.

Moore, J. W., & Vigil, D. (1993). Barrios in transition. In J. W. Moore & R. Pinderhughes (Eds.), *In the barrios: Latinos and the underclass debate* (pp. 27–49). New York: Russell Sage Foundation.

Moore, J. W., Vigil, D., & Garcia, R. (1983). Residence and territoriality in Chicano gangs. *Social Problems, 31,* 182–194.

Moore, S. (2007). Reporting while Black. *New York Times,* (September 30), Week in Review, 4.

Morash, M. (1983). Gangs, groups, and delinquency. *British Journal of Criminology, 23,* 309–335.

Morash, M., & Chesney-Lind, M. (2009). The context of girls' violence: Peer groups, families, schools, and communities. In M. Zahn (Ed.), *The delinquent girl: Findings from the girls study group* (pp. 182–206). Philadelphia: Temple University Press.

Morenoff, J. D., Sampson, R. J., & Raudenbush, S. W. (2001). Neighborhood inequality, collective efficacy, and the spatial dynamics of urban violence. *Criminology, 39*(3), 517–559.

Mulvihill, D. J., & Tumin, M. M. (Eds.). (1969). *Crimes of violence.* Washington, DC: U.S. Government Printing Office.

Muncie, J. (2004). *Youth & crime.* Los Angeles: Sage.

Murray, B. (2000). *The Old Firm: Sectarians, sport and society in Scotland.* Edinburgh, Scotland: John Donald Publishers.

National Alliance of Gang Investigators. (2005). *National gang threat assessment: 2005.* Washington, DC: Bureau of Justice Assistance, U.S. Department of Justice.

National Center on Addiction and Substance Abuse. (2010). *National Survey of American Attitudes on Substance Abuse XV: Teens and parents, 2010.* New York: Author.

National Drug Intelligence Center. (2008). *National drug threat assessment: 2009.* Washington, DC: U.S. Department of Justice, National Drug Intelligence Center.

National Gang Center. (2009). *Federal and state definitions of the terms "gang," "gang crime," and "gang member."* Tallahassee, FL: Institute for Intergovernmental Research, National Gang Center.

National Gang Center. (2010a). *Best practices to address community gang problems: OJJDP's Comprehensive Gang Model.* Washington, DC: Author.

National Gang Center. (2010b). *Highlights of gang-related legislation: Spring 2010.* Tallahassee, FL: Institute for Intergovernmental Research, National Gang Center. Retrieved from http://www.national gangcenter.gov/Legislation/Highlights

National Youth Gang Center. (2000). *1998 National Youth Gang Survey.* Washington, DC: U.S. Department of Justice, Office of Juvenile Justice and Delinquency Prevention.

Natland, S. (2006). *Volden, horen og vennskapet: En kulturanalytisk studie av unge jenter som utøvere av vold.* Doctoral dissertation submitted to the University of Bergen, Norway. Retrieved from https://bora.uib.no/bitstream/1956/2298/1/DrAvh_Natland.pdf

Ness, C. D. (2004). Why girls fight: Female youth violence in the inner city. *The Annals of the American Academy of Political and Social Science, 595,* 32–48.

Ness, C. D. (2010). *Why girls fight: Female youth violence in the inner city.* New York: New York University Press.

Nickel, M., Luley, J., Nickel, C., & Widermann, C. (2006). Bullying girls—changes after Brief Strategic Family Therapy: A randomized, prospective, controlled trial with one-year follow-up. *Psychotherapy and Psychosomatics, 75,* 47–55.

North Carolina Department of Juvenile Justice and Delinquency Prevention and Department of Public Instruction. (2008). *School violence/gang activity study* (S.L. 2008-56). Raleigh: Author.

North Carolina Department of Juvenile Justice and Delinquency Prevention. (2009). *Request for proposals: Community-based youth gang violence prevention grant project.* Raleigh: Author.

North Carolina Department of Public Instruction. (2011). *Annual report on school crime and violence, the annual study of suspensions and expulsions: 2009–2010.* Raleigh, NC: North Carolina State Board of Education, Department of Public Instruction.

Nurge, D. (2003). Liberating yet limiting: The paradox of female gang membership. In L. Kontos, D. Brotherton, & L. Barrios (Eds.), *Gangs and society: Alternative perspectives* (pp. 161–182). New York: Columbia University Press.

Obeidallah, D. A., & Earls, F. J. (1999). Adolescent girls: The role of depression in the development of delinquency. *Research Preview.* Washington, DC: U.S. Department of Justice, National Institute of Justice.

Oehme, C. G. (1997). *Gangs, groups and crime: Perceptions and responses of community organizations.* Durham, NC: Carolina Academic Press.

Office of Juvenile Justice and Delinquency Prevention. (1996). Innovative local law enforcement and community policing programs for the juvenile justice system. *Juvenile Justice Bulletin.* Washington, DC: U.S. Department of Justice, Office of Juvenile Justice and Delinquency Prevention.

Office of Juvenile Justice and Delinquency Prevention. (2009a). *OJJDP Comprehensive Gang Model: A guide to assessing a community's youth gang problems.* Washington, DC: U.S. Department of Justice, Office of Juvenile Justice and Delinquency Prevention. Retrieved from www.nationalgang center.gov/Content/Documents/Assessment-Guide/Assessment-Guide.pdf

Office of Juvenile Justice and Delinquency Prevention. (2009b). *OJJDP Comprehensive Gang Model: Planning for implementation.* Washington, DC: U.S. Department of Justice, Office of Juvenile Justice and Delinquency Prevention. Retrieved from www .nationalgangcenter.gov/Content/Documents/Implementation-Manual/Implementation-Manual.pdf

Olson, D. E., & Dooley, B. (2006). Gang membership and community corrections populations: Characteristics and recidivism rates relative to other offenders. In J. F. Short & L. A. Hughes (Eds.), *Studying youth gangs* (pp. 193–202). Lanham, MD: AltaMira Press.

Olweus, D., Limber, S., & Mihalic, S. F. (1999). *Blueprints for violence prevention, book nine: Bullying prevention program.* Boulder, CO: Center for the Study and Prevention of Violence.

Orr, L., Feins, J., Jacob, R., Beecroft, E., Sanbonmatsu, L., Katz, L., Liebman, J., & Kling, J. (2003). *Moving to Opportunity for Fair Housing Demonstration Program: Interim impacts evaluation.* Executive summary. Washington, DC: U.S. Department of Housing and Urban Development.

Osgood, D. W., & Anderson, A. L. (2004). Unstructured socializing and rates of delinquency. *Criminology, 42*, 519–550.

Osher, D., Dwyer, K., & Jackson, S. (2004). *Safe, supportive and successful schools: Step by step.* Longmont, CO: Sopris West.

Ousey, G. O., & Lee, M. R. (2004). Investigating the connections between race, illicit drug markets, and lethal violence, 1984–1997. *Journal of Research in Crime and Delinquency, 41*, 352–383.

Papachristos, A. V. (2001). *A.D., After the Disciples: The neighborhood impact of federal gang prosecution.* Peotone, IL: New Chicago Schools Press.

Papachristos, A. V. (2004). *Gangs in the global city: The impact of globalization on post-industrial street gangs.* Paper presented at the Workshop on the Sociology and Cultures of Globalization. Chicago: University of Chicago, Department of Sociology.

Papachristos, A. V. (2005a). Gang world. *Foreign Policy, 147*(March-April), 48–55.

Papachristos, A. V. (2005b). Interpreting inkblots: Deciphering and doing something about modern street gangs. *Criminology & Public Policy, 4*, 643–651.

Papachristos, A. V. (2006). Social network analysis and gang research: Theory and methods. In J. F. Short & L. A. Hughes (Eds.), *Studying youth gangs* (pp. 99–116). Lanham, MD: AltaMira Press.

Papachristos, A. V. (2009). Murder by structure: Dominance relations and the social structure of gang homicide. *American Journal of Sociology, 115*, 74–128.

Papachristos, A. V., & Kirk, D. S. (2006). Neighborhood effects on street gang behavior. In J. F. Short & L. A. Hughes (Eds.), *Studying youth gangs* (pp. 63–84). Lanham, MD: AltaMira Press.

Park, R. E. (1921). Sociology and the social sciences: The social organism and the collective mind. *The American Journal of Sociology, 27*(1), 1–21.

Park, R. E. (1936a). Human ecology. *The American Journal of Sociology, 42*(July), 1–15.

Park, R. E. (1936b). Succession: An ecological concept. *American Sociological Review, 1*(April), 171–179.

Park, R. E., & Burgess, E. W. (1921). *Introduction to the science of sociology.* Chicago: University of Chicago Press.

Park, R. E., & Burgess, E. W. (Eds.). (1925). *The city.* Chicago: University of Chicago Press.

Park, S., Morash, M., & Stevens, T. (2010). Gender differences in predictors of assaultive behavior in late adolescence. *Youth Violence and Juvenile Justice, 8*, 314–331.

Parsons, T. (1951). *The social system.* New York: The Free Press.

Pastor, M. Jr., Sadd, J., & Hipp, J. (2001). Which came first? Toxic facilities, minority move-in, and environmental justice. *Journal of Urban Affairs, 23*, 1–21.

Patterson, G. R., Capaldi, D., & Bank, L. (1991). An early starter model for predicting delinquency. In D. J. Pepler & K. H. Rubin (Eds.), *The development and treatment of childhood aggression* (pp. 139–168). Hillsdale, NJ: Lawrence Erlbaum.

Pearson, G. (1983). *Hooligan: A history of reportable fears.* London: Macmillan.

Pepler, D. J., Madsen, K., Levene, K., & Webster, C. (2004). *The development and treatment of girlhood aggression.* Hillsdale, NJ: Lawrence Erlbaum.

Perkins, U. E. (1987). *Explosion of Chicago's Black street gangs: 1900 to the present.* Chicago: Third World Press.

Petersen, R. D. (2000). Gangs and inmate subcultures of female youth. *Free Inquiry in Creative Sociology, 20*, 100–116.

Petersen, R. D. (2004). Definitions of a gang and impacts on public policy. In R.D. Petersen (Ed.), *Understanding contemporary gangs in America* (pp. 19–34). Upper Saddle River, NJ: Pearson Prentice Hall.

Petersen, R. D., & Valdez, A. (2004). Intimate partner violence among Hispanic females. *Journal of Ethnicity in Criminal Justice, 2*, 67–89.

Petersen, R. D., & Valdez, A. (2005). Using snowball-based methods in hidden populations to generate a randomized community sample. *Youth Violence and Juvenile Justice, 3*, 151–167.

Peterson, D. (Forthcoming). Girlfriends, gun-holders, and ghetto-rats? Moving beyond narrow views of girls in gangs. In S. Miller, L. D. Leve, & P. K. Kerig (Eds.), *Delinquent girls: Contexts, relationships, and adaptation.* New York: Springer.

Peterson, D., Lien, I.-L., & Van Gemert, F. (2008). Concluding remarks: The roles of migration and ethnicity in street gang formation, involvement, and response. In F. van Gemert, D. Peterson, & I.-L. Lien (Eds.), *Street gangs, migration and ethnicity* (pp. 255–272). Portland, OR: Willan Publishing.

Peterson, D., Miller, J., & Esbensen, F. (2001). The impact of sex composition on gangs and gang delinquency. *Criminology, 39,* 411–439.

Peterson, D., & Morgan, K. (2010). *Risky business: Measuring and analyzing sex differences in risk factors for gang involvement.* Paper presented at the annual meeting of the American Society of Criminology, Philadelphia.

Peterson, D., Taylor, T. J., & Esbensen, F. (2004). Gang membership and violent victimization. *Justice Quarterly, 21,* 793–815.

Peterson, R. D., & Krivo, L. J. (2009). Segregated spatial locations, race-ethnic composition, and neighborhood violent crime. *The Annals of the American Academy of Political and Social Science, 623,* 93–107.

Peterson, V. W. (1963). Chicago: Shades of Capone. *The Annals of the American Academy of Political and Social Science, 347,* 30–39.

Petrosino, A., Turpin-Petrosino, C., & Buehler, J. (2003). Scared Straight and other juvenile awareness programs for preventing juvenile delinquency: A systematic review of the randomized experimental evidence. *The Annals of the American Academy of Political and Social Science, 589,* 41–62.

Pike, L. O. (1873). *A history of crime in England.* London: Smith, Elder.

Pincus, F. L., & Ehrlich, H. J. (1999). Immigration. In F. L. Pincus & H. J. Ehrlich (Eds.), *Race and ethnic conflict* (pp. 223–228). Boulder, CO: Westview Press.

Piquero, A. R. (2008). Taking stock of developmental trajectories of criminal activity over the life course. In A. Liberman (Ed.), *The long view of crime: A synthesis of longitudinal research* (pp. 23–78). New York: Springer.

Pitts, J. (2008). *Reluctant gangsters: The changing face of youth crime.* Devon, UK: Willan Publishing.

Pizarro, J. M., & McGloin, J. M. (2006). Explaining gang homicides in Newark, New Jersey: Collective behavior or social disorganization? *Journal of Criminal Justice, 34,* 195–207.

Pogarsky, G., Lizotte, A. J., & Thornberry, T. P. (2003). The delinquency of children born to young mothers: Results from the Rochester Youth Development Study. *Criminology, 41,* 1249–1286.

Polk, K. (1999). Males and honor contest violence. *Homicide Studies, 3,* 6–29.

Popkin, S. J., Leventhal, T., & Weismann, G. (2008). Girls in the 'hood: The importance of feeling safe (Brief No. 1). Washington, DC: The Urban Institute.

Popkin, S. J., Leventhal, T., & Weismann, G. (2010). Girls in the 'hood: How safety affects the life chances of low-income girls. *Urban Affairs Review, 45,* 715–744.

Portes, A., & Rumbaut, R. G. (2005). Introduction: The second generation and the Children of Immigrants Longitudinal Study. *Ethnic and Racial Studies, 28,* 983–999.

Portes, A., & Zhou, M. (1993). The new second generation: Segmented assimilation and its variants. *The Annals of the American Academy of Political and Social Sciences, 530,* 74–96.

Pratt, T. C., & Cullen, F. T. (2005). Assessing macro-level predictors and theories of crime: A meta-analysis. In M. Tonry (Ed.), *Crime and justice: A review of research* (Vol. 32, pp. 373–450). Chicago: University of Chicago Press.

Puffer, J. A. (1912). *The boy and his gang.* Boston: Houghton Mifflin.

Pyrooz, D. C., Decker, S. H., & Webb, V. J. (Forthcoming). The ties that bind: Desistance from gangs. *Crime and Delinquency.*

Pyrooz, D. C., Fox, A. M., & Decker, S. H. (2010). Racial and ethnic heterogeneity, economic disadvantage, and gangs: A macro-level study of gang membership in urban America. *Justice Quarterly, 14,* 1–26.

Quicker, J. C. (1983). *Homegirls: Characterizing Chicano gangs.* San Pedro, CA: International University Press.

Ralph, P., Hunter, J., Marquart, W., Cuvelier, J., & Merianos, D. (1996). Exploring the differences between gang and non-gang prisoners. In C. R. Huff (Ed.), *Gangs in America* (2nd ed., pp. 241–256). Thousand Oaks, CA: Sage.

Ransford, C., Kane, C., Metzger, T., Quintana, E., & Slutkin, G. (2010). An examination of the role of CeaseFire, the Chicago police, Project Safe Neighborhoods, and displacement in the reduction in homicide in Chicago in 2004. In R. J. Chaskin (Ed.), *Youth gangs and community intervention: Research, practice, and evidence* (pp. 76–108). New York: Columbia University Press.

Reckless, W. C., Dinitz, S., & Murray, E. (1956). Self-concept as an insulator against delinquency. *American Sociological Review, 21,* 744–746.

Redfield, R. (1941). *Folk culture of Yucatan.* Chicago: University of Chicago Press.

Reeves, J. L. & Campbell, R. (1994). *Cracked coverage: Television news, the anti-cocaine crusade, and the Reagan legacy.* Durham, NC: Duke University Press.

Reiner, I. (1992). *Gangs, crime, and violence in Los Angeles.* Los Angeles: Office of the District Attorney of the County of Los Angeles.

Rennison, C. M., & Melde, C. (2009). Exploring the use of victim surveys to study gang crime: Prospects and possibilities. *Criminal Justice Review, 34,* 489–514.

Resnick, M. D., Ireland, M., & Borowsky, I. (2004). Youth violence perpetration: What protects? What predicts? Findings from the National Longitudinal Study of Adolescent Health. *Journal of Adolescent Health, 35,* 424.e1–424.e10.

Ribando, C. (2005). *Gangs in Central America.* Washington, DC: Congressional Research Service, Library of Congress.

Rice, R. (1963). A report at large: The Persian Queens. *New Yorker, 39,* 153–187.

Riis, J. A. (1892). *Children of the poor.* New York: Charles Scribner's Sons.

Riis, J. A. (1902/1969). *The battle with the slum.* Montclair, NJ: Paterson Smith.

Ro, R. (1996). *Gangsta: Merchandizing the rhymes of violence.* New York: St. Martin's Press.

Robbins, M. S., & Szapocznik, J. (2000). Brief strategic family therapy. *Juvenile Justice Bulletin.* Washington, DC: U.S. Department of Justice, Office of Juvenile Justice and Delinquency Prevention.

Robers, S., Zhang, J., Truman, J., & Snyder, T. D. (2010). *Indicators of school crime and safety, 2010.* Washington, DC: U.S. Department of Justice, National Center for Education Statistics, Bureau of Justice Statistics.

Rodgers, D. (2006). Living in the shadow of death: Gangs, violence, and social order in urban Nicaragua, 1996–2002. *Journal of Latin American Studies, 38,* 267–292.

Rosenfeld, R., Bray, T. M., & Egley, A., Jr. (1999). Facilitating violence: A comparison of gang-motivated, gang-affiliated, and nongang youth homicides. *Journal of Quantitative Criminology, 15,* 495–515.

Roush, D. W., Miesner, L. D., & Winslow, C. M. (2002). *Managing youth gang members in juvenile detention facilities.* East Lansing: Center for Research and Professional Development, National Juvenile Detention Association, School of Criminal Justice, Michigan State University.

Rubel, A. J. (1965). The Mexican American palomilla. *Anthropological Linguistics, 4,* 29–97.

Sabol, W. J., Minton, T. D., & Harrison, P. M. (2007). *Prison and jail inmates at midyear 2006.* Bulletin. Washington, DC: U.S. Department of Justice, Bureau of Justice Statistics.

Salagaev, A., Shaskin, A., Sherbakova, I., & Touriyanskiy, E. (2005). Contemporary Russian gangs: History, membership, and crime involvement. In S. H. Decker & F. M. Weerman (Eds.), *European street gangs and troublesome youth groups* (pp. 169–192). Lanham, MD: AltaMira Press.

Sampson, R. J. (2002). The community. In J. Petersilia & J. Q. Wilson (Eds.), *Crime: Public policies for crime control* (pp. 225–252). Oakland, CA: Institute for Contemporary Studies Press.

Sampson, R. J., & Graif, C. (2009). Neighborhood social capital as differential social organization: Resident and leadership dimensions. *American Behavioral Scientist, 52,* 1579–1605.

Sampson, R. J., & Groves, B. W. (1989). Community structure and crime: Testing social-disorganization theory. *American Journal of Sociology, 94,* 774–802.

Sampson, R. J., & Laub, J. H. (1993). *Crime in the making: Pathways and turning points through life.* Cambridge, MA: Harvard University Press.

Sampson, R. J., & Laub, J. H. (1997). A life-course theory of cumulative disadvantage and the stability of delinquency. In T. P. Thornberry (Ed.), *Developmental theories of crime and delinquency* (pp. 133–161). New Brunswick, NJ: Transaction Publishing.

Sampson, R. J., & Laub, J. H. (2005). A life-course view of the development of crime. *The Annals of the American Academy of Political and Social Science, 602,* 12–45.

Sampson, R. J., Morenoff, J. D., & Gannon-Rowley, T. (2002). Assessing "neighborhood effects": Social processes and new directions in research. *Annual Review of Sociology, 28,* 443–478.

Sampson, R. J., & Raudenbush, S. W. (2001). Disorder in urban neighborhoods: Does it lead to crime? *Research in Brief.* Washington, DC: U.S. Department of Justice, National Institute of Justice.

Sampson, R. J., Raudenbush, S. W., & Earls, F. (1997). Neighborhoods and violent crime: A multilevel study of collective efficacy. *Science, 277,* 918–924.

Sanchez-Jankowski, M. S. (1991). *Islands in the street: Gangs and American urban society.* Berkeley: University of California Press.

Sanchez-Jankowski, M. S. (2003). Gangs and social change. *Theoretical Criminology, 7,* 191–216.

Sanders, W. B. (1994). *Gangbangs and drive-bys: Grounded culture and juvenile gang violence.* New York: Aldine de Gruyter.

Sante, L. (1991). *Low life: Lures and snares of old New York.* New York: Vintage Books.

Sarnecki, J. (2001). *Delinquent networks: Youth co-offending in Stockholm.* Cambridge, UK: Cambridge University Press.

Saunders, B. (2011). The upshot might be no hoax. *The News and Observer* (February 12), B1.

Schalet, A., Hunt, J., & Joe-Laidler, K. (2003). Respectability and autonomy: The articulation and meaning of sexuality among the girls in the gang. *Journal of Contemporary Ethnography, 32,* 108–143.

Schlosser, E. (1998). The prison-industrial complex. *The Atlantic Monthly* (December), pp. 51–77.

Schneider, E. C. (1999). *Vampires, dragons, and Egyptian kings: Youth gangs in postwar New York.* Princeton, NJ: Princeton University Press.

Schram, P. J., & Gaines, L. K. (2005). Examining delinquent nongang members and delinquent gang members: A comparison of juvenile probationers at intake and outcomes. *Youth Violence and Juvenile Justice, 3,* 99–115.

Schweinhart, L. J., Montie, J., Xiang, Z., Barnett, W. S., Belfield, C. R., & Nores, M. (2005). *Lifetime effects: The High/Scope Perry Preschool study through age 40.* Ypsilanti, MI: The High/Scope Press.

Scott, G. (2004). "It's a sucker's outfit": How urban gangs enable and impede the integration of ex-convicts. *Ethnography, 5,* 107–140.

Scott, J. H. (1905). Social instinct and its development in boy life. Association Seminar. (June-July). New York: Young Men's Christian Associations.

Seelke, C. R. (2008). *Gangs in Central America* (CRS Report for Congress RL34112, updated October 17, 2008). Washington, DC: Congressional Research Service, Library of Congress.

Shacklady-Smith, L. (1978). Sexist assumptions and female delinquency: An empirical investigation. In C. Smart & B. Smart (Eds.), *Women, sexuality and social control* (pp. 84–88). London: Routledge & Kegan Paul.

Sharp, C., Aldridge, J., & Medina, J. (2006). *Delinquent youth groups and offending behavior: Findings from the 2004 Offending, Crime and Justice Survey, home office online report 14/06.* London, UK: Research Development and Statistics Directorate, Home Office. Retrieved from http://www.homeoffice.gov.uk/rds/pdfs06/rdsolr1406.pdf

Shaw, C. R. (1930). *The jack roller.* Chicago: University of Chicago Press.

Shaw, C. R., & McKay, H. D. (1931). Social factors in juvenile delinquency: A study of the community, the family, and the gang in relation to delinquent behavior. In *National Commission on Law Observance and Enforcement, Report on the causes of crime* (Vol. II, No. 13, Chapter 6). Washington, DC: U.S. Government Printing Office.

Shaw, C. R., & McKay, H. D. (1942). *Juvenile delinquency and urban areas.* Chicago: University of Chicago Press.

Shaw, C. R., & McKay, H. D. (1969). *Juvenile delinquency and urban areas* (2nd ed.). Chicago: University of Chicago Press.

Shaw, C. R., Zorbaugh, F. M., McKay, H. D., & Cottrell, L. S. (1929). *Delinquency areas.* Chicago: University of Chicago Press.

Shelden, R. G. (1991). A comparison of gang members and non-gang members in a prison setting. *The Prison Journal, 81*(2), 50–60.

Sheldon, H. D. (1898). The institutional activities of American children. *The American Journal of Psychology, 9,* 424–448.

Sheley, J. F., & Wright, J. D. (1993). Gun acquisition and possession in selected juvenile samples. *Research in Brief.* Washington, DC: National Institute of Justice and Office of Juvenile Justice and Delinquency Prevention.

Sheley, J. F., & Wright, J. D. (1995). *In the line of fire: Youth, guns and violence in urban America.* Hawthorne, NY: Aldine De Gruyter.

Short, J. F., Jr. (1996). *Gangs and adolescent violence.* Boulder, CO: Center for the Study of Prevention of Violence, University of Colorado.

Short, J. F., Jr. (1998). The level of explanation problem revisited—The American Society of Criminology 1997 Presidential Address. *Criminology, 36,* 3–36.

Short, J. F., Jr. (2006). Why study gangs? An intellectual journey. In J. F. Short & L. A. Hughes (Eds.), *Studying youth gangs* (pp. 1–14). Lanham, MD: AltaMira Press.

Short, J. F., Jr., & Hughes, L. A. (2006a). Moving gang research forward. In J. F. Short & L. A. Hughes (Eds.), *Studying youth gangs* (pp. 225–238). Lanham, MD: AltaMira Press.

Short, J. F., Jr., & Hughes, L. A. (Eds.). (2006b). *Studying youth gangs.* Lanham, MD: AltaMira Press.

Short, J. F., Jr. & Hughes, L. A. (2009). Urban ethnography and research integrity: Empirical and theoretical dimensions. *Ethnography, 10,* 397–415.

Short, J. F., Jr., & Hughes, L. A. (2010). Promoting research integrity in community-based intervention research. In R. J. Chaskin (Ed.), *Youth gangs and community intervention: Research, practice, and evidence* (pp. 127–151). New York: Columbia University Press.

Short, J. F., Jr., & Meier, R. F. (1981). Criminology and the study of deviance. *American Behavioral Scientist, 24,* 462–478.

Short, J. F., Jr., & Strodtbeck, F. L. (1965/1974). *Group process and gang delinquency.* Chicago: University of Chicago Press.

Shufelt, J. S., & Cocozza, J. C. (2006). *Youth with mental health disorders in the juvenile justice system.* Delmar, NY: National Center for Mental Health and Juvenile Justice.

Sikes, G. (1996). *8 ball chicks: A year in the violent world of girl gangsters.* New York: Anchor Books.

Silberman, M. (1995). *A world of violence.* Belmont, CA: Wadsworth.

Skiba, R. J., & Noam, G. G. (2001). *Zero tolerance: Can suspension and expulsion keep schools safe? New directions for youth development.* San Francisco: Jossey-Bass.

Skiba, R. J., & Peterson, R. (1999). The dark side of zero tolerance: Can punishment lead to safe schools? *Phi Delta Kappan, 80,* 372–376.

Skiba, R., Reynolds, C. R., Graham, S., Sheras, P., et al. (2006). *Are zero tolerance policies effective in the schools? An evidentiary review and recommendations.* Washington, DC: American Psychological Association.

Skogan, W. G., & Hartnett, S. M. (1997). *Community policing, Chicago style.* New York: Oxford University Press.

Skogan, W. G., Hartnett, S. M., Bump, N., & Dubois, J. (2008). *Evaluation of CeaseFire-Chicago.* Final Report to the National Institute of Justice. Chicago, IL: Northwestern University. Retrieved from http://www.ncjrs.gov/pdffiles1/nij/grants/227181.pdf

Skogan, W. G., & Steiner, L. (2004). *CAPS at ten: Community policing in Chicago.* Chicago: Illinois Criminal Justice Information Authority.

Skolnick, J. H. (1989). *Gang organization and migration.* Sacramento, CA: Office of the Attorney General of the State of California.

Skolnick, J. H. (1990). The social structure of street drug dealing. *American Journal of Police, 9,* 1–41.

Smith, D. J., & Bradshaw, P. (2005). *Gang membership and teenage offending.* Study of Youth Transitions and Crime, No. 8. Edinburgh, UK: Centre for Law and Society. Retrieved from http://www.law.ed.ac.uk/cls/esytc/findings/digest8.pdf

Snell, P. (2010). From Durkheim to the Chicago school: Against the "variables sociology" paradigm. *Journal of Classical Sociology, 10,* 51–67.

Snyder, H. N., & Sickmund, M. (1999). *Juvenile offenders and victims: 1999 national report.* Washington, DC: U.S. Department of Justice, Office of Juvenile Justice and Delinquency Prevention.

Snyder, H. N., & Sickmund, M. (2006). *Juvenile offenders and victims: 2006 national report.* Washington, DC: Office of Juvenile Justice and Delinquency Prevention.

Snyder, T. D., & Dillow, S. A. (2009). *Digest of education statistics, 2008.* Washington, DC: National Center for Education Statistics, U.S. Department of Education.

Spaulding, C. B. (1948). Cliques, gangs, and networks. *Sociology and Social Research, 32,* 928–937.

Spergel, I. A. (1964). *Racketville, Slumtown and Haulberg: An exploratory study of delinquent subcultures.* Chicago: University of Chicago Press.

Spergel, I. A. (1966). *Street gang work: Theory and practice.* Reading, MA: Addison-Wesley.

Spergel, I. A. (1984). Violent gangs in Chicago, IL: In search of social policy. *Social Service Review, 58,* 199–226.

Spergel, I. A. (1990). Youth gangs: Continuity and change. In M. Tonry & N. Morris (Eds.), *Crime and justice: A review of research* (Vol. 12, pp. 171–275). Chicago: University of Chicago.

Spergel, I. A. (1995). *The youth gang problem.* New York: Oxford University Press.

Spergel, I. A. (2007). *Reducing youth gang violence: The Little Village Gang Project in Chicago.* Lanham, MD: AltaMira Press.

Spergel, I. A. (2010). *A comprehensive, community-wide approach to the youth gang problem.* Paper presented at the 2010 Canada-U.S. Gang Summit, Toronto, March.

Spergel, I. A., & Bobrowski, L. (1989). *Minutes from the Law Enforcement Youth Gang Definitional Conference: September 25, 1989.* Rockville, MD: Juvenile Justice Clearinghouse.

Spergel, I. A., Chance, R., Ehrensaft, C., Regulus, T., Kane, C., & Laseter, R. (1992). *Technical assistance manuals: National gang suppression and intervention program.* Chicago: University of Chicago, School of Social Service Administration.

Spergel, I. A., Chance, R., Ehrensaft, C., Regulus, T., Kane, C., Laseter, R., Alexander, A., & Oh, S. (1994). *Gang suppression and intervention: Community models.* Washington, DC: U.S. Department of Justice, Office of Juvenile Justice and Delinquency Prevention.

Spergel, I. A., & Curry, G. D. (1993). The National youth gang survey: A research and development process. In A. Goldstein & C. R. Huff (Eds.), *The* gang intervention handbook (pp. 359–400). Champaign, IL: Research Press.

Spergel, I. A., Wa, K. M., & Sosa, R. V. (2006). The comprehensive, community-wide, gang program model: Success and failure. In J. F. Short & L. A. Hughes (Eds.), *Studying youth gangs* (pp. 203–224). Lanham, MD: AltaMira Press.

Starbuck, D., Howell, J. C., & Lindquist, D. J. (2001). Into the millennium: Hybrids and other modern gangs. *Juvenile Justice Bulletin. Youth Gang Series.* Washington, DC: U.S. Department of Justice, Office of Juvenile Justice and Delinquency Prevention.

Stevens, D. J. (1997, Summer). Origins and effects of prison drug gangs in North Carolina. *Journal of Gang Research, 4,* 23–35.

Stewart, E. A., Schreck, C. J., & Simons, R. L. (2006). "I ain't gonna let no one disrespect me": Does the code of the street reduce or increase violent victimization among African American adolescents? *Journal of Research in Crime and Delinquency, 43,* 427–458.

Stewart, E. A., & Simons, R. L. (2010). Race, code of the street, and violent delinquency: A multilevel investigation of neighborhood street culture and individual norms of violence. *Criminology, 48,* 569–605.

Stewart, F. H. (1994). *Honor.* Chicago: University of Chicago Press.

Stouthamer-Loeber, M., Loeber, R., Stallings, R., and Lacourse, E. (2008). Desistance from and persistence in offending. In R. Loeber, D. P. Farrington, M. Stouthamer-Loeber, et al. (Eds.), *Violence and serious theft: Development and prediction from childhood to adulthood* (pp. 269–306). New York: Routledge.

Stouthamer-Loeber, M., Loeber, R., Wei, E., Farrington, D. P., & Wikstrom, P. H. (2002). Risk and promotive effects in the explanation of persistent serious delinquency in boys. *Journal of Consulting and Clinical Psychology, 70,* 111–123.

Stouthamer-Loeber, M., & Wei, E. (1998). The precursors of youth fatherhood and its effect on the delinquency careers of teenage males. *Journal of Adolescent Health, 22,* 56–65.

Stouthamer-Loeber, M., Wei, E., Loeber, R., & Masten, A. S. (2004). Desistance from persistent serious delinquency in the transition to adulthood. *Development and Psychopathology, 16,* 897–918.

Stretesky, P. B., & Pogrebin, M. R. (2007). Gang-related gun violence: Socialization, identity, and self. *Journal of Contemporary Ethnography, 36,* 85–114.

Strom, K. J., Colwell, A., Dawes, D., & Hawkins, S. (2010). *Evaluation of the Methodist Home for Children's value-based therapeutic environment model.* Research Triangle Park, NC: Research Triangle Institute.

Sullivan, J. P. (2006). Maras morphing: Revisiting third generation gangs. *Global Crime,* August–November, 488–490.

Sullivan, M. L. (1993). Puerto Ricans in Sunset Park, Brooklyn: Poverty amidst ethnic and economic diversity. In J. W. Moore & R. Pinderhughes (Eds.), *In the barrios: Latinos and the underclass debate* (pp. 1–25). New York: Russell Sage Foundation.

Sullivan, M. L. (2005). Maybe we shouldn't study "gangs": Does reification obscure youth violence? *Journal of Contemporary Criminal Justice, 21,* 170–190.

Sullivan, M. L. (2006). Are "gang" studies dangerous? Youth violence, local context, and the problem of reification. In J. F. Short & L. A. Hughes (Eds.), *Studying youth gangs* (pp. 15–36). Lanham, MD: AltaMira Press.

Sutherland, E. H. (1947). *Principles of criminology* (4th ed.). Philadelphia: J. B. Lippincott.

Swenson, C. C., Henggeler, S. W., Taylor, I. S., & Addison, O. W. (2005). *Multisystemic therapy and neighborhood partnerships.* New York: Guilford Press.

Tasca, M., Griffin, M. L., & Rodriguez, N. (2010). The effect of importation and deprivation factors on violent misconduct: An examination of Black and Latino youth in prison. *Youth Violence and Juvenile Justice, 8,* 234–249.

Taylor, C. S. (1990a). *Dangerous society.* East Lansing: Michigan State University Press.

Taylor, C. S. (1990b). Gang imperialism. In C. R. Huff (Ed.), *Gangs in America* (pp. 103–115). Newbury Park, CA: Sage.

Taylor, C. S. (1993). *Girls, gangs, women, drugs.* East Lansing: Michigan State University Press.

Taylor, T. J. (2008). The boulevard ain't safe for your kids. . . . Youth gang membership and violent victimization. *Journal of Contemporary Criminal Justice, 24,* 125–136.

Taylor, T. J., Freng, A., Esbensen, F., & Peterson, D. (2008). Youth gang membership and serious violent victimization: The importance of lifestyles and routine activities. *Journal of Interpersonal Violence, 23,* 1441–1464.

Telles, E. E., & Ortiz, V. (2008). *Generations of exclusion.* New York: Russell Sage Foundation.

Teplin, L. A., Abram, K. M., McClelland, G. M., Dulcan, M. K., & Washburn, J. J. (2006). Psychiatric disorders of youth in detention.

Juvenile Justice Bulletin. Washington, DC: Office of Juvenile Justice and Delinquency Prevention.

Teske, S. C., & Huff, J. B. (2011). When did making adults mad become a crime? The court's role in dismantling the school-to-prison pipeline. *Juvenile and Family Justice Today,* Winter, 14–17.

Thale, G. (2007). Testimony before the House Committee on Foreign Affairs, Subcommittee on the Western Hemisphere, June 26. Washington Office on Latin America.

Thomas, W. I. (1923). *The unadjusted girl.* Boston: Little, Brown.

Thomas, W. I., & Znaniecki, F., (1920). *The polish peasant in Europe and America* (Volume IV). Boston: Gorham Press.

Thompson, C., Young, R. L., & Burns, R. (2000). Representing gangs in the news: Media constructions of criminal gangs. *Sociological Spectrum, 20,* 409–432.

Thornberry, T. P. (1987). Toward an interactional theory of delinquency. *Criminology, 25*(4), 863–891.

Thornberry, T. P. (1996). Empirical support for interactional theory: A review of the literature. In D. Hawkins (Ed.), *Delinquency and crime: Current theories* (pp. 198–235). New York: Cambridge University Press.

Thornberry, T. P. (Ed.). (1997). *Developmental theories of crime and delinquency.* New Brunswick, NJ: Transaction Publishing.

Thornberry, T. P. (1998). Membership in youth gangs and involvement in serious and violent offending. In R. Loeber & D. P. Farrington (Eds.), *Serious and violent juvenile offenders: Risk factors and successful interventions* (pp. 147–166). Thousand Oaks, CA: Sage.

Thornberry, T. P. (2005). Explaining multiple patterns of offending across the life course and across generations. *The Annals of the American Academy of Political and Social Science, 602,* 156–195.

Thornberry, T. P., & Burch, J. H. (1997). Gang members and delinquent behavior. *Juvenile Justice Bulletin. Youth Gang Series.* Washington, DC: Office of Juvenile Justice and Delinquency Prevention.

Thornberry, T. P., & Krohn, M. D. (2001). The development of delinquency: An interactional perspective. In S. O. White (Ed.), *Handbook of youth and justice* (pp. 289–305). New York: Plenum.

Thornberry, T. P., & Krohn, M. D. (2005). Applying interactional theory to the explanation of continuity and change in antisocial behavior. In D. P. Farrington (Ed.), *Integrated developmental and life-course theories of offending* (pp. 183–210). New Brunswick, NJ: Transaction Publishing.

Thornberry, T. P., Krohn, M. D., Lizotte, A. J., & Chard-Wierschem, D. (1993). The role of juvenile gangs in facilitating delinquent behavior. *Journal of Research in Crime and Delinquency, 30,* 55–87.

Thornberry, T. P., Krohn, M. D., Lizotte, A. J., Smith, C. A., & Tobin, K. (2003). *Gangs and delinquency in developmental perspective.* New York: Cambridge University Press.

Thornberry, T. P., Lizotte, A. J., Krohn, M. D., Smith, C. A., & Porter, P. K. (2003). Causes and consequences of delinquency: Findings from the Rochester Youth Development Study. In T. P. Thornberry & M. D. Krohn (Eds.), *Taking stock of delinquency: An overview of findings from contemporary longitudinal studies* (pp. 11–46). New York: Kluwer Academic/Plenum Publishers.

Thornberry, T. P., & Porter, P. (2001). Advantages of longitudinal research designs in studying gang behavior. In M. W. Klein, H.-J. Kerner, C. L. Maxson, & E. G. M. Weitekamp (Eds.), *The Eurogang paradox* (pp. 59–78). Boston: Kluwer Academic Publishers.

Thrasher, F. M. 1927/2000. *The gang—A study of 1,313 gangs in Chicago.* Chicago: New Chicago School Press.

Tita, G. E. (1999). *An ecological study of violent urban street gangs and their crime.* Unpublished dissertation. Pittsburgh, PA: Carnegie Mellon University.

Tita, G. E., & Cohen, J. (2004). Measuring spatial diffusion of shots fired activity across city neighborhoods. In M. F. Goodchild & D. G. Janelle (Eds.), *Spatially integrated social science* (pp. 171–204). New York: Oxford University Press.

Tita, G. E., Cohen, J., & Engberg, J. (2005). An ecological study of the location of gang "set space." *Social Problems, 52,* 272–299.

Tita, G. E., & Griffiths, E. (2005). Traveling to violence: The case for a mobility-based spatial typology of homicide. *Journal of Research in Crime and Delinquency, 42,* 275–308.

Tita, G. E., & Ridgeway, G. (2007). The impact of gang formation on local patterns of crime. *Journal of Research in Crime and Delinquency, 44,* 208–237.

Tita, G. E., Riley, K. J., & Greenwood, P. (2003). From Boston to Boyle Heights: The process and prospects of a "pulling levers" strategy in a Los Angeles barrio. In S. H. Decker (Ed.), *Policing gangs and youth violence* (pp. 102–130). Belmont, CA: Wadsworth/Thompson Learning.

Tita, G. E., Riley, K. J., & Greenwood, P. (2005). *Reducing gun violence: Operation Ceasefire in Los Angeles.* Washington, DC: National Institute of Justice.

Tita, G. E., Riley, K. J., Ridgeway, G., Grammich, C., Abrahamse, A., & Greenwood, P. W. (2003). *Reducing gun violence: Results from an intervention in East Los Angeles.* Santa Monica, CA: RAND.

Tobin, K. (2008). *Gangs: An individual and group perspective.* Upper Saddle River, NJ: Prentice Hall.

Toch, H. (2007). Sequestering gang members, burning witches, and subverting due process. *Criminal Justice and Behavior, 32,* 274–288.

Toch, H., & Adams, K. (1988). *Coping, maladaptation in prison.* New Brunswick, NJ: Transaction Publishing.

Tofi, M. M., Farrington, D. P., & Baldry, A. (2008). *Effectiveness of programs to reduce school bullying: A systematic review.* Stockholm: Swedish National Council for Crime Prevention.

Tolan, P. H., & Gorman-Smith, D. (1998). Development of serious and violent offending careers. In R. Loeber & D. P. Farrington (Eds.), *Serious and violent juvenile offenders: Risk factors and successful interventions* (pp. 68–85). Thousand Oaks, CA: Sage.

Tolan, P. H., Gorman-Smith, D., & Henry, D. (2004). Supporting families in a high-risk setting: Proximal effects of the SAFEChildren Preventive Intervention. *Journal of Consulting and Clinical Psychology, 72*(5), 855–869.

Tolan, P. H., Gorman-Smith, D., & Loeber, R. (2000). Developmental timing of onsets of disruptive behaviors and later delinquency of inner-city youth. *Journal of Child and Family Studies, 9,* 203–330.

Tonry, M. (2009). Explanations of American punishment policies: A national history. *Punishment and Society, 11,* 377–394.

Tremblay, R. E. (2003). Why socialization fails: The case of chronic physical aggression. In B. B. Lahey, T. E. Moffitt, & A. Caspi (Eds.), *Causes of conduct disorder and juvenile delinquency* (pp. 182–224). New York: Guilford Press.

Tremblay, R. E., Masse, L., Pagani, L., & Vitaro, F. (1996). From childhood physical aggression to adolescent maladjustment: The Montreal Prevention Experiment. In R. D. Peters & R. J. McMahon (Eds.), *Preventing childhood disorders, substance abuse, and delinquency* (pp. 268–298). Thousand Oaks, CA: Sage.

Turner, J. H. (1978). *The structure of sociological theory.* Homewood, IL: Dorsey.

Tyler, K. A., & Bersani, B. E. (2008). A longitudinal study of early adolescent precursors to running away. *Journal of Early Adolescence, 28,* 230–251.

Tyler, K. A., Hoyt, D. R., & Whitbeck, L. B. (2000). The effects of early sexual abuse on later sexual victimization among female homeless and runaway adolescents. *Journal of Interpersonal Violence, 15,* 235–250.

United Nations Office on Drugs and Crime. (2007). *Crime and development in Central America: Caught in the crossfire.* New York: Author.

U.S. Agency for International Development. (2006). *Central America and Mexico Gang Assessment.* Washington, DC: Bureau for Latin American and Caribbean Affairs, U.S. Agency for International Development.

U.S. General Accounting Office. (1989). *Nontraditional organized crime.* Washington, DC: U.S. Government Printing Office.

U.S. Government Accountability Office. (1996). *Violent crime: Federal law enforcement assistance in fighting Los Angeles gang violence.* Washington, DC: U.S. Government Printing Office.

U.S. Government Accountability Office. (2010). *Combating gangs: Federal agencies have implemented a Central American gang strategy, but could strengthen oversight and measurement of efforts.* Washington, DC: U.S. Government Accountability Office.

Valdés, D. N. (1999). *Region, nation, and world-system: Perspectives on midwestern Chicana/o history.* JSRI Occasional Paper #20. East Lansing: Michigan State University, The Julian Samora Research Institute.

Valdez, A. (2007a). *Gangs: A guide to understanding street gangs* (5th ed.). San Clemente, CA: LawTech Publishing.

Valdez, A. (2003). Toward a typology of contemporary Mexican American youth gangs. In L. Kontos, D. Brotherton, & L. Barrios (Eds.), *Gangs and society: Alternative perspectives* (pp. 12–40). New York: Columbia University Press.

Valdez, A. (2005). Mexican American youth and adult prison gangs in a changing heroin market. *Journal of Drug Issues, 35,* 843–867.

Valdez, A. (2002). *Mexican American girls and gang violence: Beyond risk.* New York: Palgrave Macmillan.

Valdez, A., Cepeda, A., & Kaplan, C. (2009). Homicidal events among Mexican American street gangs: A situational analysis. *Homicide Studies, 13,* 288–306.

Valdez, A., Kaplan, C. D., & Codina, E. (2000). Psychopathy among Mexican American gang members: A comparative study. *International*

Journal of Offender Therapy and Comparative Criminology, 44, 46–58.

Valdez, A., & Sifaneck, S. J. (2004). "Getting high and getting by": Dimensions of drug selling behaviors among Mexican gang members in south Texas. *Journal of Research in Crime and Delinquency, 41*(1), 82–105.

Van Gemert, F., Lien, I., & Peterson, D. (2008). Introduction. In F. Gemert, D. Peterson, & I.-L. Lien (Eds.), *Street gangs, migration and ethnicity* (pp. 3–14). Portland, OR: Willan Publishing.

Van Gemert, F., Peterson, D., & Lien, I.-L. (Eds.). (2008). *Street gangs, migration and ethnicity.* Portland, OR: Willan Publishing.

Vance County Gang Assessment Project. (2010). *An assessment of gang activity in Vance County, NC.* A report prepared for the Vance County Juvenile Crime Prevention Council. Henderson, NC: Vance County Juvenile Crime Prevention Council.

Vaughn, M. G., DeLisi, M., Beaver, K. M., & Wright, J. P. (2009). Identification of latent classes of behavioral risk based on early childhood manifestations of self-control. *Youth Violence and Juvenile Justice, 7,* 16–31.

Veal, R. T. (1919). *Classified bibliography of boy life and organized work with boys.* New York: Association Press.

Venkatesh, S. A. (1996). The gang and the community. In C. R. Huff (Ed.), *Gangs in America* (2nd ed., pp. 241–256). Thousand Oaks, CA: Sage.

Venkatesh, S. A. (2000). *American project: The rise and fall of a modern ghetto.* Cambridge, MA: Harvard University Press.

Venkatesh, S. A. (2008). *Gang leader for a day: A rogue sociologist takes to the streets.* New York: Penguin Press.

Vericker, T., Pergamit, M., Macomber, J., & Kuehn, D. (2009). *Vulnerable youth and the transition to adulthood: Second-generation Latinos connecting to school and work* (ASPE Research Brief, July). Washington, DC: Office of Human Services Policy, U.S. Department of Education.

Vigil, J. D. (1979). Adaptation strategies and cultural life styles of Mexican American adolescents. *Hispanic Journal of Behavioral Sciences, 1,* 375–392.

Vigil, J. D. (1988). *Barrio gangs: Street life and identity in Southern California.* Austin: University of Texas Press.

Vigil, J. D. (1990). Cholos and gangs: Culture change and street youth in Los Angeles. In C. R. Huff (Ed.), *Gangs in America* (pp. 116–128). Newbury Park, CA: Sage.

Vigil, J. D. (1993). The established gang. In S. Cummings & D. J. Monti (Eds.), *Gangs: The origins and impact of contemporary youth gangs in the United States* (pp. 95–112). Albany: State University of New York Press.

Vigil, J. D. (1998). *From Indians to Chicanos: The dynamics of Mexican-American culture* (2nd ed.). Prospect Heights, IL: Waveland.

Vigil, J. D. (2002). *A rainbow of gangs: Street cultures in the mega-city.* Austin: University of Texas Press.

Vigil, J. D. (2004). Street baptism: Chicano gang initiation. In F-A. Esbensen, S. G. Tibbetts, & L. Gaines (Eds.), *American youth gangs at the millennium* (pp. 218–228). Long Grove, IL: Waveland Press.

Vigil, J. D. (2006). A multiple marginality framework of gangs. In A. Egley, C. L. Maxson, J. Miller, and M. W. Klein (Eds.), *The modern gang reader* (3rd ed., pp. 20–29). Los Angeles: Roxbury.

Vigil, J. D. (2007). *The projects: Gang and non-gang families in East Los Angeles.* Thousand Oaks, CA: Sage.

Vigil, J. D. (2008). Mexican migrants in gangs: A second-generation history. In F. van Gemert, D. Peterson, & I.-L. Lien (Eds.), *Street gangs, migration and ethnicity* (pp. 49–62). Portland, OR: Willan Publishing.

Vigil, J. D. (2010). *Gang redux: A balanced anti-gang strategy.* Long Grove, IL: Waveland Press.

Vigil, J. D., & Long, J. M. (1990). Emic and etic perspectives on gang culture. In C. R. Huff (Ed.), *Gangs in America* (pp. 55–70). Newbury Park, CA: Sage.

Vigil, J. D., & Yun, S. C. (1990). Vietnamese youth gangs in Southern California. In C. R. Huff (Ed.), *Gangs in America* (pp. 146–162). Newbury Park, CA: Sage.

Vogel, C. (2007). Gang lite? A new kind of gang in Texas and Houston promises protection but no lifelong commitment. *HoustonPress.com.* Retrieved from http://www.houstonpress.com/2007–08–16/news/gang-lite/

Wacquant, L. (2007). *Urban outcasts: A comparative sociology of advanced marginality.* Cambridge, UK: Polity Press.

Walker, M. L., & Schmidt, L. M. (1996). Gang reduction efforts by the Task Force on Violent Crime in Cleveland, Ohio. In C. R. Huff (Ed.), *Gangs in America* (2nd ed., pp. 263–269). Thousand Oaks, CA: Sage.

Warr, M. (1996). Organization and instigation in delinquent groups. *Criminology, 34,* 11–37.

Warr, M. (2002). *Companions in crime: The social aspects of criminal conduct.* New York: Cambridge University Press.

Washington Office on Latin America. (2010). *Executive summary: Transnational youth gangs in Central America, Mexico and the United States.* Washington, DC: Author.

Wasserman, G. A., & Seracini, A. M. (2001). Family risk factors and interventions. In R. Loeber & D. P. Farrington (Eds.), *Child delinquents: Development, interventions, and service needs* (pp. 165–189). Thousand Oaks, CA: Sage.

Waters, T. (1999). *Crime and immigrant youth.* Thousand Oaks, CA: Sage.

Watkins, A., Huebner, B., & Decker, S. (2008). Patterns of gun acquisition, carrying and use among juvenile and adult arrestees: Evidence from a high-crime city. *Justice Quarterly, 25,* 674–700.

Webb, G. (1999). *Dark alliance: The CIA, the contras, and the crack cocaine explosion.* New York: Seven Stories Press.

Weerman, F. M., & Decker, S. H. (Eds.). (2005a). *European street gangs and troublesome youth groups.* Lanham, MD: AltaMira Press.

Weerman, F. M., & Decker, S. H. (2005b). European street gangs and troublesome youth groups: Findings from the Eurogang research program. In S. H. Decker & F. M. Weerman (Eds.), *European street gangs and troublesome youth groups* (pp. 287–310). Lanham, MD: AltaMira Press.

Weerman, F. M., & Esbensen, F. (2005). A cross-national comparison of youth gangs: The United States and the Netherlands. In S. H. Decker & F. M. Weerman (Eds.), *European street gangs and troublesome youth groups* (pp. 219–255). Lanham, MD: AltaMira Press.

Wei, E., Hipwell, A., Pardini, D., Beyers, J. M., & Loeber, R. (2005). Block observations of neighbourhood physical disorder are associated with neighbourhood crime, firearm injuries, deaths, and teen births. *Journal of Epidemiology and Community Health, 59,* 904–908.

Weisburd, D., Bushway, S., Lum, C., & Yang, S. (2004). Trajectories of crime at places: A longitudinal study of street segments on the city of Seattle. *Criminology, 42,* 283–321.

Weisel, D. L. (2002a). *Contemporary gangs: An organizational analysis.* New York: LFB Scholarly Publishing.

Weisel, D. L. (2002b). The evolution of street gangs: An examination of form and variation. In W. Reed & S. Decker (Eds.), *Responding to gangs: Evaluation and research* (pp. 25–65). Washington, DC: U.S. Department of Justice, National Institute of Justice.

Weisel, D. L. (2004). Graffiti. *Problem-Oriented Guides for Police. Guide No. 9.* Washington, DC: Office of Community Oriented Policing Services.

Weisel, D. L., & Howell, J. C. (2007). *Comprehensive gang assessment: A report to the Durham Police Department and Durham County Sheriff's Office.* Durham, NC: Durham Police Department.

Welch, M., Price, E. A., & Yankey, N. (2002). Moral panic over youth violence: Wilding and the manufacture of menace in the media. *Youth & Society, 34,* 3–30.

Wellford, C., Pepper, J. V., & Petrie, C. (2005). *Firearms and violence: A critical review.* Washington, DC: National Academies Press.

Wells, L. E., & Weisheit, R. A. (2001). Gang problems in nonmetropolitan areas: A longitudinal assessment. *Justice Quarterly, 18,* 791–823.

Welsh, B. C., & Farrington, D. P. (2006). *Preventing crime: What works for children, offenders, victims, and places.* Dordrecht, The Netherlands: Springer.

Welsh, B. C., & Farrington, D. P. (2007). Save children from a life of crime. *Criminology and Public Policy, 6,* 871–879.

Welsh, B. C., Sullivan, C. J., & Olds, D. L. (2010). When early crime prevention goes to scale: A new look at the evidence. *Prevention Science, 11,* 115–125.

West, C. (1993). *Race matters.* Boston: Beacon.

Whitbeck, L. B., Hoyt, D. R., & Yoder, K. A. (1999). A risk-amplification model of victimization and depressive symptoms among runaway and homeless adolescents. *American Journal of Community Psychology, 27,* 273–296

White, N. A., & Loeber, R. (2008). Bullying and special education as predictors of serious delinquency. *Journal of Research in Crime and Delinquency, 45,* 380–397.

White, R. (2008). Disputed definitions and fluid identities: The limitations of social profiling in relation to ethnic youth gangs. *Youth Justice, 8,* 149–161.

Whyte, W. F. (1941). Corner boys: A study of clique behavior. *The American Journal of Sociology, 46,* 647–664.

Whyte, W. F. (1943a). Social organization in the slums. *The American Sociological Review, 8,* 34–39.

Whyte, W. F. (1943b). *Street corner society: The social structure of an Italian slum.* Chicago: University of Chicago Press.

Wiebe, D. (1998). *Targeting and gang crime: Assessing the impacts of a multi-agency suppression strategy in Orange County, California.* Paper presented at the annual meeting of the American Society of Criminology, Washington, DC, November.

Wiebe, D. J., Meeker, J. W., & Vila, B. (1999). *Hourly trends of gang crime incidents, 1995-1998.* University of California, Irvine: Focused Research Group on Gangs.

Wiist, W. H., Jackson, R. H., & Jackson, K. W. (1996). Peer and community leader education to prevent youth violence. *American Journal of Preventive Medicine,* Suppl., *12*(5), 60.

Wikstrom, P. H., & Loeber, R. (2000). Do disadvantaged neighborhoods cause well-adjusted children to become adolescent delinquents? A study of male juvenile serious offending, individual risk and protective factors and neighborhood context. *Criminology, 38,* 1109–1142.

Wikstrom, P. H., & Treiber, K. (2009). Violence as situational action. *International Journal of Conflict and Violence, 3,* 75–96.

Wilkerson, D. (1962). *The cross and the switchblade.* New York: Penguin Books.

Williams, K., Curry, G. D., & Cohen, M. (2002). Gang prevention programs for female adolescents: An evaluation. In W. L. Reed & S. H. Decker (Eds.), *Responding to gangs: Evaluation and research* (pp. 225–263). Washington, DC: U.S. Department of Justice, National Institute of Justice.

Wilson, J. J., & Howell, J. C. (1993). *A comprehensive strategy for serious, violent and chronic juvenile offenders.* Washington, DC: U.S. Department of Justice, Office of Juvenile Justice and Delinquency Prevention.

Wilson, J. Q. (1995). Crime and public policy. In J. Q. Wilson & J. Petersilia (Eds.), *Crime* (pp. 489–507). San Francisco: ICS Press.

Wilson, S. J., & Lipsey, M. W. (2007). School-based interventions for aggressive and disruptive behavior: Update of a meta-analysis. *American Journal of Preventive Medicine, 33* (Supplement), S130-S143.

Wilson, S. J., Lipsey, M. W., & Derzon, J. H. (2003). The effects of school-based intervention programs on aggressive behavior: A meta-analysis. *Journal of Consulting and Clinical Psychology, 71,* 136–149.

Wilson, W. J. (1987). *The truly disadvantaged: The inner city, the underclass, and public policy.* Chicago: University of Chicago Press.

Wilson, W. J. (1999). Societal changes and vulnerable neighborhoods. In F. L. Pincus & H. J. Ehrlich (Eds.), *Race and ethnic conflict* (pp. 110–119). Boulder, CO: Westview Press.

Winfree L. T., Fuller, K., Vigil, T., & Mays, G. L. (1992). The definition and measurement of "gang status": Policy implications for juvenile justice. *Juvenile and Family Court Journal, 43,* 29–38.

Winfree, L. T., Weitekamp, E., Kerner, H. J., Weerman, F., Medina, J., Aldridge, J., & Maljevic A. (2007). *Youth gangs in five nations.* Paper presented at the annual meeting of the American Society of Criminology, Atlanta, November.

Winton, A. (2007). Using "participatory" methods with young people in contexts of violence: Reflections from Guatemala. *Bulletin of Latin American Research, 26,* 497–515.

Wirth, L. (1928). *The ghetto.* Chicago: University of Chicago Press.

Wolfgang, M. E., & Ferracuti, F. (1967). *The subculture of violence.* London: Tavistock Publications.

Wortley, S., & Tanner, J. (2006). Immigration, social disadvantage and urban youth gangs: Results of a Toronto-area study. *Canadian Journal of Urban Research, 15,* 1–20.

Wortley, S., & Tanner, J. (2008). Respect, friendship, and racial justice: Justifying gang membership in a Canadian city. In F. van Gemert, D. Peterson, & I.-L. Lien (Eds.), *Street gangs, migration and ethnicity* (pp. 192–208). Portland, OR: Willan Publishing.

Wright, J. D. (2005). The constitutional failure of gang databases. *Stanford Journal of Civil Rights and Civil Liberties, 1,* 115–142.

Wyrick, P. A. (2000). *Vietnamese youth gang involvement* (Fact Sheet No. 2000–01). Washington, DC: Office of Juvenile Justice and Delinquency Prevention.

Wyrick, P. A. (2006). Gang prevention: How to make the "front end" of your anti-gang effort work. *United States Attorneys' Bulletin, 54,* 52–60.

Yablonsky, L. (1959). The delinquent gang as a near-group. *Social Problems, 7,* 108–117.

Yablonsky, L. (1967). *The violent gang* (Rev. ed.). New York: Penguin.

Yoder, K. A., Whitbeck, L. B., & Hoyt, D. R. (2003). Gang involvement and membership among homeless and runaway youth. *Youth and Society, 34,* 441–467.

Zahn, M. A., Day, J. C., Mihalic, S. F., & Tichavsky, L. (2009). Determining what works for girls in the juvenile justice system: A summary of evaluation evidence. *Crime and Delinquency, 55,* 266–293.

Zatz, M. S. (1987). Chicano youth gangs and crime: The creation of moral panic. *Contemporary Crises, 11,* 129–158.

Zatz, M. S., & Portillos, E. L. (2000). Voices from the barrio: Chicano/a gangs, families, and communities. *Criminology, 38,* 369–401.

Zhou, M., Lee, J., Vallejo, J. A., Tafora-Estrada, R., & Xiong, Y. S. (2008). Success attained, deterred, and denied: Divergent pathways to social mobility in Los Angeles's new second generation. *The Annals of the American Academy of Political and Social Science, 620,* 37–61.

Zorbaugh, H. (1929). *Gold coast and slum.* Chicago: University of Chicago Press.

Index

About the Author

Dr. James C. (Buddy) Howell is a Senior Research Associate with the National Gang Center, in Tallahassee, Florida, where he has worked for 15 years. He formerly worked at the U.S. Department of Justice for 23 years, mostly as director of research and program development in the Office of Juvenile Justice and Delinquency Prevention. He has published 40 works on youth and street gangs, and a similar number on juvenile justice and delinquency prevention, including four books. His gang publication topics include street gang history; gang homicides; drug trafficking; gangs in schools; hybrid gangs; myths about gangs; risk factors; gang problem trends; and what works in preventing gang activity, combating gangs, and reducing gang crime. He is very active in helping states and localities reform their juvenile justice systems and use evidence-based programs and in working with these entities to address youth gang problems in a balanced approach.

SAGE Research Methods Online

The essential tool for researchers

**Sign up now at
www.sagepub.com/srmo
for more information.**

An expert research tool

- An **expertly designed taxonomy** with more than 1,400 unique terms for social and behavioral science research methods

- **Visual and hierarchical search tools** to help you discover material and link to related methods

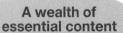

- Easy-to-use navigation tools
- Content organized by complexity
- Tools for citing, printing, and downloading content with ease
- Regularly updated content and features

A wealth of essential content

- The most comprehensive picture of quantitative, qualitative, and mixed methods available today

- More than **100,000 pages of SAGE book and reference material** on research methods as well as editorially selected material from SAGE journals

- More than **600 books** available in their entirety online

Launching 2011!

⑤SAGE research methods online